Vortex Flow in Nature and Technology

Vortex Flow in Nature and Technology

HANS J. LUGT
David W. Taylor Naval Ship Research and Development Center, Bethesda, Maryland

A Wiley-Interscience Publication
JOHN WILEY & SONS
New York · Chichester · Brisbane · Toronto · Singapore

The German version of this book was published by G. Braun, Karlsruhe, 1979.

Copyright © 1983 by John Wiley & Sons, Inc.

All rights reserved. Published simultaneously in Canada.

Reproduction or translation of any part of this work beyond that permitted by Section 107 or 108 of the 1976 United States Copyright Act without the permission of the copyright owner is unlawful. Requests for permission or further information should be addressed to the Permissions Department, John Wiley & Sons, Inc.

Library of Congress Cataloging in Publication Data:

Lugt, Hans J.
   Vortex flow in nature and technology.

   Translation of: Wirbelströmung in Natur und Technik.
   "A Wiley-Interscience publication."
   Includes bibliographical references and indexes.
   1. Vortex-motion.   2. Dynamic meteorology.
3. Ocean-atmosphere interactions.   I. Title.

QC159.L8413 1983   532'.0595   82-23903
ISBN 0-471-86925-2

Printed in the United States of America

10 9 8 7 6 5 4 3 2 1

*TO ANNELIESE,
CHRISTIAN, and BRIGITTE*

TO MY MOTHER and

TO THE MEMORY OF MY FATHER

Preface to the German Edition

Every movement of matter, except for a few special cases, may be considered as vortical. This book is an attempt to give a unified and comprehensive account of the far-reaching significance of the vortex concept in science and technology. The basic and characteristic properties of vortices have been extracted from the abundance of observational material and illustrated with typical examples. It is unavoidable that in this effort the limits of the author's own special field have been exceeded.

The book consists of two parts. In the first six chapters it is assumed that the earth's rotation and the density changes of air and water have no influence on vortex motion. The second part (Chapters 7 through 12) treats vortices in rotating systems and those under the influence of stratification. The historical significance of the vortex concept in science is elucidated in the first chapter. Additional historical and philosophical remarks are contained in all other chapters.

Mathematical derivations and estimates have not been included, since the approximate solutions of the basic physical equations, except for a few simple cases, are extremely complicated and are often obtainable only by means of large computers. The general circulation of the atmosphere may be mentioned as an example. Without such a rigorous mathematical treatment, the explanations and interpretations of the vortex phenomena are based only on descriptions. The author believes this approach will reach a larger audience. Quantitative information is provided by means of dimensionless parameters (like the Reynolds number), graphic displays, and computer-generated pictures.

References had to be restricted to essential sources, reviews, and recent work. Two examples may give an idea of the overwhelming number of publications: More than 1000 references are known to the author which deal with Kármán-vortex streets, and more than a dozen papers on the Presocratic vortex concept "dine." However, the 500 references were selected in such a way that they may be used for further studies and for finding additional references. Because of the many English literature citations, a brief dictionary of English technical words is added.

For students of fluid dynamics Chapters 4 through 6 are intended to be an unorthodox introduction, based on vorticity dynamics, to the theory of viscous fluids. The author's many years of experiments with computer calculations have convinced him that the theory of vorticity dynamics should be used as an introduction to fluid dynamics. The first course on this subject was given by the author in 1969 at the American University, Washington, D.C.

Friends and colleagues have helped in the preparation of the manuscript with discussions, advice, and stimulation. I thank in particular Professor Dr. J. Zierep, Karlsruhe, for valuable remarks and for reading the manuscript. I owe thanks to Professor Dr. A. Walz, Emmendingen, for much encouragement and for pointing out the forgotten anecdote on Meerwein, the first glider from Emmendingen (Section 4.7). Professor G. von Ihering, Washington, D.C., has critically read the first part of the book and provided the nice remarks on J. S. Bach's "turbae" (Chapter 1) and on the art of sculling boats (Chapter 4). With Dr. L. Schmid, NASA Goddard, I had stimulating discussions on the subjects of Chapters 7 and 12. Finally, I owe sincere thanks to my wife for reading part of the manuscript and above all for her continuous encouragement.

HANS J. LUGT

Potomac, Maryland
December 1978

Preface to the English Edition

Since the publication of the German edition, the text has been improved, updated, and considerably enlarged during the translation process. In particular, Chapters 5 and 6 were expanded to reflect the rapid progress in the study of three-dimensional vortex shedding, of flow-induced vibration, and of turbulent vortices.

Readers have suggested the inclusion of a mathematical treatment of the subject matter, an idea that was abandoned in the original version. Such inclusion would far exceed the original intent of this book (which is not a textbook) and, because of the wide range of vortex phenomena covered, would call for the treatment of either a limited area of vortex motion or a multivolume and multiauthor effort. A compromise was attempted by including a mathematical supplement that demonstrates analytically the essential properties of vortices.

The author owes thanks to Mrs. J. Schot of DTNSRDC who went over most parts of the book and helped to clarify and improve the technical content. H. Cheng of DTNSRDC read the mathematical supplement, D. Rockwell of Lehigh University read Chapter 6, and F. H. Busse of the University of California, Los Angeles, read Chapter 8.

A successful translation would not have been possible without the diligent and cheerful assistance of Mrs. A. Phillips of DTNSRDC who improved considerably the readability of the text.

Potomac, Maryland
February 1983

HANS J. LUGT

Contents

Notation   xv

PART ONE   1

1. Historical Survey   3

1.1. Myth and Prehistory, 3
1.2. Antiquity and the Middle Ages, 6
1.3. The Renaissance, 10
1.4. The Cartesian Vortex Theory, 11
1.5. The Era of Classical Mechanics, 13
1.6. Significance of the Vortex Concept in the 20th Century, 15

2. Basic Concepts and Kinematic Considerations   18

2.1. Two Definitions, 18
2.2. Rotation and Conservation Law of Matter, 19
2.3. Streamlines and Pathlines, 21
2.4. Are Vortices Visible?, 24
2.5. The Spectrum of Vortices, 26

3. Properties of Simple Vortices   28

3.1. State Variables and Basic Laws, 28
3.2. The Two Basic Types of Plane Vortices, 30
3.3. The Bucket Experiment, 34
3.4. Intake and Discharge Vortices, 34
3.5. The Motion of Several Point Vortices, 38
3.6. Curved Vortex Tubes, 40

4. Vorticity   45

4.1. Generation and Spreading of Vorticity, 45
4.2. Dynamic Similarity, 46
4.3. Numerical Experiments, 47
4.4. Fluid Motion at Very Small Reynolds Numbers, 48
4.5. Boundary Layer and Shear Flow, 52
4.6. The Classical Theory of Vorticity, 54
4.7. The Lanchester–Prandtl Hypothesis of Flying, 56
4.8. Wheel, Propeller, and Boomerang, 59

5. Separation   67

5.1. Flow Separation and Vortex Formation, 67
5.2. Vortices Behind an Edge, 71
5.3. Vortex Separation, 74
5.4. Note on the Evolution of Fast-Swimming Fish, 77
5.5. Three-Dimensional Vortices Behind Bodies, 81
5.6. Swirling Flows, 87
5.7. Free Vortex Rings, 91

6. Instability and Turbulence   97

6.1. What is Instability?, 97
6.2. Instability of Shear Flows and Boundary Layers, 98
6.3. Periodic Vortex Shedding, 101
6.4. Flow-Induced Vibration, 106

xii   Contents

- 6.5. Rotating Bodies, 112
- 6.6. Some Remarks on Turbulence, 119
- 6.7. Examples of Turbulent Vortices, 121
- 6.8. Flow Drag and its Control, 123

PART TWO   127

7. Fluid Flow in a Rotating System   129

- 7.1. Absolute Rotation and Mach's Principle, 129
- 7.2. Centrifugal Force and Coriolis Force, 131
- 7.3. Dynamic Similarity in a Rotating System, 134
- 7.4. Hyperbolicity, 135
- 7.5. Circulation in Rotating Vessels, 138
- 7.6. Instability of Rotating Fluids, 145

8. Stratification in Ocean and Atmosphere   150

- 8.1. The Earth's Air and Water Envelope, 150
- 8.2. Upward and Downward Motions, 152
- 8.3. Cellular Convections, 158
- 8.4. Vortices and Vorticity in Stratified Flows, 162
- 8.5. Analogy Between Rotating and Stratified Fluids, 164
- 8.6. Rotation in Stratified Fluids, 165

9. Circulations in Atmosphere, Ocean, and Earth   169

- 9.1. The General Circulation of the Atmosphere, 169
- 9.2. Ocean Circulations, 175
- 9.3. The Gulf Stream, 178
- 9.4. Laboratory Experiments and Computer Calculations, 181
- 9.5. An Excursion into the Earth's History, 183

10. Single Vortices in Atmosphere and Ocean   187

- 10.1. Preferred Frequency Ranges in the Spectrum of Vortices, 187
- 10.2. Cold Front and Squall Line as Origins of Local Vortices, 188
- 10.3. Tornadoes and Waterspouts, 190
- 10.4. Dust Devils, 196
- 10.5. Langmuir Vortices, 197
- 10.6. Tidal Vortices, 198

11. Hurricanes   203

- 11.1. Historical Notes, 203
- 11.2. Hurricanes in the 20th Century, 204
- 11.3. Conditions for the Development of Hurricanes, 206
- 11.4. The Hurricane as a Heat Engine, 207

12. Extraterrestrial Vortices   212

- 12.1. Periodicity as an Ordering Principle, 212
- 12.2. The Solar System, 212
- 12.3. Circulations and Vortices in Planetary Atmospheres, 215
- 12.4. Rotating Stars, 218
- 12.5. Exotic Vortex Stars, 220
- 12.6. Galactic Vortices, 222
- 12.7. The Limits of Human Perception, 224

References   227

MATHEMATICAL SUPPLEMENT   243

Notation for the Mathematical Supplement   245

1M. Some Definitions and Kinematical Aspects   247

- 1.1M. Some Definitions, 247
- 1.2M. Rotation and Conservation Law of Matter, 248
- 1.3M. Streamlines and Pathlines, 248
- 1.4M. Lines, Tubes, and Filaments, 249

2M.	**Conservation Laws of Viscous Fluids**	**250**	5.3M. Spiral Discontinuity Lines, 269	

2M. Conservation Laws of Viscous Fluids — 250

2.1M. The Navier–Stokes Equations, 250
2.2M. The Vorticity-Transport Equation, 251
2.3M. Analogy Between Vorticity and Heat Transfer, 251
2.4M. Dimensionless Form of the Vorticity-Transport Equation, 252

3M. Vortices in Viscous Fluid Flows — 253

3.1M. Some Closed-Form Solutions, 253
3.2M. Note on the Occurrence of Extremal Values Inside a Vorticity Field, 257
3.3M. Similarity Solutions, 257
3.4M. Boundary-Layer Flows, 259

4M. Vorticity and Circulation Theorems of Inviscid Fluids — 263

4.1M. Helmholtz's Theorems, 263
4.2M. Kelvin's Circulation Theorem, 264
4.3M. Extensions, 264
4.4M. Some Solutions, 265

5M. Vortices in Potential Flow — 267

5.1M. Point Vortices, 267
5.2M. Point Vortices as a Hamiltonian System, 269
5.3M. Spiral Discontinuity Lines, 269
5.4M. Vortex Lines and Tubes, 270
5.5M. Vortex Concept in Wing Theory, 271

6M. Vortices in Slow Motion — 273

6.1M. Plane Vortices, 273
6.2M. Axisymmetric Vortices, 274

7M. Rotating Fluid Systems — 276

7.1M. Centrifugal and Coriolis Forces, 276
7.2M. Hyperbolicity and Taylor–Proudman Theorem, 276
7.3M. Geostrophic Vortices, 277
7.4M. Gradient, Inertia, and Cyclostrophic Motions, 278
7.5M. The Ekman Layer, 279

8M. Vortices in Stratified Fluids — 280

8.1M. The Boussinesq Approximation, 280
8.2M. Viscous-Flow Solutions, 280
8.3M. Inviscid-Flow Solutions, 281

References for the Mathematical Supplement — 283

Name Index — 285

Subject Index — 291

Notation

a	thermal diffusivity
A	area
b	expansion coefficient
c_D	drag coefficient
c_L	lift coefficient
D	diameter, drag
Ek	Ekman number
f	frequency, Coriolis parameter
Fr	Froude number
g	gravitational constant, gram
Gr	Grashof number
h	hour
Hz	Hertz
l	half a wavelength
L	length, lift
mb	millibar
N	Brunt–Väisälä frequency
p	pressure, roll parameter
Pr	Prandtl number
r	coordinate in the radial direction
R	radius, resultant
Ra	Rayleigh number
Re	Reynolds number
Ro	Rossby number
s	second
St	Strouhal number
t	dimensionless time
T	temperature
U	wave or phase velocity
V	parallel-flow velocity
U, V	sometimes parallel and tangential velocity components, respectively
v_r, v_φ, v_z	velocity components in r, φ, z coordinates
W	axial velocity
y	coordinate in the northern direction
z	coordinate perpendicular to the earth's surface
α	angle of attack
β	glide angle, coefficient in the change of the Coriolis force
Γ	circulation
λ	wavelength, friction factor
ν	kinematic viscosity
ρ	density
σ	swirl angle
ϕ	latitude
φ	angle in polar coordinate system r, φ
ω	vorticity component perpendicular to the plane of rotation
Ω	angular velocity

References are cited in square brackets in text, for example, [3] and superscript numbers refer the reader to Remarks at the end of each chapter.

Vortex Flow in Nature and Technology

PART ONE

1. Historical Survey

From the earliest humans to their present-day descendants, vortices have fascinated mankind. Everyone has noted at some time whirling leaves on a stormy autumn day or eddies at the surface of a river. What is the attraction of this phenomenon? Is it the unknown force behind the fast swirling movement of matter? Until recently, humans had thought that supernatural forces were the causes of vortices. In legends and fairy tales, nymphs and monsters appear in whirlpools and rapid ocean currents; in the Homeric epic, Odysseus barely escapes the fangs of Charybdis; and in the far north, the legendary Maelstrom draws everything close to it into the depths of the sea. Again and again the danger and mystery of vortical motion have excited fantasy and superstition. Today, however, no student of science believes in supernatural causes or meditates on the cryptic meaning of vortices, but both danger and mystery have remained characteristic of many vortex flows. Typhoons and hurricanes still devastate the shores of Asia and America, and many people die in tornadoes every year. The circulation patterns of the deep sea are as mysterious as ever, and the great astronomical enigmas, such as the origin of the solar system and the structure of galaxies, are unsolved vortex problems.

Obviously vortices play a much more powerful role in nature than merely whirling leaves or eddies in a river suggest. Indeed, vortices are not rare caprices of nature but are essential for the movement of matter. Thus it is not surprising that the vortex concept is of central importance in the history of science and philosophy. Since prehistoric times vortex motion has been used to explain basic phenomena in nature. In the realm of the macrocosmos conclusions by analogy were drawn from the properties of commonly observed vortices in attempts to solve cosmological problems. In particular, the evolution of the universe and the motion of heavenly bodies were related to vortical flow. In the microcosmos vortices served as models to describe the atomic structure of matter. These attempts to understand nature by means of vortices were invalidated with the arrival of modern physics at the beginning of this century. On the other hand, the development of a vortex theory within the realm of classical mechanics has led to a great triumph of the vortex concept, as will be shown further.

The following text provides a brief historical survey of the significance of the vortex concept. Additional historical information on special vortex motions is included in each chapter.

1.1. MYTH AND PREHISTORY

Observations and speculations on vortices and their meanings go back to prehistoric times. According to Mackenzie [1], the spiral motif plays a decisive role in their study. In the paleolithic period, the hunting era of mankind, the view of whorled shells motivated the use of the spiral as a magicoreligious symbol of life. Early humans believed that life developed in cavities, inferred from the formation of human and animal life in the womb. It might be that hollow spaces were assumed to be formed by spiral movement and the curling in of folds.[1]

A turning point in the history of mankind occurred in the mesolithic age when hunters became farmers and herdsmen. No longer did humans depend on the accidental prey obtained by hunting but produced their food themselves. During this new phase humans became more aware of the forces and cyclic events in nature. Sun, wind, and water were the life-giving sources of energy. The cycles of growth and decay, day and night, full moon and new

4 Historical Survey

Figure 1.1. Ornament from the Stone Age. Development of animals from vortices (from C. Schuchhardt [3]).

moon, the change of the seasons revealed eternal laws. Humans at that time regarded these events as supernatural because they did not have any rational explanation. The concept of the hereafter developed in this context [2]. The inescapable cycles in nature, so it was believed, originated from the interventions of gods and ghosts, and earthbound humans were at their complete mercy. This development probably started in the Near East or perhaps in Southeast Asia. It also modified the symbolic meaning of the spiral. Whereas until then the spiral had represented a kind of static state or solidified movement, which consisted of cavities in organic matter, it now became identified with movement in the form of vortex flow, again as a symbol for life, growth, and energy (Fig. 1.1). The spiral was also a symbol for transmigration of the soul and the paths of gods and demons. At that time one of the greatest inventions of mankind was made, that of the wheel,[2] which has become a symbol for periodicity.

Vortices were observed in water, in air, and in the sky, and the different forms were often related to each other. The significance ascribed to them in the magical and mythical sense is summarized here from many sources.

Life started in the water of the primeval vortex. In whirlpools, which were also considered gates to the nether world, lurked danger. In whirlwinds gods, spirits, and witches traveled through the air. The heavenly vortex, which humans envisioned in the revolution of the celestial bodies around the north star, was the source of all energy. These ideas can be found in various versions all over the earth.[3]

The oldest evidence for this concept is found in the Egyptian pyramid texts. There a direct relationship is mentioned between stellar revolution and the annual inundations of the Nile—a periodicity that is extremely important for agriculture in the Nile delta. At another place in the pyramid texts it is said: "King Unis goes to the sky, on the wind." Here, the writer is apparently referring to the updraft in a whirlwind that can lift objects skyward with sudden speed [1].

Remarks on whirlwinds can also be found in the Bible, one of the most fertile sources for the history of the Near East [4]. Whirlwinds are mechanisms which God uses to come in contact with mortals:

Then the Lord answered Job out of the whirlwind
(Job 38.1);
... the Lord hath his way in the whirlwind and in the storm, and the clouds are the dust of his feet
(Nahum 1.3);
... and Elijah went up by a whirlwind into heaven.
[King (2) 2.11].

According to the oldest sacred script of the Hindu, the revelations of *Rigveda*, the embryo lies in the depth of the primeval vortex. A similar idea is found in the tales of the Zuni Indians: The world mother created life by stirring water in an earthen bowl [1]. In the Minoan culture spiral motifs are widespread. The Cretan labyrinth of King Minos is a spiral symbol for the entrance to the underworld [5]. Decorations on the bathtub at the palace of Knossos on Crete depict cuttlefish and spirals (ca. 1400 BC), but whether these ornaments had a cultic meaning is not known.

In contrast, the written documents from ancient Greece have a unique meaning. The oldest transmitted description of a tidal vortex is given by Homer (end of 8th century BC). The returning heroes of the *Odyssey* had to face the danger of the giant whirlpool Charybdis [Reprinted from Homer: *The Odyssey*, translated by E. V. Rieu (Penguin Classics, 1946), p. 195. Copyright © the Estate of E. V. Rieu, 1946. By permission of Penguin Books Ltd.]:

For on the one side we had Scylla, while on the other the mysterious Charybdis sucked down the salt sea water in her dreadful way. When she vomited it up, she was stirred to her depths and seethed over like a cauldron on a blazing fire; and the spray she flung on high rained down on the tops of the crags at either side. But when she swallowed the salt water down, the whole interior of her vortex was exposed, the rocks re-echoed to her fearful roar, and the dark sands of the sea bottom came into view.

Hesiod (ca. 700 BC) narrates in his *Theogony* how the heaven of the gods and the earth were created. It goes like this:

To Ocean Tethys brought the rivers forth
in whirlpool waters roll'd Eridanus
deep-eddied, and Alpheus, and the Nile:
and the divine Scamander.

The whirlpool Eridanus, by the way, was considered to be one which flowed through the underworld.[4] A similar circular flow is mentioned in Plato's *Phaidon*[5] as the mechanism for reincarnation: "After winding (the river Pyriphlegeton) about many times underground, it flows into Tartarus at a lower level." Aphrodite, the "sea-nourished" goddess of love, originated in whirling water.

According to a Finnish legend, the whirlpool goes through the whole globe [6]. The Polynesians also thought the whirlpool was the entrance to the world of the dead. The Maori of New Zealand believe that the souls of the dead go to heaven in a whirlwind [1]. There are tribes in India that perform dances on spiral paths because demons use such roads [7].

In China and Japan whirlwinds and whirlpools are related to dragons, and the spiral is the symbol for the curled dragon. According to a Chinese legend a dragonlike serpent-god rules the universe in the form of a celestial vortex. In the Chinese Yin–Yang doctrine, which assumes two primeval forces acting on each other, the spiral is closely related to Yang, the source of life and energy. The straits of Naruto in Japan, in which probably the largest tidal vortices on earth occur, are considered as the "eastern gate of the dragon palace." A Japanese serpent-god causes whirlwinds. This idea can also be found among the Sumerians, who probably venerated a goddess of whirlpool: Is-ḫa-ra of Ur. She was connected with the sea serpent [1].

In the Celtic and Teutonic cultures, vortex motifs and legends about vortices are also widespread. They too believed that life was created in the whirlpool. The pagan spiral motif for life has been adopted even in the Christian religion. Figure 1.2 shows a Scottish cross on which the figure of the crucified as a symbol of new life is replaced by a spiral. In Scandinavian mythology, the starry sky is turned by the "world mill" (this expression refers to the importance of the corn mill in the early period of agriculture). The Maelstrom also originated in the world mill. This connection between cosmic whirl and the Maelstrom is elucidated by de Santillana and von Dechend in a revealing book called *Hamlet's Mill* [6]. According to these authors the mythical origin of the Malestrom is related to the precession of the equinoxes,[6] which defines a world age. In the old Norse poetry of the *Edda*, one can read that the world mill ground gold, peace, and happiness in the "golden age." Later, the mill was grinding salt one day when the millstone unhinged and sank into the ocean. Water came out of the hole

Figure 1.2. Scottish cross with spiral as a symbol for life (after Stuart in Reference 1).

of the millstone and formed the Maelstrom. That is how the ocean became salt. In Reference 1 is cited from Rydberg: "Of the mill it is said that it is dangerous to men, dangerous to fleets and to crews, and that it causes the Maelstrom (svelgr) when the water of the ocean rushes down through the eye of the millstone." The Maelstrom goes into earth, an idea which previously cited examples show to be worldwide. Another name for Malestrom is "Navel of the sea" [6].

Spirals that represent vortices without cultic or mythological meaning are quite rare. Interesting examples are found on reliefs of the neo-Assyrian era (11th through 7th centuries BC). These representations are very realistic and their details are astonishing. However, even these reliefs cannot be considered as *l'art pour l'art*. They were made to glorify the might of the king; and this might was received from the gods. Figure 1.3 shows a relief from the 9th century BC. In addition to vortex spirals in the river, one can see a spiral at the corner of the

Figure 1.3. Assyrian relief from the palace of King Ashur-nasir-pal II in Nimrod, 9th century BC. Corner vortex at the front of the fortress (by courtesy of the Trustees of the British Museum).

fortress. This spiral is probably the oldest likeness of a corner vortex [8].

1.2. ANTIQUITY AND THE MIDDLE AGES

All things come into being by necessity, the cause of the coming into being of all things being the vortex, which Democritus calls "law of nature."
Diogenes Laertius

One way to achieve human orientation in the world is to draw conclusions from local observations and apply them to distant phenomena. This procedure was used in mythological times to unravel the secrets of the universe through anthropomorphic analogies. As an example from Greek culture Hesiod, in his *Theogony,* describes the gods as humanized beings, invested with power and immortality, who determined the events in the world.

A decisive turn in thinking occurred in Greek Asia Minor during the 7th and 6th centuries BC, when the Ionic natural philosophers expressed a principally new idea: They replaced the anthropomorphic analogies with impersonal and mechanically based arguments [9]. Everyday observations of the movement of bodies in nature and at home were exploited for discussing profound philosophical problems. The central theme was adopted from Hesiod: the question of the origin of the universe and of the world order. However, the way gods acted was no longer used an an explaining principle. Instead, physically oriented questions were asked: What are the elements the world is composed of? How was the cosmos created out of chaos, or more precisely, how were the elements separated from the chaotic primeval state and formed into a world order? The formulation of these problems (more than the answers proper) is one of the great achievements of the early Greek thinkers and may be considered the birth of science.

Of the Ionic cities, which had a close contact with the cultures of the Near East and were open to new ideas, Miletus was the most famous. The names of three Milesians are known: Thales (624–546), Anaximander (610–545), and Anaximenes (585–528). They believed that only one simple primary substance exists in the world and everything else originates from it. Actually, from Thales it is only known that he considered water as the primary substance of all being. Anaximander abstracted the concept of the primary substance to something that is not a special substance, but a general and undetermined one *(apeiron).* He made the transition from the perceptual to the conceptual. But Anaximenes had already abandoned this idea and proposed air as the primary substance.

Anaximander pioneered in answering the question of how matter was separated at the beginning of the world [9]. The explaining principle was the vortex (Greek: δύνη or δῖνος). This thought was derived without doubt from the mythological concept of the "cosmic whirl." Anaximander believed in an evolution of the universe and in the decisive role of the vortex in transporting heavier material into the vortex center to form the earth there. The lighter matter would settle around the earth as ocean and sky. This idea was taken up by Anaximenes and later by Anaxagoras (500–428), Empedocles (492–

432), and Democritus (460–370). However, except for Empedocles and Democritus, very little is known about the teachings of these men.

The opinions of the philosophers diverged with regard to the form and size of the earth relative to the sky and with regard to the way the earth is held in place in the cosmos. Anaximander considered a cylindrical earth around which lighter rings of matter revolved like wheels. Anaxagoras believed that the earth is held in the center of the cosmic whirl through its disklike form and that it hovers there like a lid over a pillow of air. The radius of the earth-disk was assumed to be almost as large as that of the firmament [10].

The opinion of Empedocles, who lived in Sicily, differed considerably from that of the Milesians. He did not believe in a single primary substance but in four: earth, water, air, and fire. He also recognized that matter does not move by itself, but requires an impulse, something which today would be called a force. He distinguished between attracting and repelling "forces" called "love" and "strife." With these concepts he explained vortex motions, of which he discussed two types. The first may be related to what is called in modern terminology the centrifugal effect. Empedocles demonstrated this effect with a revolving, liquid-filled ladle. During the rotating motion around the end of the ladle the liquid remains inside the ladle without spilling (Fig. 1.4). Empedocles used this effect to explain the positions of the celestial bodies in the firmament. The second type of vortex motion he inferred from the observation of whirlwinds and of liquids stirred in a container. When a fluid rotates over a solid surface, a secondary circulation in the direction of the center is generated because of the adherence of the fluid at the surface (Fig. 1.5). Solid particles within the fluid migrate toward the center and accumulate there as long as they are heavy enough to withstand the updraft. This observation is called today the "tea-cup phenomenon." Tea leaves collect in the center of a cup when the liquid is stirred. The ancient Greeks observed that sediments were deposited when wine was stirred. Empedocles assumed that near the bottom "love" would bring the liquid together; at the surface, however, "strife" would separate it. Aristotle generalized this property of rotating fluids in the sense that heavier particles always tend to migrate toward the center of rotation (even without a solid bottom). This conclusion is not correct. Empedocles assumed that the earth was formed in the center of the primeval vortex by means of the "tea-

Figure 1.4. Bronze ladle from Greece.

cup effect." He also believed that the earth was much smaller than the firmament. Thus he rejected Anaxagoras' "lid theory." Empedocles thought that the earth hovers in the updraft of the primeval vortex [11].

Anaxagoras, a contemporary of Empedocles, did not believe in the structure of matter based on the four elements. Rather, he thought of a kind of continuum. In his view, matter does not change, regardless of how small it is cut. A piece of bone crushed to powder remains bone. Anaxagoras envisioned the driving force as being in the mind, an idea which differs from the mechanistic perception of Empedocles. Of importance, moreover, is the larger role which Anaxagoras attributes to the vortex concept. While the vortex concept so far had been restricted to cosmological events (although the tendency toward a generalization is clearly recognizable in Empedocles), Anaxagoras used the vortex as the explaining principle for the total world process [9].

End and climax in the presocratic development of the vortex concept were reached in the teachings of the first atomists, Leucippus and his student Democritus. They considered matter to be composed of very small, indivisible particles, and they interpreted the accumulation and order of those "atoms" by means of vortices. This idea received its greatest attention in the 19th century (Section 1.5). In contrast to Anaxagoras, whose "mind" controls the rotation of matter, Democritus assumed that the unordered motion of the atoms is changed to an or-

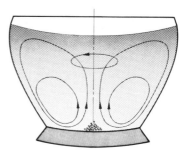

Figure 1.5. Meridional flow due to stirring of a fluid in a container. Solid particles, which are heavier than the fluid, migrate toward the center of the bottom and accumulate there.

dered one by a random process (if one waits long enough). Democritus considered the vortex motion so basic that he saw in it simply the general law of nature. He equated the vortex concept δίνη to the law of nature ἀνάγκη (see the citation at the beginning of Section 1.2). This was probably the first attempt in the history of science to formulate a "unified theory of physics."

The dominating role of the vortex concept was evident not only in the mechanics of the micro- and macrocosmos, but also in medicine. According to Theophrastus, the most eminent student of Aristotle, an attack of vertigo is caused by a disturbance of the equilibrium in the head. This disturbance is generated by vortical motion. One has only to turn around several times to observe this effect. Also, according to the famous Greek physician Hippocrates, a contemporary of Democritus, bladder stones are formed by deposits from unclean urine. The explanation for this was again the tea-cup phenomenon, Fig. 1.5 [9].

The change from the mythological conception of the world to the mechanistic view during the Ionic enlightenment also included a critique of the Olympian gods. Since physical arguments questioned the power of the gods, even their very existence became questionable. At least their part in the creation and order of the world was restricted. However, the Ionic philosophers themselves were still so embedded in the mythological tradition that in the new teachings they did not see a conflict with the belief in gods but a wonderful complement.

This situation changed 100 years later. Empedocles and the atomists Leucippus and Democritus believed merely in a pure mechanism. Here, without doubt, the roots of today's ideological materialism must be sought. The dominance of this idea at the end of the 4th century BC can be read in Aristophanes' comedy *The Clouds* from 423 BC: "The Whirlwind has driven out Zeus and is king now." This spirit of the age included also the critique of humans. Athens, in the meantime, had become the spiritual center of Greece (after the fall of the cities in Asia Minor to the Persians). Here, the sophists taught from about 450 BC. They questioned the ability of human perception and the validity of social norms. The role of the sophists in history has been tarnished because of their sharp rejection by Socrates (469–399 BC) and Plato (427–347 BC). The great sophist Protagoras (480–411 BC) taught that the sensual perceptions of humans are the only source of knowledge. This idea has played an important role in natural philosophy until the present time. Unfortunately, sophism was misapplied to develop frivolous fictitious arguments, which discredited the whole teaching.

The reaction of the establishment to the critique of gods, morals, and society did not fail to appear. Every pursuit of natural philosophy was summarily considered atheism by the Athenian authorities and was persecuted. The most prominent victim of this early inquisition was Socrates (for which, by the way, Aristophanes was not completely innocent). Unjustly, he was accused of atheism and of corrupting the youth, condemned to death, and in 399 BC took the poisoned cup.

Plato's idealism was the philosophical answer to the materialism of Empedocles and Democritus [12]. The vortex theory of the micro- and macrocosmos, which contained an element of chance, was rejected in favor of a teleological view by Plato and his disciple Aristotle (384–322 BC). Although Epicurus took over the ideas of the first atomists on the atomic structure of matter, he rejected the role of vortices in the interplay of the atoms. The Roman natural philosopher and poet Lucretius (100–55 BC), who was strongly influenced by Epicurus, does not mention vortices at all in his description of the atoms.

Aristotle, whose writings have had a tremendous influence on western culture, did not share (as mentioned before) the opinions of the presocratic natural philosophers nor did he accept the Pythagorean belief in an oscillating universe. Influenced by Plato's *Timaeus* he envisioned a world that consisted of two parts: the space beyond the moon, which was in an eternal unchanged state, and the sublunar space, which was subject to changes. The heavenly bodies in the first space moved in circular orbits because they were considered perfect and eternal[7] (Pythagorean influence) [13]. Aristotle was probably the first to use a minimum principle as an argument for the simplicity and perfection of the circle [14] (The circle has the shortest perimeter for a given area). This idea was certainly pleasing, although additional assumptions in the form of composed circular orbits had to be made to explain the epicyclic motion of the planets. Aristotle thought that immaterial spirits maintained the motion of the celestial bodies. Later during the scholastic period this belief was abandoned by Buridan and replaced by his impetus theory (ca. 1350 AD) [15]. However, scientists held to the idea of circular orbits until Kepler's discovery of the elliptic planetary orbits in 1609.

In addition to speculative philosophical work on cosmology, Aristotle also dealt with purely physical problems. He was probably the first to separate the study of natural events from their mythological and philosophical aspects. In his book *Meteorologica* and in the pseudo-Aristotelian scripts *Problemata*, *Mechanica*, and *De Mundo*, vortices are described for their own sake. The cause, occurrence, and motion of whirlwinds and tidal vortices are discussed:

A whirlwind thus arises when a storm that has been produced is unable to free itself from the cloud: it is caused by the resistance of the vortex, and occurs when the spiral sinks to the earth and carries with it the cloud from which it is unable to free itself.
(Meteorologica 371a9, The Loeb Classical Library)

Why is it that sometimes craft traveling on the sea in fine weather are sunk and completely disappear, so that no wreckage even comes to the surface? Is it because when a spot in the ground underneath the sea breaks and forms a hollow, they follow the movement of the air down into the sea and into the hollow? Similarly the sea traveling in a circle in every direction is carried below. This is a whirlpool.
(Problemata XXIII, 5, The Loeb Classical Library)

The Roman authors followed the Greek tradition. Descriptions of meteorological vortices by Lucretius, Seneca (0–65 AD), and Pliny (23–79 AD) have been handed down. Seneca tells about the cause of whirlpools:

At this point, if you like, we may ask why a whirlwind occurs. In the case of rivers it usually happens that as long as the rivers move along without obstruction their channels are uniform and straight. When they run into some boulder projecting on the side of the bank they are forced back and their waters twist in a circle with no way out. Thus they are swirled around, sucked into themselves, and form a whirlpool.
(Seneca, Natural Questions V,13,1, The Loeb Classical Library)

Other Latin sources, which do not deal expressly with meteorological phenomena, also reveal a knowledge of whirlwinds. A remark from Lucan (39–65 AD) on dust devils is cited in Section 10.4.

In the Middle Ages the ideas of antiquity were adopted, partly just as they were, partly reviewed critically. For example, Dante's tale of Charybdis in the seventh book of the *Inferno* is a story which goes back through Virgil to Homer. Dante's Odysseus, however, meets a more dramatic end than that encountered in the original: Odysseus goes to hell through a whirlpool to be punished for his alleged evil deeds in the Trojan war.

In the late Middle Ages a critical examination of the teachings of Aristotle took place. During this process new ideas were developed that led directly into the Renaissance. First steps toward a scientific terminology are discernible [16]. In this book the following partial aspects are of interest [17]:

1. The kinematics of rotating bodies was taken up by Gerard of Brussels (13th century) in his *Liber de motu*. Parts that are farther away from the center of rotation move faster than those closer to the center. He apparently thought of solid-body rotation.

2. Buridan rejected the teachings of Aristotle on the dynamics of moving bodies. Aristotle taught that a body thrown into the air is carried forward by the air itself, since air had to flow in the rear of the body in the direction of motion because there could not be a vacuum. Buridan offered instead his impetus theory. He claimed that the body is driven by the force that acts on it (impetus or vis motiva), since air does not propel but resists the movement. This idea was also applied by Buridan, as mentioned before, to the circular motion of translunar heavenly bodies, which according to Aristotle are propelled by immaterial beings. (Buridan, however, was not the first to question the Aristotelian theory. He adopted the criticism from Avicenna and he again from Philoponus [18].) At this point the "questions on the four books of the Aristotelian *Meteorologica*" may be mentioned which Themon, the "Son of the Jew" (ca. 1349–1361) has asked. In the tenth question of the second book is cited: "We investigate whether a whirlwind or cyclone is generated by hot and dry exhalation." Unfortunately, no answer is given in Reference 17.

3. Nicole Oresme (ca. 1325–1382) studied the commensurability of rotating celestial bodies and its cosmological consequences. In other words: Is the ratio of the periods of two celestial bodies a ra-

10 Historical Survey

tional number or not? Oresme distinguished between "circulation" and "revolution." The return of a single body along a circular orbit from an arbitrary point to the same point is a circulation. The return of several bodies from an arbitrary initial state to exactly the same state (commensurable) or to a similar state (incommensurable) is a revolution. In antiquity and the Middle Ages it was generally assumed that revolution is commensurable (Pythagoras' great world year, see Plato, *Timaeus* 39D). In his work *De commensurabilitate vel incommensurabilitate motuum celi*, Oresme probably assumed (he left the question open) that the celestial motions were incommensurable, for only with this assumption could he discredit astrology [17] (see also Reference 19).

1.3. THE RENAISSANCE

Since earliest times humans have used forms of art to express their response to experience and the essence of their ideas. These forms include perception of vortices. In most documents of antiquity and the Middle Ages, however, vortices played only a secondary role in the sense that they were not considered for their own sake. In these records, a vortex may have been used as (a) a symbol for mythological and philosophical ideas (such as the spiral motif and the idea of the cosmic whirl), (b) part of a tale or a picture and only loosely connected with the theme of the work proper (Homer's description of Charybdis in the *Odyssey*), or (c) an artistic subject in itself (Dante's whirlpools and tornadoes to illustrate the horrors of the *Inferno*).

It has already been pointed out in Section 1.2 that Aristotle was probably the first to describe natural events for their own sake, among them vortices. In the Roman era Lucretius, Seneca, and Pliny produced descriptions of nature. The climax, however, in the use of art as a means for describing vortex motion was reached in the Renaissance. Here, the undisputed master was Leonardo da Vinci (1452–1519).

Leonardo's work represents a link between the Middle Ages and modern times. He had outgrown the thinking of late Scholasticism but did not yet have available the powerful resources of rational mechanics and mathematics; nor had the idea of scientific experiments fully matured during his lifetime. But the artistic description of nature, which has in common with scientific experiment the sin-

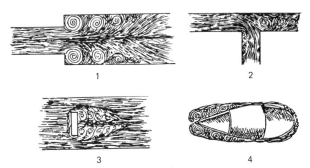

Figure 1.6. Sketches by Leonardo (from *Handbuch der Experimentalphysik* [20]).

gling out of a particular phenomenon from a total happening in nature, reached its culmination in Leonardo.

Even in his early years, Leonardo perceived the wave and the vortex as the manifestation of motion and power, and he applied wavy and spiral lines as stylistic elements in his paintings.[8] With advancing age he turned more and more to the scientific mastery of these forms of motion. His sketch pads and manuscripts testify to his deep understanding of relationships in nature, which included vortex motion. He was the first to describe turbulent motion, to recognize the difference between potential vortex and solid-body rotation, and to study vortical motions in channels and in the wake of obstacles. Figure 1.6 shows some sketches from Reference 20, borrowed from a selection of manuscripts republished in 1923 by Carusi and Favaro [21]. The first three figures are from the fourth book of this edition under the title *De retrosi dell' acqua* (on vortices in water); the fourth figure is from the seventh book.

Leonardo also applied his knowledge of vortical motion to anatomical studies [22]. Some years ago scientists confirmed Leonardo's theory that vortices inside the aortic valve [23] are essential for the control and efficiency of this valve (Fig. 1.7). Leonardo discovered this relation without knowing that blood circulates.

At the end of his life Leonardo revived the age-old belief in the cosmic whirl. The more intensively he studied natural science, geology in particular, the more he recognized how insignificant and helpless humans are within the happenings of nature. Such thoughts might have caused his tendency to brooding and melancholy (Fig. 1.8). They certainly led to his mystic and apocalyptic perception of the destructive power of nature. Both the beginning and the end of the world appear to him delineated by natural catastrophes of cosmic dimensions. How

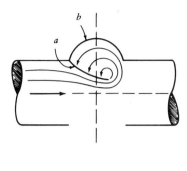

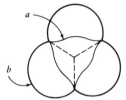

Figure 1.7. Schematic presentation of the flow in the aortic valve. The heart generates a pulsating stream of blood in the aorta. The return flow of the blood is prevented by means of three leaflets (*a*). Development and decay of the vortex inside the three protrusions (*b*) guarantee the uniform and effective work of the valve. The leaflets open and close once a second and have to last a whole human life. The function of the protrusions and the task of the vortex were interpreted correctly by Leonardo (adapted from Bellhouse and Talbot, *Journal of Fluid Mechanics,* Cambridge University Press [23]).

otherwise could he describe the primordial power of the Deluge in the Old Testament as the vision of a vortex of water that destroys everything (Fig. 1.9).

It is probably no coincidence that the same idea about the diluvial vortex was used by another painter of the Renaissance. Dürer, a contemporary of Leonardo, was influenced by a nightmare he experienced in 1525 to draw a landscape which is threatened by a large column of water. Whirlwinds in the distance forebode the disaster [24].

1.4. THE CARTESIAN VORTEX THEORY

... que tous les mouvements qui se font au Monde sont en quelque façon circulaire: c'est à dire que, quand un corps quitte sa place, il entre toujours en celle d'un autre, & celui-ci en celle d'un autre, & ainsi de suite jusques au dernier, qui occupe au même instant le lieu délaissé par le premier.
                           Descartes, Le Monde

During the 16th and 17th centuries the Aristotelian–scholastic world view was shaken by the new ideas of Copernicus, Galileo, and Kepler. Coperni-

Figure 1.8. Leonardo: *Old man and Vortices;* probably a self-portrait (Windsor Castle, Royal Library, copyright reserved).

Figure 1.9. Leonardo: *The Deluge* (Windsor Castle, Royal Library, copyright reserved).

cus moved the center of the universe from the earth to the sun. Galileo discovered the law of free fall, which contributed to the clarification of the concept of inertia. Finally, Kepler found that the orbits of the planets are not circular, but elliptic. With these discoveries the traditional conceptions had to be completely changed. When the center of the universe was moved from the earth to the sun, the Aristotelian partition of heavenly and earthly regions lost its meaning. When Galileo saw a rough lunar surface with the aid of his newly developed telescope, Aristotle's view that the heavenly bodies were mathematically exact spheres could be maintained no longer. However, Galileo still was convinced that the orbits of the planets were circular. On the basis of his experiments about the free fall of bodies, he developed the idea that the circle is an inertial path. Kepler's discovery of the elliptic orbit made it necessary to look for a new explanation of the planetary motions. Although Kepler's laws of the planetary orbits are correct, he could not detach himself from the idea of a circular orbit as a natural path, and he imagined the elliptic orbit as a perturbation of the circle. Kepler believed that the "species immateriata" kept the planets on their circular paths, but, perpendicular to it, a disturbance in the form of magnetic attraction and repulsion caused the "elliptic" orbits. Kepler believed that a magnetic vortex caused by the rotation of the sun was the reason for the disturbance. This idea already contained the essence of the Cartesian vortex theory [25].

Modern natural philosophy, based on these new ideas of the 16th century, starts with Descartes (1596–1650). He developed on the foundations of mathematics a unified philosophical system, whose significance for modern times can hardly be overestimated. This tremendous work, published in 1644, also contains a general theory of physical phenomena, Descartes's famous vortex theory. This work triggered a struggle among the most illustrious intellects of that time, a struggle which was to last for more than 100 years [25, 26].

The basic idea of the Cartesian vortex theory rests on the assumption that matter has extension and is identical with space. Consequently, there is no vacuum and bodies can interact with each other only by direct contact. If a body moves, the surrounding fluid particles are induced to a circular motion. Although this conception is correct for a continuum (Section 2.2), Descartes also applied the idea to celestial bodies, which he thought were suspended in etheric vortices. In addition, the dynamical aspects of his theory—the transfer of force by means of pressure and impact—are not correct and had to collide sooner or later with Newton's gravitational theory. For instance, the Cartesian vortices could not describe the elliptic orbits of the planets, which are derived from the Keplerian laws. Descartes' successors modified and improved the vortex theory. For instance, Huygens in 1669 tried to explain gravitation with the vortex theory. In 1689 Leibniz developed a complicated model, the harmonic vortex theory, to include Kepler's elliptic orbits.

In 1687 Newton published the *Principia*, in which he advocated the theory that material particles interact with each other over a distance by attraction. This theory did not remain unchallenged. The idea that forces act over a certain distance was difficult to comprehend for scientists at that time, all the more so as Newton believed he needed divine intervention to keep the heavenly bodies in motion. Scientists like Leibniz, Huygens, and the Bernoullis, therefore, generally accepted the basic Cartesian vortex concept, although improvements appeared to be necessary. Another grave objection was raised against Newton's postulate of an "absolute space" with the property of being able to accelerate a body without the presence of other matter. A contemporary of Newton, Berkeley, argued in 1710 that the movement of a body makes sense only in relation to another body, and that absolute space is an unacceptable abstraction (Section 7.1). Berkeley's ideas were shared by Leibniz and were later taken up by Mach and Einstein. However, to this day the problem has still not been entirely resolved [27].

About the middle of the 18th century, after a short period of reconciliation, the struggle was lost for the Cartesians. The success of Newtonian mechanics in describing planetary motion was overwhelming. Another consequence of the Cartesian vortex theory was that the earth should be flatter at the equator than at the poles, whereas the opposite should be true according to Newton's gravitational theory. Maupertuis' famous Lapland expedition in 1736–1737 also resolved this controversy in favor of Newton.

1.5. THE ERA OF CLASSICAL MECHANICS

Based on Newton's theory, mechanics was the first physical theory that was studied systematically. In

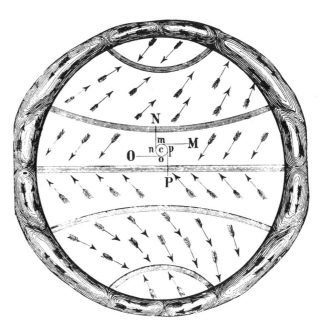

Figure 1.10. Ferrel's model of the general circulation from 1856 (from E. N. Lorenz [28]).

this environment bold ideas also surfaced on the interpretation of natural events that are much more complicated than planetary motions. In 1735 Hadley published his theory on the general circulation of the atmosphere. Although his model must be rejected from today's point of view, it is considered the beginning of a long and painful struggle to develop a satisfactory explanation of atmospheric motions [28]. Figure 1.10 shows a sketch by Ferrel from 1856. In his model Ferrel applied for the first time the force in a rotating system, which was discovered by Coriolis in 1835 and named after him. Shortly after Hadley, in 1749, Boscovich published the first comprehensive study on tornadoes [29]. A far-reaching hypothesis on the origin of the solar system was put forward by Kant (1755) and Laplace (1796). According to this hypothesis, the sun and planets developed out of a rotating gaseous cloud, an idea which is still basically accepted today (Section 12.2). Here, the age-old idea of the "cosmic whirl" is resurrected within the framework of Newtonian mechanics. In contrast to these grandiose speculations, laboratory experiments on vortices were modest, more in the spirit of rococo. Modifications of the tea-cup experiment (Fig. 1.5) were made by Saulmon in 1714 to study gravitation and by Wilcke in 1780 to investigate atmospheric vortices (Fig. 1.11) [30].

In this period, in 1786, the hydraulic engineer DuBuat gave the first explicit evidence that a fluid

Figure 1.11. Wilcke's apparatus from 1780 to produce a vortex.

adheres to a solid surface (nonslip condition). This phenomenon, which is so important in understanding flow separation and the formation of vortices (Chapter 5), was subsequently the subject of a century-long investigation of the nature of nonslip and the possible existence of slip [31, 32]. Other hydraulicians were less successful in lifting the veil from the secret of flow separation. Figure 1.12 shows a sketch from Venturoli's book *Elementi d' idraulica*, first printed in 1807 [33] that reveals a misconception (provided the lines are interpreted as streak lines, otherwise, as streamlines, they would contain even more errors): The separating streak lines in front of the plate are incorrectly drawn. The lines should branch out on the plate (Figs. 2.15 and 5.1).

The slight condescension that is sometimes connected with the word rococo is certainly not justified for the development of the rigorous mathematical and mechanical disciplines during the 18th century. Indeed, it was a heroic age for mechanics, and the names of the giants like Leibniz, d'Alembert, Bernoulli, Euler, Lagrange, and Laplace speak for themselves. The Cartesian and Newtonian systems,

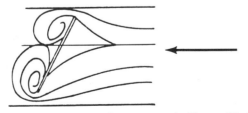

Figure 1.12. Erroneous flow patterns in Venturoli's book from 1807.

however, were not suited to the description of fluid motions. Euler deserves credit for having laid the foundation of fluid dynamics [34]. He derived the equations of motion for an inviscid fluid that today bear his name. He also was the first—together with d'Alembert—to have used a mathematical term that later became known as "vorticity vector." The kinematic significance of this important concept was first recognized by Cauchy in 1841 and Stokes in 1845. Stokes and a few years earlier Navier in 1827, Poisson in 1831, and St. Venant in 1843 derived under various hypotheses the viscous term in the equations of motion that are called today "Navier–Stokes equations" [35].

In 1858 Helmholtz derived the famous vorticity theorems for the vorticity field of an inviscid fluid. These theorems say in essence that vorticity in an inviscid fluid can neither be generated nor destroyed. Eleven years later Lord Kelvin formulated another important version of Helmholtz's vorticity laws in the form of the circulation theorem named after him. These discoveries laid the basis for a modern vortex theory. They also inspired the last philosophical attempt (after the Cartesian vortex theory had failed) to use the vortex concept for a unified physical theory within the realm of the microcosmos.

In 1861 Maxwell used vortices as a mechanical model for his electromagnetic theory. He justified his idea with the argument that magnetic fields are rotatory by nature. In this argument, magnetic fields can be replaced by electric fields, an idea which was brought up by Boltzmann in 1891 [26].

In 1867 Lord Kelvin developed a theory that the properties of atoms and molecules may be interpreted by vortex rings. He imagined knotted and linked vortex rings, and the variety of combinations would reflect the multitude of the elements and compounds (Fig. 1.13). Such vortex rings can vibrate and can have kinetic energy just as atoms and molecules do. Details and changes in this theory were made by Tait, Thomson (the discoverer of the electron), and Hicks. Pearson and FitzGerald tried to develop a theory of ether with the aid of vortices [26, 36–38].

Toward the end of the century Lord Kelvin had to admit that the natural fate of vortices was instability and decay. These properties are incompatible with an atomic model. Thus, with the advent of modern physics died the last dream of interpreting the structure of the physical world with a mechanical vortex model.[9] One would, however, do injustice

Figure 1.13. Lord Kelvin's knotted and linked vortex rings as mechanical models for atoms and molecules (from W. Thomson [36]). These complicated vortices so far have never been realized experimentally (Section 3.6).

to the past to regard the tremendous effort to develop a unified vortex theory as a mere episode in the history of science. The ideas and labors of these scientists were important stepping stones to today's accumulated knowledge of humankind (The historian Ed. Meyer remarked: "Historical is what is or has been efficacious.").

1.6. SIGNIFICANCE OF THE VORTEX CONCEPT IN THE 20TH CENTURY

Vortices are ordered structures of fluid motion, which nature prefers over chaos in many situations.

Although the ambitious philosophical attempt to formulate a unified vortex theory failed, the development of the vortex concept and its engineering applications within the framework of classical mechanics have had great success. Its effects on fluid dynamics are so numerous that they can only be suggested in this historical survey. Descriptions of various types of vortices and their properties are the substance of this book.

Over the years the vortex theorems of Helmholtz, which are valid only for incompressible, ideal fluid flow, have been generalized by Lord Kelvin (1869), Bjerknes (1898), Crocco (1937), Ertel (1942), Oswatitsch (1943), and Vazsonyi (1945) [39]. Rotating fluid systems are vortices by definition. Here, the formulation of the Taylor–Proudman theorem was an important theoretical contribution [40].

The study of various vortex configurations, especially those named for Bénard, von Kármán, G. I. Taylor, and Görtler, was helpful in understanding instability, which was first investigated by Helmholtz, Lord Rayleigh, and Lord Kelvin. Since instability leads in most cases to turbulence, the model of "spectral eddy distribution" is useful here too. Early research on turbulence is connected with the names of Reynolds, Boussinesq, Prandtl, G. I. Taylor, Heisenberg, and Kolmogorov [41, 42]. A basic theory on turbulence, however, is still missing. More recently, the work on this subject has received a new impetus by the recognition of the existence of coherent structures in the form of "large-scale eddies" [43]. Prigogine [44] explained the preference of such ordered structures in nature over a chaotic state in many situations by means of nonequilibrium thermodynamics.

In aero- and hydrodynamics the theory of lift (based on the bound-vortex concept) by Lanchester, Prandtl, Kutta, and Joukowsky initiated the rigorous development of airplanes, propellers, and turbines [45, 46]. Between 1904 and 1918 Riabouchinsky [47] performed fundamental experiments on vortices and rotating plates in his private laboratory at Koutchino. Other examples of the application of the vortex concept include dust cyclones, combustion chambers, and centrifuges in the chemical and nuclear industries, and rotating gauge and control elements in cybernetics [48, 49].

Solutions of the basic equations in fluid dynamics, which describe vortices, have been sought for since the formulation of the Navier–Stokes equations [50]. After World War II a new chapter in the history of fluid dynamics was opened with the development of electronic computers. For the first time it became possible to solve numerically the basic equations of motion for certain difficult problems. The vorticity-transport equation was the key to understanding unsteady flows [51].

In meteorology Bjerknes' circulation theorem and Rossby's solution of the (simplified) vorticity-transport equation became the basis for theoretical weather prediction. In 1926 Jeffreys discovered that knowledge of the occurrence and migration of cyclonic and anticyclonic air masses in midlatitudes is essential in understanding global atmospheric circulation [28].

In acoustics the theory of aerodynamic sound generation is based on vortical motion in an unsteady flow. In 1908 Bénard related the vortex concept to sound for the first time. More recent research was carried out by Yudin in 1944 and by Lighthill and Powell [52, 53]. Closely related to vortical sound generation is vortex-induced vibration of bodies [54].

In astronomy the development of the solar system and the galaxies, the motion of planetary atmospheres, and the convection flows inside stars are more or less unsolved vortex problems [55, 56]. Hubble identified spiral nebulas as galaxies in the mid-1920s, and in 1943 von Weizsäcker developed a vortex theory for the generation of the planetary system. During the last few decades studies have been made on exotic vortex stars whose physical properties are scarcely imaginable. Neutron stars with incredible angular velocity have been postulated, and it is believed that they have been discovered. Still more fantastic are the characteristics of rotating black holes, whose description requires the theory of relativity and quantum mechanics (Section 12.5) [57]. The spacecraft Voyagers 1 and 2 transmitted beautiful pictures of atmospheric vortex arrays from the planet Jupiter [58].

Even in quantum theory, which replaced Lord Kelvin's theory of vortex atoms, the vortex concept has a useful role. In their study of liquid helium Landau and Feynman introduced the concept of "roton" (quantum vortex) [59]. Also, since vortices are described by solutions of the field equations [60], the topological characteristics of vortices have recently become of interest in the physics of elementary particles.

Modern science has extended human understanding of vortices and their role in nature in two ways: It has clarified the basic significance of vortical motion in mechanics, and it has extended the spatial horizon of humans so that they can observe large-scale vortices and interpret them. The ancient Greeks could see only local vortices like whirlwinds and whirlpools. Today one can follow on satellite photographs the movements of hurricanes. Space probes are sent on interplanetary missions to signal back pictures of the wonders of the planets. Astronomers use modern equipment to study the spiral structure of galaxies as they ponder the questions of when, where, and how these cosmic whirls were created in the primeval hour of the universe.

Remarks on Chapter 1

1. The spiral, as an imitative symbol, which is obtained by direct observation of nature, stands for life, evolution, and a source of energy. In particular, the spiral motif represents the fundamental process of biological creation. In Goethe's book on the metamorphosis of plants is written: "This spiral tendency, as a basic law of life...." Many

Figure 1.14. The three-spiral figure from Newgrange, Ireland, in a neolithic cemetery (ca. 3000 BC).

human and animal organs exhibit not only spiral patterns during their formation but require them for functioning in the developed stage. Examples are the intestines and the cochlea of the ear [61, 62]. The spiral structure of matter is demonstrated in DNA, which is the basic substance of all body cells and has the form of a double helix. This "creative" spiral must be distinguished from the "destructive" spiral exemplified by a whirlwind [63]. The spiral as a symbol for energy, in particular for cyclic energy, is found in such diverse places as Yucatan (Maya) and Ireland (Fig. 1.14).

2. A history of the wheel is given in Reference 64 in the context of a general outline on the significance of rotatory movements for the development of tools. One must distinguish in rotatory movement whether it occurs continuously in the same direction or whether it changes its direction alternately. The development of the latter type of rotation, as that used in drilling and fire making, occurred much earlier than the discovery of the wheel proper. The mechanism of drilling and fire making probably goes back to the to and fro rubbing movement of the human hand, and it may be as old as the oldest tool making, that is, perhaps 500,000 years. In comparison, the use of the wheel as a pottery disk and a cart wheel goes back only to about 3500–3000 BC. It was probably first developed in the Tigris valley or in southern Russia.

3. The important archeological question of whether symbols and tools spread out from a single place all over the earth (along trade routes of ores) or whether they appeared independently at different places on earth will not be discussed in this book. However, it may be mentioned that accord-

ing to Reference 64 it is unlikely that the oldest and most widespread wheel, the tripartite disk, was discovered independently in several cultural centers. The same is true for the spiral motif, which has been found more widely than any other symbol all over the world.

4. Eridanus, a constellation of stars in the southern sky, is also considered a whirlpool in the sky. The mythological connection between the astronomical and the earthly Eridanus is described in Reference 6.

5. The relation between transmigration of souls and circulation (in the sense of return) is a thought of Plato, which he adopted from the Pythagoreans and they, in turn, from the Orphics. The Orphics were probably influenced from the Orient. In any case, at that time the ideas of reincarnation were already known in Egypt and India [65].

6. The precession of the equinoxes refers to the wobbling of the earth as it spins on its axis like a spinning top. The rotation around the ecliptical pole takes about 26,000 years. This is considered a "world age." According to Reference 6, the precession was discovered in prehistoric times and is considered the factual background for cosmic myths all over the world. Of particular interest for this book is the relation between the heavenly whirlpool and the Maelstrom described briefly in Section 1.1.

7. The circle and spiral may be interpreted symbolically. The points of a circle form a curved line on which no point is favored. Since constant movement on circular orbits is periodic and purely stationary, the circle symbolizes eternal, uniform rhythm without progress. Spiral motion too is periodic, but, in contrast to the circle, the spiral has a preferred point, the center, and it is rotation without repetition. The spiral thus symbolizes growth and evolution. Visually, the different kinds of symmetry that occur in the circle and spiral give the illusion of static and dynamic characteristics, respectively. Figures with various symmetry properties are exemplified in Fig. 1.15. The flywheel with its sixfold axial symmetry gives the impression of rotation, whereas the snowflake

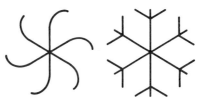

Figure 1.15. Flywheels and snowflakes are examples of figures which give the illusion of dynamic and static properties.

with the additional sixfold plane symmetry appears static [66]. The circle has infinitely many symmetry planes; the spiral has none.

8. Van Gogh expressed his inner strife and turmoil in landscape paintings in which the whirling and pulsating element of nature prevails. These paintings were made in the last year of his life: *Starry night* (1889) and *Street with Cypresses* (1890) are special examples. In modern art the vortex motif is met frequently; for instance, *Whirlpool* by Dali and Escher, *Wave of Naruto* by Hiroshige, *The Onyx of Electra* by Matta, *Inland Sea* by Pasmore, and *Vortex* by Korpi. The vortex concept has also penetrated music where it is most clearly visible in the composition form of the circle canon. The most obvious analogy, which is even documented by name, is found in the short, highly dramatic, mostly fugued choral sections in the passions by J. S. Bach, which express the fury and indignation of agitated crowds and were designated "turbae" because of their whirling process. A spiritual essay in vortex motions with many beautiful photographs was written by Schwenk [67].

9. The idea of attributing to vortices a general role in nature was taken up once more in 1936 by Krafft [68]. He developed the working hypothesis that life is a vortex phenomenon. The process of life is primarily the generation and maintenance of protoplasmic vortices. This idea reminds one of the remark of the German poet Goethe in his *Maximen und Reflexionen:* "The highest, what we have received from God and nature, is life, the rotating motion of the monas around itself, which knows neither rest nor calm." See also Remark 1 on the spiral structure of living substances.

2. Basic Concepts and Kinematic Considerations

2.1. TWO DEFINITIONS

What is a vortex? The answer to this question is neither simple nor unique [1]. Without any doubt, vortex motion is connected with rotating matter. However, a distinction must be made between a single particle rotating around its axis and one revolving with many other particles about a common center. For the time being, the rotation of individual particles around their axes will not be considered. Rather their paths will be compared with those of neighboring particles, and the following definition will then apply:

Definition 1
   A vortex is the rotating motion of a multitude of material particles around a common center.

The paths of the individual particles do not have to be circular, but may also be asymmetrical as shown in Fig. 2.1.

The representation of vortices in a plane such as those in Fig. 2.1 is justified if the vortices are cylindrical (Fig. 2.2). Then the paths are the same in every plane normal to the axis of rotation, and it is sufficient to draw only one plane. Cylindrical vortices, thus, are simply called "plane vortices." Most vortices in nature, however, have a "spatial" structure, that is, the pathlines are not perpendicular to the axis of rotation but have a component parallel to it. These paths are not closed as are those in Fig. 2.1. An example of a spatial vortex is the spiral vortex in Fig. 2.3.

A measure for the spatial structure of a vortex is the ratio of vortex diameter to length of axis. Large-scale vortices on earth (> 100 km) are planarlike because the layer of the atmosphere and of the oceans is thin in relation to their extensions. Vortices whose ratio of diameter to axis is very large will be called "disklike" (Fig. 2.4a). For instance, hurricanes have diameters of about 1000 km and a height of 10 km. The ratio is thus 100 to 1. In contrast, local vortices are "columnar" (Fig. 2.4b). Dust devils, for example, have diameters of 10 m and heights of 1000 m. In this case the ratio is 1 to 100.

Vortex Definition 1 is based on the pathlines of material particles. However, the flow behavior at a point in space may also be used for a vortex definition. The classical definition of Cauchy and Stokes [Reference 39, Chapter 1] calls the angular velocity of a fluid at a point in space "vorticity." In this general sense vortical motion is a basic mode of motion along with translational motion and motion due to deformation (Fig. 2.5).

A flow in which the fluid does not rotate at every point in space is called "irrotational" or "potential" flow. In general, there is no potential flow in nature, but many flow regions can be considered irrotational in an approximate way.

The statement that every motion at a point in space and time can be divided into rotation, translation, and deformation can be proven in a mathematically exact manner [Reference 39, Chapter 1]. The necessary assumption that the space filled with matter can be considered a "continuum" means that the state variables of matter, such as temperature, density, and pressure, and the laws that determine these state variables do not lose their meaning

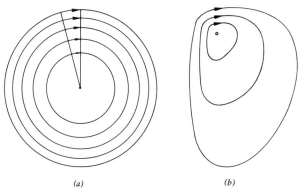

Figure 2.1. (a) Concentric circular vortex, (b) asymmetrical vortex.

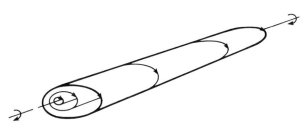

Figure 2.2. Cylindrical vortex, seen in perspective.

Figure 2.3. Spiral vortex.

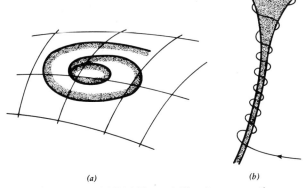

Figure 2.4. (a) Disklike and (b) columnar vortices.

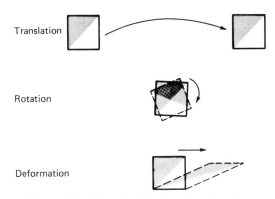

Figure 2.5. The three basic modes of motion.

if the volume of the material particle is arbitrarily decreased (Section 3.1). For instance, the density of a fluid is the ratio of its mass to the volume the mass occupies. If the volume under consideration collapses to a point, it is still meaningful for macroscopic investigations to talk about density at a point. Values for the state variables can be associated with every point in space. The totality of these values forms a "field." The fact that in reality matter cannot be divided into arbitrarily small parts, because the concept of continuum contradicts the quantum structure of matter, does not restrict the usefulness of this abstraction for describing macroscopic events. The vorticity concept of Cauchy and Stokes means that a certain angular velocity can be assigned to every point in space.[1]

Definition 2

The angular velocity of matter at a point in continuum space is called vorticity.[2]

The concept of vortex itself is used in the meaning of Definition 1. Synonyms for the word "vortex" like whirl, eddy, circulation, cyclone, or swirl occur in the literature.[3] They differ only slightly in their physical or engineering meaning and will not be uniquely defined.

Although there is no vortex without vorticity (Section 3.2), a vorticity field does not have to represent a vortex. For instance, the parallel shear flow in Fig. 2.6 has vorticity but is not a vortex [2].

2.2. ROTATION AND CONSERVATION LAW OF MATTER

... of which the largest and farthest stream flowing around the earth is the so-called "Okeanos."
Plato, Phaidon

20 Basic Concepts and Kinematic Considerations

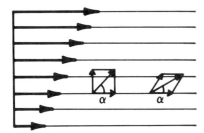

Figure 2.6. The parallel shear flow is not a vortex. Every single fluid particle, however, rotates. The flow, therefore, has vorticity. The change of the angle α is a measure for the rotation of a particle (which also deforms).

In the conception of the first Greek thinkers the earth was a finite disk surrounded by a huge whirlpool, the Okeanos. The heavenly bodies moved on spherical or cylindrical shells around the earth. This view of the world originated in the immediate and unreflected observation of the stars and in the experience of the ancient seafarers. Behind this concept, however, lies the almost trivial statement that every movement in a finite, closed space must obviously lead to a rotation. This follows from the conservation law of matter, namely, matter cannot be generated and cannot be destroyed, at least not within the framework of classical physics.[4] Every movement of a material particle in a continuum must influence its neighboring particle either by displacing it or by attracting it to follow. This idea was clearly expressed by Descartes (see the quotation in Section 1.4).

The purely kinematic statement that rotation follows from the conservation law of matter for a finite and closed space is for the time being restricted to steady motions, that is, the paths of the particles do not change with time. The restriction "kinematic" means that the forces that cause the motion or maintain it are not of concern. Some examples will illustrate the relationship between rotation and the conservation law of matter.

If one stirs coffee in a cup, rotation in some form is always generated, although it is usually manifested in rather complicated form. The simplest stirring movement is circular. Blowing into a cup toward the center generates symmetrical vortices, which at the surface look similar to the pattern depicted in Fig. 2.7.

It is not necessary to consider a finite space surrounded by solid walls to realize the necessity for vortical motion. Such motion is also possible in infinite space, provided one assumes that the fluid far away from the source of motion is at rest. For in-

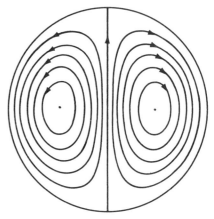

Figure 2.7. Two vortices in a finite circular region.

stance, a camp fire on a field creates a vortex ring above it, which is generated by the buoyancy of the heated (and thus lighter) air. The conservation law of matter requires that the heated air must be replaced by the cooler surrounding air (Fig. 2.8).

If "flow separation" or "instability" creates a closed region within the fluid, one or more vortices must be generated. The physical explanation for their occurrence goes beyond the kinematic considerations of this chapter and is reserved for later study (Chapters 5 and 6). Therefore, in the following examples the existence of separation and instability is assumed.

When a moving fluid encounters an interior corner, some particles near the wall avoid the corner and shorten the path by moving away from the wall. A separated area is generated in which other particles are trapped and forced to revolve around a center (Fig. 2.9).

The instability of a flow and the vortex formation connected with it are illustrated in the following example. Imagine fluid situated between two horizontal heat conducting plates. If the lower plate is uniformly heated, the fluid layer next to the plate will be heated, too, and the fluid becomes unstable since

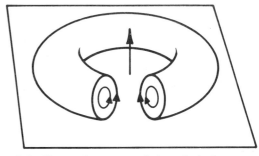

Figure 2.8. Vortex ring generated through the buoyancy of the locally heated air.

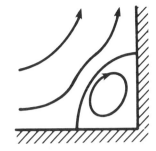

Figure 2.9. Flow in a corner.

Figure 2.11. Direction field of velocities and streamlines.

the heated fluid is lighter than the cooler fluid above it (except for water in the range of anomaly). A small disturbance of the fluid, therefore, is sufficient for the heavier fluid to slide below the lighter one due to the force of gravity. However, because of the viscosity, a certain temperature difference between the two plates must exist before instability occurs. If this critical temperature difference is reached, a small disturbance is sufficient to change the unstable state. The warmer fluid particles migrate upward, the cooler ones downward. When they arrive at their new place, the warmer particles are cooled by the upper plate and the cooler particles are heated by the lower plate, and the process of exchange is repeated. After a while a flow develops that (provided the viscosity of the fluid changes with temperature) consists of vortex rings in hexagonal cell-like spaces (Fig. 2.10). These fluid cells are called "Bénard cells." As will be described later in more detail, they play an important role in meteorology and geophysics. For instance, a "mackerel sky" indicates an unstable layer of air in the atmosphere. Humid air condenses while streaming upward within Bénard cells and is visible in the form of clouds.

2.3. STREAMLINES AND PATHLINES

So far, only time-independent flows have been considered, where the pathlines do not change with time. In order to study time-dependent flows, the

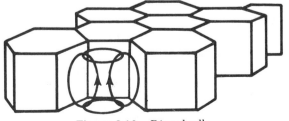

Figure 2.10. Bénard cells.

concept of "streamline" is introduced. If the velocities of individual particles at a certain time in a plane are drawn as arrows, a "direction field" is obtained (Fig. 2.11). The lines tangential to these arrows are called "streamlines." Since the particles move in the direction of the streamlines, there is no motion perpendicular to the streamlines. Therefore, the mass flux, that is, the mass that flows per unit time between two streamlines, remains constant. When the streamlines are drawn in distances of equal mass flux, the resulting picture gives information about regions of high and low velocities. Closely spaced streamlines indicate relatively high velocities; streamlines lying farther apart represent low velocities. Pictures of streamlines describe the instantaneous state of a flow, because they indicate the direction of motion of all particles at a given time.

In contrast, the pathline of a particle is obtained if the positions of the particle are determined over a specific time interval. Thus, if the velocity field changes in time, one obtains pathlines that obviously no longer coincide with the streamlines. Only for time-independent flows do the particles move along streamlines, and it follows that pathlines and streamlines coincide only in a time-independent flow.

For time-dependent flows the situation is illustrated in the following example: Fig. 2.12 shows a vortex that spreads in time. Two streamlines at time 0 may look at times 1 and 2 as they are sketched in Fig. 2.12a. Two particles, which are marked by a cross and a circle, migrate during these two time intervals along spiral paths (Fig. 2.12b).

Sometimes a time-dependent flow can be made time independent by changing the position of the observer with respect to the motion of the body. For instance, Fig. 2.13a shows the constant movement of a sphere through a fluid. The position of the observer is fixed in space, and, as the sphere moves to the left, it passes by the observer. The sphere

22 Basic Concepts and Kinematic Considerations

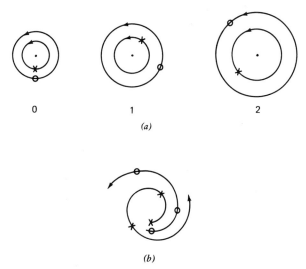

Figure 2.12. (a) Streamlines and (b) pathlines of a vortex spreading in time.

pushes the fluid particles in front of it ahead as it moves, whereas it drags the particles in the rear along with it. The instantaneous streamline patterns consist of closed curves, since far away from the sphere the fluid is at rest. A moment later the sphere is moved to the left. At this instant the streamlines look as before, but they are at a different location. Therefore, the streamlines change in space with time, and the flow is time dependent (the flow in this example is considered very viscous). The pathlines are loops near the sphere (Fig. 2.13b). However, if the observer moves with the sphere, or what is the same, if a parallel flow is superposed, the flow becomes time independent (Fig. 2.13c). The closed streamlines have vanished. Instead, they are parallel far away from the sphere. Of course, the conservation law of matter is valid. The space in this case in not closed any more. Fluid enters from the left and leaves at the right. Since the flow is time independent, pathlines and streamlines coincide.

Figure 2.13 reveals something strange: Although the physical process of the moving sphere does not change if the position of observation is altered, the streamlines (and pathlines) are completely different in different "reference frames." (In physics reference frames that move relative to each other with constant translational velocity are called "inertial systems.") Thus the following theorem is true: Streamlines and pathlines are not invariant when changing the inertial system.

Something else in Fig. 2.13 is astonishing. Since the fluid far away from the sphere in Fig. 2.13a is at rest, the streamlines must be closed according to the

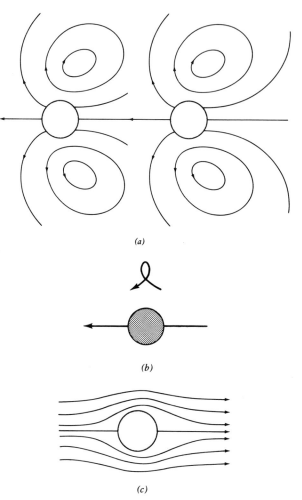

Figure 2.13. (a) A sphere moves in a fluid at rest from right to left. The reference frame is fixed in the fluid at rest, that is, the sphere passes by the observer. (b) For case (a) the pathlines near the sphere form loops. They do not coincide with the streamlines. (c) The flow in case (a) becomes time-independent if the reference frame is fixed with the sphere, or what is the same, if a parallel flow from left to right is superposed in such a way that the sphere does not move relative to the observer.

conservation law of matter. However, they do not exhibit a vortex in the sense of Definition 1, because the particles do not rotate about a common center (Fig. 2.13b). The time-independent case (Fig. 2.13c) demonstrates more clearly that no vortex exists in the flow. Descartes' remark in the citation of Section 1.4, therefore, does not necessarily describe a vortex in the sense of Definition 1 [1].

On the other hand, a vortex is not always characterized by closed streamlines. Figure 2.14a shows a vortex whose axis of rotation is at rest relative to an observer. If a parallel flow is superposed or if the observer moves relative to the center of the vortex,

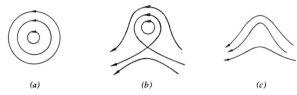

Figure 2.14. On a circular vortex (a) a parallel flow is superposed. In (b) the velocity of the parallel flow is small with respect to the tangential velocity of the vortex; in (c) the velocity of the parallel flow is large.

the streamlines open up to a wave (Figs. 2.14b and 2.14c).

Figure 2.15 shows a situation in which closed and wavy streamlines occur simultaneously. Behind a plate obliquely placed in a stream vortices form with alternating direction of rotation (vortex street, Section 6.3). The observer moves with the plate so that the streamlines far away from the plate are parallel, and the surface of the plate itself is a streamline. The closed separation region behind the front edge of the plate is a vortex since body and separation region do not move relative to the reference frame. The flow inside the separation region, however, is so weak that no streamline other than the one which marks the boundary is computed. To the right is a strong vortex with four closed streamlines. One recognizes from the form of the neighboring streamlines that this vortex rotates counterclockwise and has a small translational motion relative to the plate (Fig. 2.15b). This vortex originated at the trailing edge of the plate. The third vortex rotates clockwise and has already moved away from the front edge of the plate.

The difficulty of defining a vortex is evident.[5] Note that Definition 1 has the great disadvantage that it is valid only in a reference frame that does not move relative to the center of the vortex. In addition, Fig. 2.13 tells us that closed streamlines do not necessarily represent a vortex, and Fig. 2.14, that wavy streamlines may manifest a vortex. It will be shown in Chapter 8 that in stratified flows wavy streamlines can exist that have nothing to do with vortices. The instantaneous streamline patterns are, therefore, not suited in general to define and identify a vortex (in contradistinction to time-independent flows).

Pathlines, like streamlines, are not invariant in general when the reference frame is changed. However, a reference frame can be sought in which closed or spiral pathlines are revealed (if they do in fact exist). Then, the very nature of pathlines, as obtained by a time-integration process (in contrast to instantaneous streamlines) ensures that the flow is observed in the reference frame fixed to the vortex over the period under consideration. Such closed or spiralling pathlines reveal the existence of vortices according to Definition 1.

Alternatively, a time sequence of instantaneous streamline patterns can be used in the following way: All possible inertial frames are checked to locate regions with closed streamlines. Then at another instant, the closed streamline regions in their

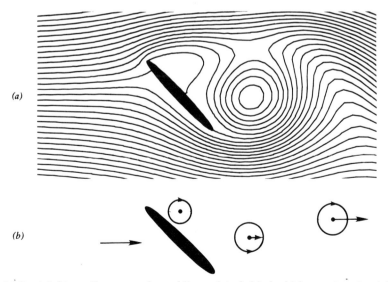

Figure 2.15. (a) Streamlines around an oblique plate behind which a vortex street develops (computer-generated picture). In this time-dependent flow three vortices can be recognized, whose movement relative to the plate is indicated in (b).

24 Basic Concepts and Kinematic Considerations

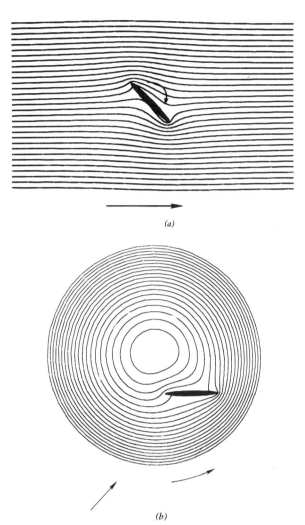

Figure 2.16. (a) Plate moving in the horizontal direction to the left and simultaneously rotating. The observer moves forward with the plate so that the plate seems to be surrounded by a parallel flow. However, the plate rotates relative to the observer. The computer-generated picture shows instantaneous streamline patterns. (b) The observer now moves with the plate, that is, the reference frame is fixed to the plate and rotates with it. Therefore, the streamlines far away from the plate are circular. If the plate did not move in a translational direction and only rotated, the center of the circles of streamlines would coincide with the center of the plate (from H. J. Lugt [4]).

respective reference frames are examined to see if their centers have moved. When the centers of such regions do not change with time, these regions represent vortices [1, 3].

In summary: The identification of a vortex in an unsteady flow requires the knowledge of the history of the motion.

So far only vortices in inertial systems have been considered. A system that rotates relative to an inertial system is, in itself, already a vortex. It is often useful to introduce a rotating system if events near a rotating body are to be studied. The events on the rotating earth, for instance, can be described more easily if the reference frame rotates with the earth. This will be described in more detail in Section 7.1. Figure 2.16 shows, for a rotating plate that simultaneously moves with constant forward speed, how streamlines behave in an inertial frame (Fig. 2.16a) and in a rotating system (Fig. 2.16b). The streamlines in both figures are time dependent since there is no reference frame at all for this flow problem in which the flow would be time independent (as contrasted with the flow in Fig. 2.13). For simplicity a potential flow has been chosen. Streamline patterns for real fluids are presented in Section 6.5.

A final word will stress the profoundness of the streamline concept. Solid-body mechanics deals with the movements of individual bodies, that is, their relation in space and time; fluid dynamics is concerned with the velocity field of the fluid at fixed points in space, a field that can be displayed in streamline patterns, which exhibit a certain permanence. The "steadiness" (or "unsteadiness") of streamlines is an intrinsic characteristic of this method of description. The idea of using the velocity field to describe the movement of matter goes back to Leonhard Euler, who introduced the mathematical formalism for this method and with it laid the basis of theoretical fluid dynamics. The difference between the "Eulerian" description of a flow field and the "Lagrangian" method of tracing individual fluid elements is illustrated by a burning candle. The visible part of the flame, its shape, is maintained (although it may flicker), whereas the journeys of the individual molecules and their chemical transformation in the flame are fleeting episodes.

More than two millenia before Euler the preservation of form in an environment of flux fascinated Heraclitus. Arius Didymus quotes him: "Upon those who step into the same rivers different and different waters flow" (Reference 9, Chapter 1). More familiar but less accurate is Plato's comment on Heraclitus: " . . . he has said that you could not step into the same river twice."

2.4. ARE VORTICES VISIBLE?

From the discussion so far, the question of whether vortices can be seen seems superfluous. In reality, however, almost all vortices in nature and technology are invisible. Who has seen vortices behind a falling leaf or behind a fish, unless it swims very close to the surface? On the other hand, rotating leaves can be observed clearly, and one infers from this the presence of a vortex. The difference is clear: Rotation of a fluid cannot be seen unless the flow has somehow been made visible. Such a marking can be made with dust, ink, clusters of leaves, droplets of a liquid, smoke, clouds, or (in outer space) with stars [5].

In water experiments, fine dust is often used as an indicator. The paths of individual dust grains observed over a certain time span reveal the pathlines of the motion. When cars are photographed at night with the shutter of the camera kept open, every headlight generates a band of light on the picture. These bands are simply pathlines. However, opening the shutter of the camera only briefly gives approximate streamlines (or exactly, short pathlines that yield streamlines when put together).[6]

When observing vortices in an experiment or in nature, one does not usually see either streamlines or pathlines. Streamlines cannot be made visible easily, and it is not always possible to follow the path of an individual particle. What, then, is actually being observed when milk is poured into a cup of coffee and beautiful spiral vortices form? What are the spiral cloud bands on satellite photographs of the earth? If paint is injected into water over an interval of time at precisely the same spot, or if smoke is blown into flowing air, paint and smoke streaks are formed. These streaks are not pathlines but "streak lines" [6]. The difference between pathlines and streak lines is understood most easily if the development of the streak line is studied at various times. In Fig. 2.17 a paint particle, which is marked by a cross, is brought into the flow from a nozzle at time 0. At time 1 this particle has already traveled a certain distance, and a new paint particle, marked by a small circle, leaves the injection nozzle. At time 2 this particle also has left the nozzle and has covered a certain distance. Its path is different from that of the first particle at time 1 because the flow has changed in the meantime. The pathlines of the individual paint particles are, therefore, different. If one takes an instantaneous picture, for instance at time 3, one does not see the pathlines (dashed) but only the paint particles at time 3, which put together form a streak line (solid line). In time-independent flows the pathlines do not change, and thus the statement can be made: In

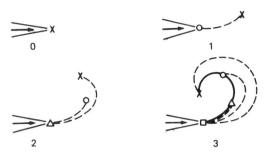

Figure 2.17. A sketch for explaining the definition of a streak line (see text).

time-independent flows pathlines, streamlines, and streak lines coincide.

The streak lines of rotating fluids are spirals in general (in special cases circles). They are directed from the outer edge toward the interior of the rotating fluid mass in the case of vortex generation and vortex stretching, as in the bathtub vortex (Fig. 2.3). Because streak lines are usually the only means of revealing the existence of a vortex, the spiral has been used since time immemorial to represent and to symbolize the vortex (Section 1.1).

In nature streak lines often develop in a more complicated way. First, the location at which the flow is made visible need not be restricted to a single point but can consist of an area, and, second, this area may change with time. For instance, upward moving humid air in the atmosphere condenses from a certain height onward and forms a stratum of cloud. The wind may cause the place of condensation to change. It is also possible that during the development of a streak line no more fluid particles will be marked after a certain time and that the existing streak line will deform (Fig. 2.18). The photographs of streak lines or streak areas do not give enough information on the temporal development of the flow unless a series of pictures at different times is taken. For example, astronomers have made beautiful photographs of spiral galaxies, which consist of billions of stars, but these instantaneous pictures do not reveal any direct clues on the evolution of these systems.

So far it has been assumed that the substance that makes the fluid flow visible has the same or at least approximately the same density as the fluid itself. This is true for ink and water, but sand and air differ considerably in their densities, although a sandstorm in the desert proves that air can carry sand with it. In this case, however, gravity and centrifugal force act differently on the "carrier fluid" and the "foreign particles." The paths of the fluid

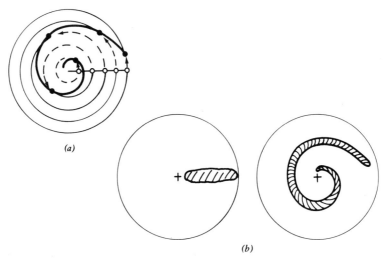

Figure 2.18. (a) Streak line marked by circles. If the particles near the axis rotate faster than those farther away, the initially straight streak line stretches to a spiral (solid points). (b) A cloud may roll up spirally in a whirlwind.

particles and the foreign particles are therefore different. An insect, which is heavier than air, is smashed against the windshield of a moving car. Because of its larger density the insect can only approximately follow the curvature of the pathlines on which it would be carried safely past the car. This small deviation in trajectory leads to its fate (Fig. 2.19). In the milk centrifuge the fact that the lighter, fatter milk and the heavier skimmed milk move along different paths is used to separate them.

2.5. THE SPECTRUM OF VORTICES

If manifested in these circles were the cosmic order of the universe, I should be well content.
　　　　　　　　　　　Dante, Divine Comedy

Vortices of all magnitudes can be found in nature. From the quantized vortices in superfluid liquid helium to the rotation of galactic systems in the universe, vortex motion is present. An idea of the multitude and variety of vortices is provided by the following list, ordered according to the spatial dimensions of the vortices:

	Diameter
Quantized vortices in liquid helium	10^{-8} cm
Smallest turbulent eddies	0.1 cm
Vortices generated by insects	
Vortices behind leaves	0.1–10 cm
Vortex rings of squids	
Dust whirls on the street	
Whirlpools in tidal currents	1–10 m
Dust devils	
Vortex rings in volcanic eruptions	
Whirlwinds and waterspouts	100–1000 m
Convection clouds	
Vortices shed from the Gulf Stream	
Hurricanes	100–2000 km
High- and low-pressure systems	
Ocean circulations	
General circulation of the atmosphere	2000–5000 km
Convection cells inside the earth	

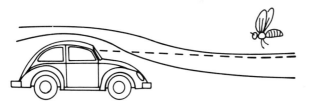

Figure 2.19. An insect does not follow the pathlines of the air around the car, but hits the windshield.

	Diameter
Planetary atmospheres	
Great Red Spot of Jupiter	$5000-10^5$ km
Rings of the planet Saturn	
Sun spots	
Rotation inside of stars	Depending on size of stars
Galaxies	Order of light years

Vortices may form in air, water, gas, plasma,' and the earth. They may also consist of accumulations of solid bodies. Therefore, the existence of vortices does not depend on the kind of medium. For this reason the word "fluid" is used in the following discussions.

Vortices can be very small or can have the dimensions of galaxies. They are generated through differences in density and temperature, through friction, through the action of electrodynamic forces, or through gravity. In spite of the variety in generation, size, angular velocity, and kind of medium, vortices have common characteristics and structures. The purpose of this book is to find such structures and to describe them. In the first half of this book the fundamentals of vortical flow are given; then, the generation and properties of those vortices that are not influenced by the rotation of the earth or by the difference in densities of the oceans and the atmosphere (local vortices) are discussed. The second half of this book describes vortices whose generation and properties are essentially determined by rotation and stratification. Finally, extraterrestrial vortices, that is, vortices that occur in outer space, are briefly discussed.

Remarks on Chapter 2

1. The rotation of a fluid at a point in continuum space is, of course, not related to the spin of elementary particles like that of electrons and protons.
2. To equate vorticity with the angular velocity of a fluid element in a continuum is not quite exact, since deformation of the element also takes place (Fig. 3.11). The two concepts of vorticity and angular velocity are mathematically equivalent up to second order. Vorticity is a state variable that is related to a point in space and not to a fluid particle.
3. Different applications of the individual concepts may be demonstrated with the word "circulation." In mathematical fluid dynamics circulation is uniquely defined as the area integral over the vorticity. In meteorology and oceanography large-scale vortices are designated circulations. In biophysics the flow in closed tubes is called circulation, for instance, blood circulation, in contrast to the fluid flow in the ureter and urethra.
4. For most of the vortices described in this book the assumptions of classical physics, in particular the model of a continuum, are sufficiently exact. Relativistic effects, for which the conservation law of matter in its classical form does not hold, occur in vortices of cosmic dimensions (Section 12.5). In the atomic region, which is not treated in this book, relativistic and quantum mechanical assumptions must also be made.
5. There are other definitions of a vortex in the literature, all of which are unsatisfactory for a viscous fluid [1]. For instance, Lagerstrom [7] says, based on Prandtl's analysis of flow inside a closed domain: "a maximum set of nested closed streamlines." Saffman and Baker [8] restrict themselves to an ideal fluid: "A vortex is a finite volume of rotational fluid, bounded by irrotational fluid or solid walls." Extrema of the vorticity field are also used to signify vortices [9] (Section 5.3).
6. Streak lines can be made visible for certain time periods by means of flow visualization through chemical time reactions [10]:

    Color appears instantly:

    $$I_2 + 2\,Na_2S_2O_3 \rightarrow Na_2S_4O_6 + 2NaI.$$

    Color appears slowly:

    $$2KI + Na_2S_2O_8 \rightarrow Na_2SO_4 + K_2SO_4 + I_2.$$

7. "Plasma" is defined in physics as a gas that consists of free electrons, ions, and neutral particles. In addition to "solid, liquid, and gaseous," the state of plasma is often denoted as the fourth state of aggregation of matter. Plasma occurs in gas discharges.

3. Properties of Simple Vortices

3.1. STATE VARIABLES AND BASIC LAWS

An event in nature is described physically by certain basic quantities which uniquely determine the momentary state. "State variables"[1] of a fluid flow are, for instance, velocity, pressure, vorticity, density, and temperature. Each point in a flow regime, which is assumed to be a continuum, can be characterized by a number (or three numbers for vectors in three-dimensional space) for each state variable. The totality of all numbers of one state variable in a region forms the "field" of that state variable. Under certain circumstances still other variables must also be considered such as those that occur in the presence of electric and magnetic fields. Mixtures of several fluids, such as water and air, must be described by additional state variables—the humidity, for example. In the following discussion it is assumed for the time being that the fluid has the same density and temperature everywhere. Then the flow of that fluid is uniquely determined by the velocity and pressure fields, or by velocity and vorticity (Chapters 4 and 5).

State variables can be measured with instruments and the state of a flow can thus be determined. Meteorologists make use of this procedure by obtaining measurements of wind velocity, pressure, temperature, humidity, and precipitation at as many geographic locations as possible. The closer together these locations are, the more accurately the field is described. The evaluation of the measured data, which are recorded on maps, is facilitated when points of equal temperature, equal pressure, or any other state variable are connected. Such lines of equal temperature are called "isotherms," lines of equal pressure "isobars," and so forth.

For a scientist the mere collection of measured data is not very satisfying; he seeks natural laws to explain the relationships among the state variables and their changes in space and time. These natural laws can be reduced to only a few basic laws (axioms), which are found by evaluating experiments in connection with simplifying notions (models) in an inductive way. State variables together with axioms describing their relationships define a theory. The axioms of fluid dynamics can be categorized into two groups: the conservation laws and the constitutive equations (Reference 39, Chapter 1). The conservation laws assert that certain physical quantities cannot be generated nor destroyed in a closed system.[2] Such conservation laws exist for matter, energy, momentum, and angular momentum. The constitutive equations deal with the properties of matter and indicate, for instance, whether a body is elastic, rigid, liquid, plastic, or gaseous. As an example, "rigid" means that all parts of a body remain unchanged in their locations within that body. This does not happen in reality, not even for the hardest metal or rock, but the concept of "rigid" is often useful for describing complex phenomena in nature, in which compressibility and deformation of a body can be neglected.

Matter cannot be created nor can it vanish (in the sense of classical physics). This is the law of the conservation of matter. For energy, too, a conservation law holds. It says that the sum of all types of energy, such as kinetic energy, pressure energy, heat energy, or electromagnetic energy, cannot change in a closed system. The conservation law of momen-

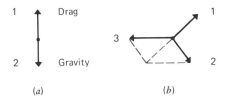

Figure 3.1. (a) The forces 1 and 2 are in equilibrium; (b) the forces 1, 2, and 3 are in equilibrium.

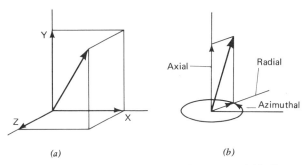

Figure 3.2. The three components of a vector **v** (a) in Cartesian coordinates and (b) in cylindrical coordinates.

tum can be expressed in such a way that the inertia (defined as mass times acceleration) of a particle is equal to the sum of all forces acting on the particle. If one considers inertia in a formal way as a force (force of inertia), then the forces acting on a particle must be in equilibrium. For instance, if a body is falling constantly in the gravitational field of the earth, resistance is equal to gravity, although with opposite sign (Fig. 3.1a). It may be recalled that forces are vectors. Vectors have a certain absolute value and a certain direction at each point in space and are added geometrically (Fig. 3.1b).

A vector in three-dimensional space can be decomposed into three components and is, therefore, determined by three numbers (Fig. 3.2a). In describing vortical motions it is often advantageous to use cylindrical coordinates (Fig. 3.2b). The resultant of all forces that act on a body can be decomposed into a component parallel to the flow, called the "drag," and into two components normal to it, the side forces. The side force opposite to gravity is called "lift" (Fig. 3.3a) (which should not be confused with the drag opposite to gravity in Fig. 3.1a). In Fig. 3.3a the component perpendicular to the page is not drawn. The resultant is kept in equilibrium at point A by an equal force acting in the opposite direction. If the resultant force does not act at the center of gravity and is not on a line through the center of gravity parallel to the resultant, the flow causes a "torque" on the body. This torque tries to rotate the body and, therefore, must be balanced by a torque in the opposite direction if the body is to remain in equilibrium (Fig. 3.3b).

In addition to these forces, velocity and vorticity are also vectors. Vorticity may be interpreted as the angular velocity of the fluid at a point in space. Angular velocity is a vector normal to the plane of rotation (thus it coincides with the axis of rotation), and its positive direction may be agreed upon in the way shown in Fig. 3.4 ("right-hand rule"). The same is true for the vorticity vector. In a plane flow all vorticity vectors are normal to the plane of motion.

The angular momentum of a closed system is also a constant. The angular momentum of a particle is proportional to its velocity and its distance from the center about which it is rotating. If this distance is diminished, the velocity must increase correspondingly. A well-known example is the velocity of the earth around the sun. According to Kepler's second law (which is merely the conservation law of angular momentum) the earth's velocity increases whenever the earth comes closer to the sun in such a way that the areas traversed by the radius vector at equal time intervals are equal (area law, Fig. 3.5).

The conservation law of angular momentum explains why the high angular velocities of vortices can occur. A clarifying example is given in Section 3.4 on the intake vortex.

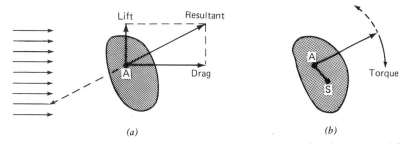

Figure 3.3. On the definition of drag, lift, and torque: A is the aerodynamic center of the forces and S is the center of gravity.

30 Properties of Simple Vortices

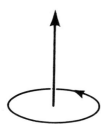

Figure 3.4. The vector of the angular velocity is normal to the plane of rotation.

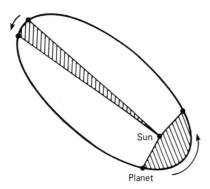

Figure 3.5. Kepler's second law as an example of the conservation law of angular momentum. The shaded areas, which are traversed by the radius vectors at equal time intervals, are equal. Hence, the earth must move faster in the vicinity of the sun than farther away.

3.2. THE TWO BASIC TYPES OF PLANE VORTICES

Among the vortices is one which is slower at the center than at the sides, another faster at the center than at the sides.

Leonardo da Vinci

The properties of the simplest plane vortex, the circular vortex, will now be discussed in more detail. There are two basic types of velocity distribution which, for a time-independent flow, represent the only possible rotations. Two simple experiments illustrate these rotations.

In the first experiment a solid body such as a disk rotates steadily around an axis through its center (Fig. 3.6a). For such a motion it is well known that the velocity of points on the disk increases linearly with their distance from the center (Figs. 2.1 and 3.6b). Imagine now that the disk is hollow like a box and completely filled with a fluid. When the experiment is repeated, the fluid will, after a transient period, rotate like a solid body because of its adherence to the walls. Hence, the velocity of the fluid particles also increases linearly with the distance

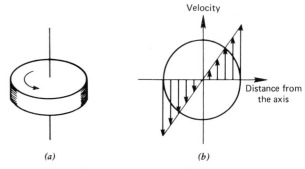

Figure 3.6. (a) Every rotating solid body is a vortex. (b) Velocity distribution in such a vortex.

from the rotation center, and the angular velocity is constant everywhere in the fluid. This rotary fluid motion is called "solid-body rotation," and it occurs only when the fluid motion is steady (time independent). In the transient phase from the start of the motion to the steady state, the fluid may oscillate or may move in a complicated way (Section 7.5). This difference between a fluid and a solid body provides a test for whether an egg is boiled or not. In the raw, unboiled state the egg cannot rotate like a top, but it is able to do so in the boiled state when the egg is solid.

A uniform translation can be superimposed on the solid-body rotation without changing the tangential velocity profile. When the motion parallel to the axis is constant, the particles travel along a helical path (Fig. 3.7).[3]

Even in the steady state a fluid can perform rotational motions which a solid body cannot, a fact which Leonardo wondered about. The following experiment, described by Newton, demonstrates this difference. A long, circular rod rotates in a fluid with constant velocity around its axis (Fig. 3.8). The fluid velocity is highest and equal to the velocity of the

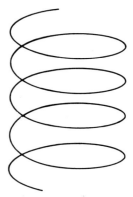

Figure 3.7. If a vortex with solid-body rotation moves in the axial direction, the pathlines are helical lines.

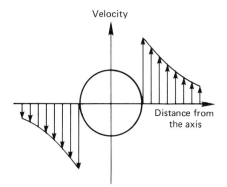

Figure 3.8. Velocity distribution in a potential vortex.

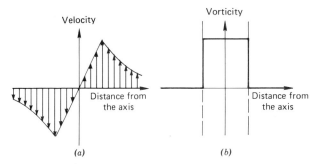

Figure 3.10. Rankine vortex: (a) velocity distribution and (b) vorticity distribution.

rod at the rod's surface, since the fluid adheres to the body of the rod. With increasing distance from the rod the velocity diminishes in inverse proportion to the distance. Such a fluid motion is called a "potential vortex" because, as will be shown later, the fluid has no vorticity except at the center of rotation.

The two types of vortical motion, the solid-body rotation and the potential vortex, are essentially different. In solid-body rotation the fluid particles may be compared with the horses on a merry-go-round, which are tightly fastened to the rotating disk and, therefore, turn once around the axis at each revolution of the disk[4] (Fig. 3.9a). Constant angular velocity in solid-body rotation also means constant vorticity in the entire fluid:

> Angular velocity and vorticity are constant in a solid-body rotation.

If, on the other hand, a potential vortex is produced in a container filled with water, and if tiny rods are floating on the surface of the water, these little rods do not change their direction on their circular paths (except at the center of rotation). These rods do not rotate around their own axes but move with a circular translational motion[5] (Fig. 3.9b):

> The potential vortex does not have vorticity outside the core.

The center, however, has vorticity. This can be seen immediately if one remembers that the poten-

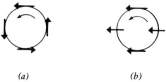

Figure 3.9. The movement of a rod (a) in solid-body rotation and (b) in a potential vortex.

tial vortex is generated by a rod which itself rotates as a solid body. The rod, of course, may also be replaced by a hollow cylinder filled with fluid. The velocity distribution inside the cylinder would not change (Fig. 3.10). Fluid motion composed of a potential vortex and a solid-body rotation is called a "Rankine vortex" after the fluid dynamicist Rankine. For a steady, circular motion without a velocity component normal to the plane of rotation (which was assumed), the Rankine vortex is the only possible vortex whose velocity is zero at the center as well as far away from it.

The Rankine vortex is useful in defining the diameter of a vortex. The diameter of the vortex core can be exactly determined, but the definition of the diameter of the vortex is quite arbitrary. It may be agreed that the radius of the vortex is that distance from the center at which the velocity has decreased to a certain fraction of the maximum velocity, say, for instance, to 5%.

If the radius of the solid core of the vortex shrinks to a point, one obtains a "vortex line." The whole flow field is free of vorticity except at the axis of rotation, where the vorticity is infinite, but the circulation remains finite.[6] The vortex line is a special vorticity line, that is, an isolated vorticity line with infinite vorticity. If the velocity vectors in Fig. 2.11 are thought of as vorticity vectors, the streamlines become vorticity lines. A bundle of vorticity lines is enclosed in a "vorticity tube." An isolated vorticity tube is called a "vortex tube," which encloses a "vortex filament." The vorticity lines in the bundle need not have infinite values. Accordingly, a Rankine vortex is a vortex filament.

The two basic types of plane vortex flows differ from each other in another respect. Whereas all particles in solid-body rotation are at rest for an observer rotating with the fluid, in a potential vortex concentric fluid layers rub against each other (Fig.

32   Properties of Simple Vortices

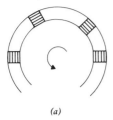 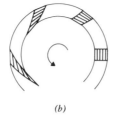

Figure 3.11. Fluid particles in a solid-body rotation are at rest relative to each other (*a*), whereas in a potential vortex fluid layers rub against each other (*b*) [1].

3.11). This may be demonstrated with the following experiment. On a slowly rotating surface of water a square is marked with powder. In a solid-body rotation the square does not change its form. However, in a potential vortex the shear forces act in such a way that the square is deformed to a kind of a parallelopiped until it is completely pulled apart. Another nice experiment was devised by G. I. Taylor [2]: If a potential vortex is produced by a rod slowly rotating in a viscous fluid, and if a square is similarly marked, the square spreads until it is no longer recognizable. If the flow direction is reversed at this moment, the original square is reformed exactly as if a movie were reversed.

Since no shear forces occur in solid-body rotation, no energy is needed to maintain its motion. In contrast, the rod, which is driven for instance by a motor, must constantly provide kinetic energy to the potential vortex in order to make the shearing of the particles possible. The kinetic energy is transformed inside the fluid into heat. This process is called "dissipation." Thus in a Rankine vortex the potential vortex demands that work to be done by the hollow cylinder. If the solid-body rotation takes place under a vacuum, rotation is maintained without a supply of energy once the motion has been started. Therefore, solid, liquid, and gaseous heavenly bodies, artificial satellites, and space ships continue to rotate from the moment at which they first receive their angular momentum.

If the rod or the hollow cylinder is suddenly pulled out of the fluid, the motion will slow down in time through friction of the fluid particles against each other. The vortex gradually decays since no more energy is transmitted from the rod (Fig. 3.12) (Reference 35, Chapter 1). Since this changing rotation is time dependent, it is not one of the two basic types, that is, it is neither solid-body rotation nor a potential vortex.

Are there vortices of the type of Fig. 3.12 that are time independent and, therefore, that continuously

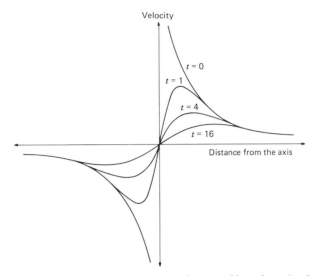

Figure 3.12. Decay of a potential vortex. Near the axis of rotation the flow behaves like a rotating solid body. The time t is given in dimensionless units.

obtain kinetic energy from somewhere? Such a vortex can exist only if it has a radial velocity component and, because of the conservation law of matter, also an axial velocity component. In this case energy is carried from the outside into the vortex. An example is the bathtub vortex, which is a spiral vortex (Fig. 2.3).

The complete description of rotation requires the pressure distribution as well as the velocity profile, provided the other state variables are constant. The different kinds of media, by the way, have no influence on the characteristic properties of vortical motions. For instance, in solid-body rotation the increase in velocity with distance from the center is linear for water as well as for air. The differences in medium can be described by the material constants that determine the magnitude of the occurring forces or the time of vortex decay. For instance, the viscosity of water is about 100 times larger than that of air. Therefore, in order to maintain a potential vortex at constant speed in water, 100 times more work must be done in water than in air. However, only the typical properties of vortices and their structure are of concern in this book. The common characteristics make it possible simultaneously to study vortices of such completely different sizes as those in a lake and in a teacup. Certain "flow parameters" will be defined later that measure these common characteristics.

The pressure in a vortex is not uniformly constant. The pressure in a solid-body rotation, for instance in a rotating box, measured and plotted over

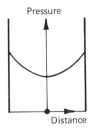

Figure 3.13. Pressure distribution in a solid-body rotation.

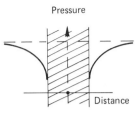

Figure 3.15. Pressure distribution in a potential vortex.

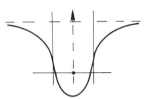

Figure 3.16. Pressure distribution in a Rankine vortex.

distance from the center, shows a parabola (Fig. 3.13). The pressure increases toward the wall of the box because the centrifugal force is larger the farther away the particles are from the center. This is easy to imagine in the following way: If one ties a ball on a string and whirls it around, the force pulling at the string is larger at constant angular velocity the longer the string happens to be. Against this centrifugal force an equally large centripetal force (the string) must act to maintain equilibrium (Fig. 3.14). If the string is released, the ball will fly away tangentially to the circular path. The movement of the moon and satellites around the earth also illustrate gravity and centrifugal force in equilibrium. If the satellite is braked by rockets and the centrifugal force thus reduced, gravity dominates and the satellite either enters a smaller orbit or drops to the earth. In a rotating fluid it is the wall of the box that prevents the fluid from spilling out. This equilibrium exists inside the fluid as long as the fluid has the same density throughout. If, however, there are particles in the fluid that are heavier or lighter than the rest, for instance oil drops in water or dust in air, then the equilibrium of the forces is disturbed. The heavier particles will fly toward the wall of the box, the lighter ones will collect at the center. The principle of centrifuges and dust cyclones is based on this phenomenon. In the milk centrifuge the cream, which is fat and therefore light, is separated from the heavier skim milk. In dust cyclones gas and air are cleaned by being separated from the heavier dirt particles. In nature separation of heavier and lighter material takes place on a gigantic scale in rotating stars.

The pressure in a potential vortex also increases with the distance from the center. In contrast to solid-body rotation, however, the increase is largest near the center (Fig. 3.15). Combining the curves of the two types of vortices gives the pressure distribution of a Rankine vortex (Fig. 3.16). Here it is assumed that the pressure is equal at the outer and inner walls of the hollow cylinder, which is not necessarily the case. For a decaying vortex (Fig. 3.12) the pressure distribution is similar to that shown in Fig. 3.16.

The pressure at the center of a vortex in water can decrease so much that the gases dissolved in the water separate or that water evaporates even at room temperature. This is called "cavitation." Around the rotation axis a cavitation of gas or air ("hollow core") is generated, which is easy to recognize (Fig. 4.35). In a whirlpool at a free water surface, the air–water interface passes over to the hollow core of the vortex (Fig. 3.18).

Sometimes wedge-shaped waves are observed on a free surface of water without the presence of a ship or a fish to cause such a disturbance. These "ghost wakes" are generated by invisible whirlwinds in sudden gusts. A whirlwind bulges the free surface at its center, and this disturbance causes wedge-shaped "ship waves" when it moves over the water [3].

At this point two basic concepts of fluid dynamics must be introduced. So far it has been assumed that fluid particles travel on their prescribed paths all the time. However, it is possible (and this is the

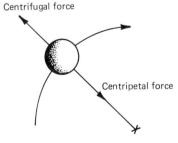

Figure 3.14. Forces that act on a revolving sphere.

most common case in nature and technology) that particles fluctuate around an average path in an irregular way and move along the exact pathlines only on the average. Such a flow is called "turbulent," as distinguished from the "laminar" motion considered so far. In the statistical sense the considerations for streamlines and pathlines are valid for turbulent motions also. Turbulent flow properties, however, differ from those of laminar flow as explained in more detail in Section 6.6. For instance, a turbulent vortex filament will decay faster than the laminar one in Fig. 3.12, and the velocity distribution outside the core is flatter [4, 5].

3.3. THE BUCKET EXPERIMENT

The experiment with the rotating box is now changed in such a way that the box is opened at the top and half emptied. This may be visualized as a bucket half filled with water and placed on a revolving disk, which causes the water to rotate. The faster the bucket rotates, the more water is pressed against the wall, forcing it to ascend (Fig. 3.17). The reason for this phenomenon is again the centrifugal force, exactly as in the rotating box completely filled with water. The change in the free surface can be explained in the following way: The surface of the water is always perpendicular to the force that acts on the water. Otherwise the water would move in the direction of the horizontal component of the force until this component reached zero. In still water only gravity is present, and the free surface of the water is, therefore, parallel to the earth's surface. In a rotating fluid the centrifugal force must be added to gravity so that the resultant of the two forces determines the new surface (Fig. 3.17), which has the form of a paraboloid [6]. In a horizontal plane the pressure, which is proportional to the height of the water, again shows the well-known parabolic pressure distribution of the solid-body rotation. In an analogous way the free water surface of a potential vortex is curved, as can be seen in Fig. 3.15.[7]

The bucket experiment played an important role in the history of science. Newton argued that the curvature of the free surface is a consequence of absolute rotation and proves the existence of an absolute space. This means that space itself need not be related to a reference frame and that absolute rotation is a motion relative to this absolute space. Berkeley rejected Newton's thesis and held that, on the contrary, the bucket experiment demonstrated the relativity of rotational motion, since the bucket moved in relation to the total matter of the universe. Details of this argument are described in Section 7.1.

3.4. INTAKE AND DISCHARGE VORTICES

In addition to the two basic types of steady circular flows described in Section 3.2, there are other time-independent rotary motions that have an azimuthal velocity component as well as radial and axial components. The spiral vortex in Fig. 2.3 is an example. If one stirs water in a container and permits it to flow out an opening at the center of the base, vortices with free surfaces are produced whose shapes depend, among other parameters, on the strength of the rotation and on the height of the water column from the bottom, the submergence. If the rotation is small or the submergence large, the water surface forms only a dimple (Fig. 3.18a). For stronger rotation or less submergence the dimple develops to an air core as depicted in Fig. 3.18b. If the air core stretches to the bottom of the tank, the "critical submergence" is reached. From then on, air is entrained into the discharging flow (Fig. 3.18c).

Such "air-entraining vortices" play an important part in hydraulics. Water from rivers and reservoirs is drawn through various types of intake pipes, either for flood control (spillway), irrigation, domestic and industrial supply, or electric power generation[8] [7–9]. Since water intakes are usually located near the water surface for economic reasons and to avoid drawing in sediments, strong vortices can form with air entrainment (Fig. 3.19). This entrainment reduces the discharging capacity considerably and may result in a dam overflowing, a reduction in the efficiency of pumps, and the

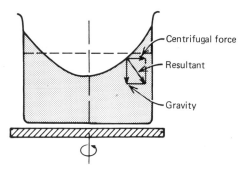

Figure 3.17. Distribution of water in a rotating bucket.

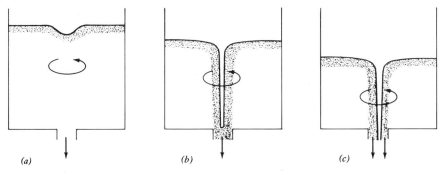

Figure 3.18. Discharge of rotating liquid through a hole in the bottom of a tank. (*a*) The vortex is weak, and the surface forms a dimple; (*b*) the air core of the vortex reaches the bottom (critical submergence); (*c*) strong, air-entraining vortex.

generation of vibrations and noise, to cite a few examples. Even more serious is the possible failure of the cooling system in a nuclear reactor through air entrainment [11].

The pipe intakes for these purposes, whether horizontal or vertical, may be constructed with or without bellmouths (Fig. 3.20). Bellmouths increase the discharge under otherwise identical conditions.

When the water at the intake is not through a pipe but through a hole in the bottom of a tank (Fig. 3.18*c*), the developing swirling jet together with the entrained air will have a shape near the orifice as shown in Fig. 3.21 according to Binnie and Davidson [12]. If the swirl is large, the jet farther down will disintegrate (spray); if it is small, it will converge again and form an annular jet.

Figure 3.19. Vortex at a turbine intake at Arapuni in New Zealand (from E. N. da C. Andrade, *New Scientist* [10]).

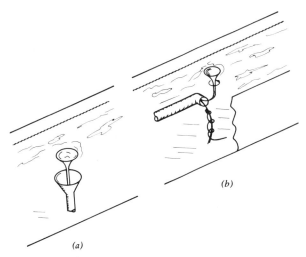

Figure 3.20. Vortices at (a) a vertical intake pipe with bellmouth and (b) a horizontal intake pipe.

A twin vortex, similar to that in Fig. 3.20b, occurs also when a jet emerges from a hole in a plate and is deflected by a fluid stream parallel to the plate [13].

How do strong vortices like those in pipe intakes form? The explanation is contained in the conservation law of angular momentum (Section 3.1). For simplicity, consider a single fluid particle with a tiny angular velocity, located about 30 cm from the outlet. As this particle is drawn toward the opening, it must increase its azimuthal velocity at a rate inversely proportional to the distance from the opening in order to preserve its angular momentum. At a distance of 0.3 cm from the center of rotation the particle will have increased its velocity 100-fold. Since the angular velocity increases with the square of the distance, the corresponding angular velocity at 0.3 cm distance will have increased 10,000-fold. This may also be expressed in the following way: The number of rotations per second is 10,000 times

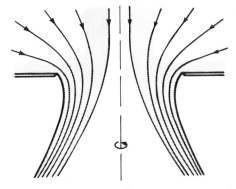

Figure 3.21. Shape of the swirling jet with entrained air at the outlet of Fig. 3.18c.

Figure 3.22. A coaxial jet in a rotating fluid generates a concentrated vortex.

larger at a distance of 0.3 cm than at 30 cm. This simple although crude calculation shows that a small initial rotation in the fluid is sufficient to create a concentrated vortex.

Another experiment will once more demonstrate the generation of a spiral vortex. If one stirs hot coffee in a cup and afterward pours cold milk into the center, the coffee starts rotating around the jet of milk with relatively high velocity. Without the jet the fluid rotates almost like a solid body. Because of the jet the fluid particles are drawn toward the center and then downward. They must rotate faster to preserve their angular momentum (Fig. 3.22). A concentrated vortex has thus been developed. If the coffee had not been stirred before pouring in the milk, that is, if the fluid had no rotation, the addition of milk would not result in a vortex. Because cold milk is heavier than hot coffee, the downward motion of the jet of milk is maintained longer than if this density difference did not exist.

The following experiment, which elucidates the generation of a vortex stretching from a solid surface into a pipe intake, has been described by Blanchette [14]. One needs a fan, a vacuum cleaner, and a table. When air is sucked into the tube of the vacuum cleaner without disturbing the air flow, an axisymmetric "sink flow" is produced at the opening of the tube (Fig. 3.23a). This symmetry is destroyed in the presence of the table top near the tube, and the dashed line, which is the axis of symmetry in Fig. 3.23a, is bent toward the table in Fig. 3.23b. If one removes the table and arranges the fan's air stream at an angle to the axis of symmetry, an intake vortex is created with the axis of symmetry as the rotation axis (Fig. 3.23c). If the table is again brought near the tube, the rotation axis bends down toward the table as shown in Fig. 3.23d. The vortex thus created can be made visible by putting a shallow bowl of water on the table. The low pressure at

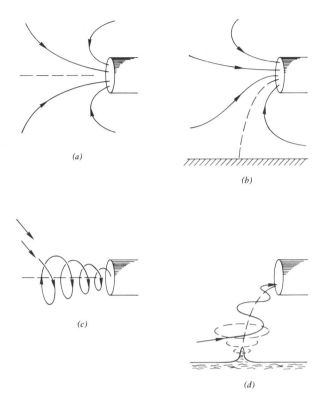

Figure 3.23. Experiment for the generation of an intake vortex according to Blanchette.

the axis of the vortex causes the free surface of the water to arch, and the ascending air pulls water drops with it, creating a miniature waterspout (Section 10.3).

This effect can also be observed near jet aircraft standing on an airfield with engines running. The jets suck in air, which is rotated at high speed. Vortex filaments can be generated, which extend from the leading edge of the jet engine down to the earth. They are strong enough to pull any nearby sand and stones into the jet engine, which can result in damage [15, 16] (see also Fig. 3.20b).

Intake and discharge vortices are also observed in symmetrical geometric surroundings. For instance, a vortex may develop at or near a rectangular intake (Fig. 3.24a) or behind a slanted round tube, placed on a horizontal bottom (Fig. 3.24b). The rotation can of course be in either direction.

A discharge or drainage vortex can also be observed in a bathtub and in any outlet of a water tank. Here again the question may be raised: How is the vortex generated after the plug has been lifted if no one has stirred the water beforehand? Since no vortex can be produced without rotational energy, a certain amount of angular momentum must have been already present to cause the vortex to develop. This is indeed the case, and a so-called "organization time" plays an important role in the development of the vortex [18]. The initial angular momentum could have been generated by small asymmetries of the boundary-layer flow at the walls of the bathtub (Section 4.5), by asymmetric pulling of the plug, by asymmetric temperature distribution, or even by asymmetric air motion over the water surface (in Section 9.2 it will be shown that water circulation can be created by air currents). The direction of rotation of the vortex that will be formed depends on this initial angular momentum.

For years the question was debated whether or not the earth's rotation generates the bathtub vortex. It is well known, as will be explained in more detail in Section 7.2, that a body that moves horizontally will be deflected by the earth's rotation to the right in the northern hemisphere and to the left in the southern hemisphere. Therefore, if the earth's rotation were the essential cause for the bathtub vortex, the vortex would have to rotate counterclockwise in the northern hemisphere, clockwise in the southern half, and not at all on the equator. Observations, however, do not show a preferred direction of rotation. Thus, the earth's rotation can be neglected in considering these disturbances. But this does not mean that the earth's rotation does not influence the bathtub vortex at all. Under carefully controlled conditions the rotation of the earth does determine the direction of the bathtub vortex [10]. This effect was described in 1908 by the Austrian physicist Turmlitz [19]. Char-

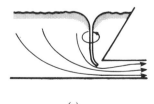
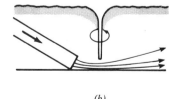

Figure 3.24. (a) Intake vortex generated at a horizontal, rectangular suction slot. (b) Discharge vortex at the opening of a tube placed on the bottom of a tank (from E. Levi [17]).

acteristically, he called his experiment a proof for the axial rotation of the earth, and indeed, the earth's rotation can be confirmed in this way. In the 1960s careful experiments were made by Shapiro in Boston and repeated at the University of Sydney in Australia [20]. The results were as expected: in Boston the bathtub vortex rotated counterclockwise, in Sydney clockwise.

3.5. THE MOTION OF SEVERAL POINT VORTICES

Sometimes two or more vortices are generated simultaneously. They influence each other and form a system that exhibits characteristic motions. The motion of only two vortices will be considered first.

If a single vortex exists in a fluid at rest far away from any wall, it does not move. Two neighboring vortices, however, act on each other in such a way that their rotating axes migrate. This can be understood easily if the motions of the two vortices are studied separately. The velocity field of the first vortex determines the velocity of the center of the second vortex, and the velocity field of the second vortex determines the motion of the first. Since the resulting velocity can be thought of as a superposition of the velocities of the two vortices, the vortex system too has a certain translational velocity. Two vortices of equal strength but opposite rotation move forward on a straight line perpendicular to a line connecting the centers of rotation (Fig. 3.25a). For simplicity, the vortices are assumed to be parallel vortex lines, that is, as in Fig. 3.25, two points (point vortices) that do not decay. If the rotational velocities of the two vortices are different, the path of the system is curved toward the stronger vortex (Fig. 3.25b) [21].

An example of such a vortex system is the motion of two vortices behind an oar or at the wing tips of an airplane (Section 4.7). This vortex pair can endanger smaller airplanes during takeoff and landing

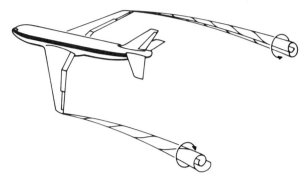

Figure 3.26. Vortex pair behind an airplane. These vortices also have an axial velocity component, which is discussed in Section 5.3.

operations of large airplanes (Fig. 3.26). These vortices persist for several minutes over the airfield and cause downdrafts of several meters per second, but they decay quickly when they reach the ground. A number of serious accidents involving smaller aircraft have been recorded, so that airport authorities are aware of the need to deal with this phenomenon [22].

If two vortices have the same direction of rotation and the same strengths, they revolve around the center of their connecting line (Fig. 3.27a). Vortices of unequal strengths also rotate around a center (Fig. 3.27b). This "center of the system" lies on the connecting line of the two centers at the point at which the velocities of both vortices are zero.

A natural drama of large dimensions is the meeting of two tropical whirlwinds. Since these atmospheric vortices are always cyclonic (Chapter 11), two neighboring vortices will have the same rotation and thus will rotate around each other. In the 1920s Fujiwara predicted this kind of motion for tropical storms. This "Fujiwara effect" has indeed been observed [23].

Another example of two vortices with the same rotational direction can occur on the flight deck of an aircraft carrier (Fig. 3.28). The two tip vortices, which are generated by the staggered flight deck,

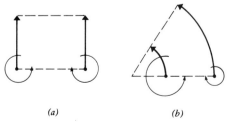

Figure 3.25. Motion of two point vortices with opposite rotations: (a) for equal strengths and (b) for unequal strengths.

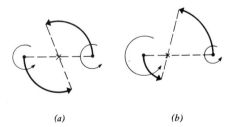

Figure 3.27. Motion of two point vortices with the same rotational directions; (a) for equal strengths and (b) for unequal strengths.

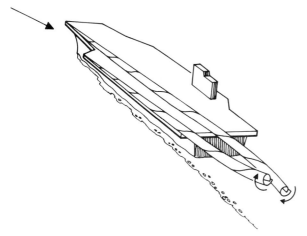

Figure 3.28. Vortex pair with the same rotational direction over and behind the flight deck of an aircraft carrier. As in Fig. 3.26 the vortices also have an axial velocity component.

are swept over the deck into the wake of the carrier and rotate around the center of the system. They can exert a dangerous force on landing airplanes.

The behavior of more than two vortices can be studied in the same way. Important for the understanding of vortex generation is the motion of a straight chain of point vortices, equally strong and rotating in the same direction (Fig. 3.29). This chain approximates a "discontinuity line," which is described in detail in Section 4.5. A small deviation of the point vortices from the straight line is sufficient to destroy this vortex configuration and produces a condition called "instability" (Section 6.1). The chain rolls up into a collection of larger vortices.

The model of a vortex chain in Fig. 3.29 is not realistic despite its intriguing simplicity. The roll-up process depends essentially on the initial disturbance and on the number of vortices per wavelength of the disturbance. In addition, viscosity is neglected (Section 4.5). A better approximation of reality can be achieved if the vortex chain is re-

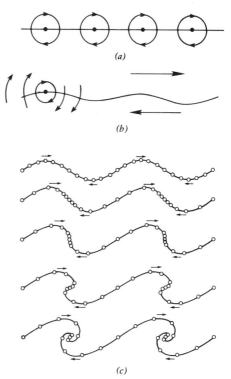

Figure 3.29. The straight vortex chain (a) represents a discontinuity line, which has a jump in the velocity (b). Such a discontinuity line is unstable since a small disturbance, which can be assumed to be wavelike, causes a further deviation from the straight line. The discontinuity line, that is, the vortex chain, rolls up to larger vortices (c). This process occurs because the individual point vortices of the chain influence each other through their velocity fields, as is demonstrated for two vortices in Fig. 3.27 (according to Rosenhead[9] [24]).

placed by a vortex band with a multitude of point vortices (Fig. 3.30).

The behavior of two vortices with the same direction of rotation will be examined once more by replacing the two point vortices in Fig. 3.27 with two vortex clusters consisting of a large number of point vortices (Fig. 3.31). When the two clusters are suf-

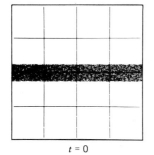 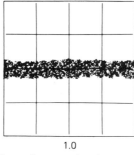 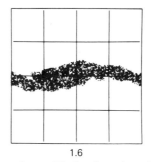 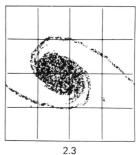

$t = 0$ 1.0   1.6   2.3

Figure 3.30. Roll-up of a vortex band to a vortex cluster. The band consists of hundreds of point vortices. The time t is given in dimensionless units (computer-generated picture from K. V. Roberts and J. P. Christiansen [25], North Holland Publishing Co.).

40 Properties of Simple Vortices

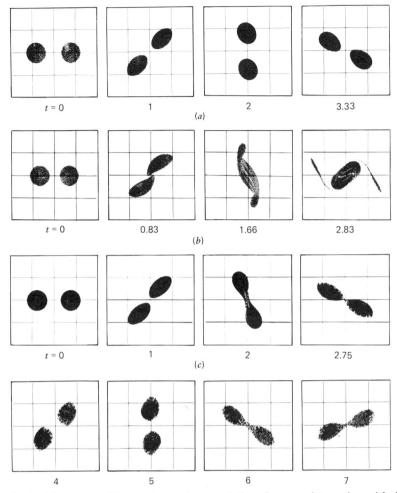

Figure 3.31. The motion of two vortex clusters consisting of many point vortices with the same direction of rotation. (*a*) The clusters are so far apart that they rotate around each other as two point vortices in Fig. 3.27. (*b*) The clusters are so close that they join each other. (*c*) Critical distance. The clusters rotate about each other and exchange point vortices (computer-generated picture from K. V. Roberts and J. P. Christiansen [25], North Holland Publishing Co.).

ficiently far away from each other, they rotate around each other as did the two point vortices in Fig. 3.27. However, the circular areas are deformed, and oscillations occur on the surfaces of the vortices (Fig. 3.31*a*). On the other hand, when the two clusters are very close, they merge (Fig. 3.31*b*). Between these two situations a critical distance exists, at which the vortex clusters alternately approach each other and draw apart; in this process they exchange point vortices (Fig. 3.31*c*).

3.6. CURVED VORTEX TUBES

In Section 3.5 only the motion of point vortices was considered, that is, the motion of vortex lines which are parallel to each other and not curved. Nonparallel straight vortex lines bend immediately, and curved vortex lines have a velocity component perpendicular to the plane of curvature. This velocity is induced by the individual elements of the vortex line in a way similar to that in which the velocity of two point vortices was related (Fig. 3.25). Unfortunately, this velocity component has an infinite value, so that the model of a curved vortex line is useless. However, if one assumes a core of nonzero diameter, which corresponds to a vortex filament in a vortex tube, the induced velocity is finite [26]. In general, curved vortex filaments change their form with time, and it is very unlikely that their initial state will ever be reached again. A turbulencelike state develops.

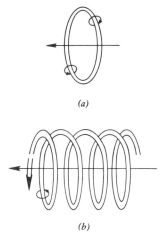

Figure 3.32. (a) Vortex ring and (b) helical vortex.

But there are exceptions, the vortex ring and the helical vortex, for example (Fig. 3.32). These vortices induce a velocity parallel to the axes of the ring and the helix, respectively, and move in space with constant speed. This traveling speed increases with decreasing radius of ring and core. The helical vortex, in addition, rotates about its own axis [27].

Figure 3.29 shows (and this can be proven in a rigorous mathematical way) that the slightest disturbance of the straight vortex chain leads to instability. The vortex chain rolls up, and the initial state will never be restored. Also, straight vortex tubes, vortex rings, and helical vortices become unstable through certain disturbances [26, 28]. However, there are also disturbances that cause the vortices to oscillate. Figure 3.33 shows a sequence of sketches (a) and a computer-generated picture (b) that demonstrate the oscillation of an elliptic vortex ring. This vortex pulsates not only in the plane of the ring but also perpendicular to it. The part of the ellipse with the larger curvature moves ahead faster than the one that is less curved. When running ahead, however, the curvature diminishes, and the piece of the ring that fell behind, now with the larger curvature, catches up. Then the cycle repeats.

Coaxial circular vortex rings of different diameters behave in a similar way (Fig. 3.34). Through induction the first vortex ring increases the diameter of the second, and the diameter of the first is diminished by the second. Simultaneously, the translational velocity of the first vortex increases in relation to the second, and it slips through the second. After that, the cycle is repeated [6].

It may be recalled in this context that the vortex property of oscillation was of utmost importance in Lord Kelvin's theory of vortex atoms (Fig. 1.13).

Disturbances can also occur in the form of periodic and solitary waves, which run along the vortex

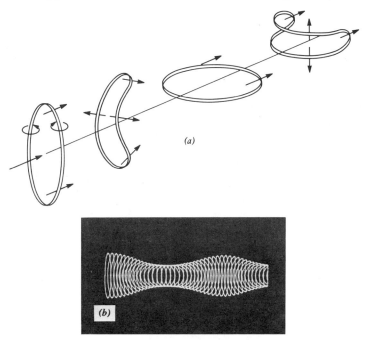

Figure 3.33. Pulsating vortex ring of elliptic cross section. The sketch (a) shows the form and direction of motion of the individual parts of the ellipse. On the computer-generated picture (b) the single phases of the oscillating vortex ring are seen from the side (H. Szu, Institute for Advanced Study, Princeton, N.J.).

42 Properties of Simple Vortices

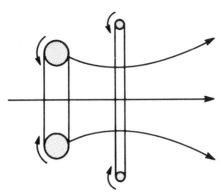

Figure 3.34. Movement of two vortex rings with the same axis and rotary direction.

tube. They are called "vortex waves" and "surges," respectively, and can be observed in the air cores of whirlpools as depicted in Figs. 3.35 and 3.36 [29–31].

Finally, it may be recalled that point vortices and vortex filaments are models in which viscosity is neglected. They must therefore be applied cautiously in describing realistic flow processes. There are, however, physical models, like those for superfluid helium and rotating neutron stars (Section 12.5), in which irrotational flow is postulated. Such fluids reveal an exotic behavior, which is briefly described here for superfluid helium-4 [32].

Liquid helium-4 becomes a superfluid below the transition temperature 2.2 K at the vapor pressure and is then called He II. This means that superfluid helium can move in apparent disregard of the laws of friction. The present view is that He II consists of two freely interpenetrating fluids, the normal fluid and the superfluid. At temperatures below 1 K the presence of the normal fluid may be neglected so that the superfluid prevails. According to Landau the superfluid is assumed to be irrotational. Thus, superfluid helium at rest cannot be made to rotate. For instance, in a rotating container of liquid helium, at below 1 K the superfluid remains at rest. If the container is speeded up, vorticity will finally appear in the form of quantized vortex lines, which are parallel to the axis of rotation and which have a circulation of $\Gamma = nh/m$, where h is Planck's constant, m is the mass of the helium atom, and n is a positive or negative integer ($h/m \approx 10^{-3}$ cm^2/s). With increasing rotation, more and more vortex lines appear, until the whole forms a solid-body rotation. Experiments have confirmed the existence of these vortex lines [33]. The core of the lines with a radius of $\approx 10^{-8}$ cm was made visible by trapping electrons

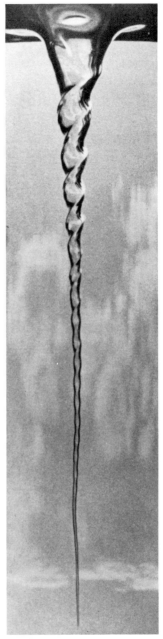

Figure 3.35. Vortex waves in the air core of a whirlpool (Photograph by T. Schwenk [Reference 67, Chapter 1]).

at the center. Since vortices cannot be created nor destroyed in the macroscopic continuum theory of irrotational flow, an explanation of the appearance of vortex lines must be sought in a microscopic quantum-mechanical process, which must also include the normal fluid part.

In a similar way quantized vortex rings (rotons) can be studied. Also, disturbances along the vortex lines (vortex waves) have been observed [34], and

Figure 3.36. Surge moving along the axis of a whirlpool (Photograph by P. E. M. Schneider, Max-Planck-Institut für Strömungsforschung, Göttingen).

turbulent flow models for superfluids have been studied [35].

Under normal conditions, however, viscosity cannot simply be neglected. For instance, the circular vortex ring in Fig. 3.32 does not move forward with constant speed but decreases in velocity and decays. In the same way the pair of vortex rings in Fig. 3.34 does not pulsate forever; rather the two vortices quickly decay. The behavior of vortices under the influence of viscous forces is discussed in Chapters 4–6.

Remarks on Chapter 3

1. The concept of a state variable is introduced here in a popular sense and not with the restrictions of exact thermodynamics.
2. A closed system is closed in the sense that the physical quantity considered can neither leave nor enter the system. For instance, matter cannot enter nor leave a closed vessel. However, heat may be exchanged with the surroundings if the vessel is not insulated.
3. The growth of coiled snail shells may be considered a kind of vortex motion (Fig. 3.37). The formation and evolution of the various types of shells have been dictated by functional adaptation to internal and external water flow [36]. It

Figure 3.37. The rotary growth of coiled shells.

may be recalled from Section 1.1, that the spiral form of fossilized shells is regarded as a "solidified movement." With this interpretation, there is no real dichotomy between the static and dynamic kinds of spiral symbolism at the beginning of neolithic time.

4. Does the moon rotate on its axis when it circles the earth? The question sounds trivial because the same part of the moon is always directed toward the earth. However, more than 100 years ago the penny paradox, which is the same problem, excited the emotions of readers in the journal *Scientific American* [37]. If one rotates a penny about the edge of a second, fixed one without slipping, the question is: Does the first penny rotate once or twice around its axis when revolving around the second penny? For an observer who looks from above on both pennies, the first penny rotates twice; for an observer on the penny, only once (Fig. 3.38). For an observer on earth the moon does not rotate; however, it rotates once in relation to the stars.
5. See Remark 2 in Chapter 2.
6. The circulation Γ, defined as the area integral over the vorticity distribution, is for a potential vortex $\Gamma = \lim_{A \to 0} \int \omega \, dA = \infty \cdot 0 = $ constant, ω is the vorticity.
7. In a fluid motion with shear flow the curvature of the free surface depends on the reaction of the fluid to stress. For instance, in air, water, and oil

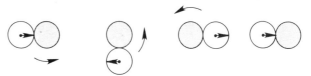

Figure 3.38. The "penny paradox."

44 Properties of Simple Vortices

Figure 3.39. The free surface of water and of dough near a rotating rod.

the particle strain is proportional to the stress. Such a fluid is called, after Newton, "Newtonian." Non-Newtonian fluids like honey, tar, and dough can behave quite differently. If one rotates a rod in oil or water, the liquid level sinks as shown in Fig. 3.39: the dough level, however, rises.

8. An unusual disaster happened in Louisiana in November 1980 when a drilling pipe on Lake Peigneur accidentally hit a salt mine underneath. The lake drained completely into the salt mine, creating a huge intake vortex which lasted for a few hours [38].

9. Rosenhead's calculation also suffered from inaccuracies, which made the results useless after a certain computing time. Fink and Soh [39] have overcome this problem by a smoothing technique (see Fig. 5.37a).

4. Vorticity

Before our eyes opens forth now the splendid prospect of three-dimensional kinematics, the mother tongue for man's perception of the changing world about him. Its peculiar and characteristic glory is the vorticity vector, for whose existence it is both requisite and sufficient that the number of dimensions be three.

C. Truesdell [1]

4.1 GENERATION AND SPREADING OF VORTICITY

In Chapter 2 the concept of vorticity was defined as the angular velocity of the fluid at a point in space. How does the fluid obtain its local angular momentum, and how is this angular momentum transmitted inside the fluid? To answer these questions, imagine a fluid completely at rest, its vorticity zero everywhere. In this fluid a body is now abruptly set in motion. The fluid is disturbed, and this disturbance spreads through the fluid. The pressure impulse travels with the speed of sound. Simultaneously, the fluid is sheared at the solid surface by the abrupt movement of the body. In this way, vorticity is generated[1] (Fig. 2.6). However, vorticity is transported much more slowly in the fluid than the pressure impulse. This spreading mechanism will be described in more detail below.

The generation of rotation may be imagined in a somewhat naive way as follows: Fluid particles do not slip along the wall but roll, owing to their adherence,[2] like balls and cylinders in a bearing. In Fig. 4.1 the flow near a blunt edge is sketched. The rotation of the particles is indicated by cylinders whose various diameters symbolize the strength of the vorticity. The stronger the shearing at the wall, the larger is the vorticity produced at the wall. Figure 4.2a shows the flow past a plate. On the centerline of the plate parallel to the edges the velocity is zero; it reaches its maximum near the edges of the plate. Correspondingly, no vorticity is created on the centerline; the edges contribute the most (Fig. 4.2b). Vorticity generated at the body surface spreads from there into the fluid through diffusion and convection [2].

In general, diffusion is the molecular exchange of physical quantities like mass, momentum, and energy from a fluid element to its neighbor. Diffusion of heat energy is designated as heat conduction, the diffusion of momentum as viscosity or friction, and the diffusion of matter in another medium as mass diffusion. Diffusion of vorticity is an exchange of momentum. In contrast to diffusion, the transport of physical quantities by the moving fluid itself is called "convection." For instance, the transport of heat energy through the flow of water in a warm-water heating system is convection, but heat conduction in a metal is a diffusion process.

The transport of vorticity is best perceived as analogous to heat convection and conduction. Convection of vorticity has the particular property that vorticity is preserved on a particle path. Thus, vorticity can be transferred to neighboring paths only by diffusion, that is, by the effect of viscosity. Immediately at the wall, to which the particles adhere, vorticity can be transferred to the fluid only by diffusion. Then, diffusion and convection together take care of the onward spreading. If diffusion is much larger than convection, vorticity expands in a way similar to a dipole distribution (Fig. 4.2b). As already mentioned, the production of vorticity is largest at the edges and zero at the centerline. Connecting the centers of the cylindrical rolls of equal radii drawn in Fig. 4.1 produces areas of equal vorticity,

46 Vorticity

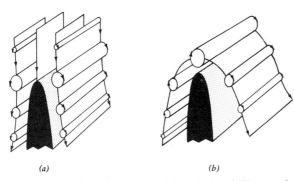

Figure 4.1. Plane slow motion (*a*) against and (*b*) around a blunt edge. The rotation of the single particles is symbolized by rolls perpendicular to the plane of motion. The larger the radius, the stronger is the vorticity.

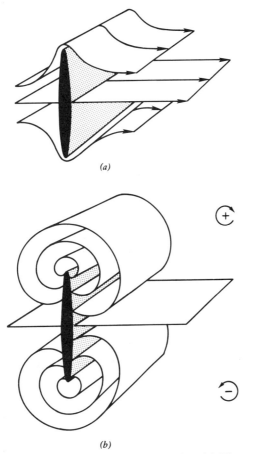

Figure 4.2. Plane slow motion against a plate (*a*). The areas of equal vorticity are dipolelike about the edges where the production of vorticity is largest (*b*).

which lie around the edges like shells. A cross section gives lines of equal vorticity, whose arrangement resembles a dipole.[3] The direction of rotation of the vorticity is opposite in the upper and lower halves.

4.2. DYNAMIC SIMILARITY

The force relation between diffusion and convection can be shown quantitatively with the concept of "dynamic similarity": Everyone understands what "geometric similarity" of two figures means. If the contour of two figures is the same but their size is not, the figures have geometric similarity. An airplane model in the toy shop is geometrically similar to the original plane on the airfield. The aerodynamicist is also interested in the similarity of air forces such as drag and lift, so that he or she may draw conclusions from measurements in the wind tunnel about the behavior of the full-scale plane to be built. The similarity of forces is called "dynamic similarity." In the same way that geometric similarity is determined by a parameter, namely by the ratio of model size to the original, dynamic similarity can be expressed quantitatively by a parameter. This will be shown in the following paragraph.

It has already been explained that vorticity spreads through diffusion and convection. The forces that cause these two processes are the viscous and inertial forces. If no other forces than these act in the fluid, dynamic similarity exists between two fluid flows as long as the ratio of the two forces is the same in both flows. The ratio of inertial force to viscous force is the "Reynolds number," named after the fluid dynamicist Osborne Reynolds [Reference 35, Chapter 1]. The mathematical expressions for the inertial and viscous forces show that the Reynolds number Re is

$$\mathrm{Re} = \frac{LV}{\nu},$$

where L, V, and ν are the characteristic length, velocity, and the kinematic viscosity, respectively. The width or length of a body may be chosen for the characteristic length, for instance, the chord of an airfoil (Fig. 4.3) or the diameter or radius of the pipe in a pipe flow. It is here irrelevant which geometric line is used as the characteristic length. The same argument is true for the characteristic velocity. An example will illustrate the application of the Reynolds number. For tests in the wind tunnel an airfoil might be built with the ratio 1:50. Since air is used for the model as well as for the full-scale design, the kinematic viscosity is the same in both systems. In order to keep the Reynolds number constant, that is, to preserve dynamic similarity, the velocities of

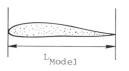 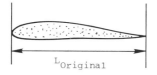

Figure 4.3. Sketch to explain Reynolds number (see text). L is the chord length.

model and full-scale design must have the ratio 50:1.

The introduction of flow parameters like the Reynolds number simplifies the study of flow phenomena considerably. For instance, if the pressure loss of an incompressible fluid in a smooth pipe of circular cross section were, in fact, to be determined experimentally, the pipe diameter, velocity, and kinematic viscosity would have to be varied individually, and the measured data plotted parametrically in a set of curves. By means of the Reynolds number, however, only one curve for the pressure loss is needed to analyze all possible variations of pipe diameter, velocity, and kinematic viscosity (Section 6.8).

4.3. NUMERICAL EXPERIMENTS

The computer is one of the most important of the many inventions that have spurred the tremendous growth of science and technology since World War II. Volumes have been written about the magnitude of data processing capabilities, the fantastic speed of computation, and the numerous possibilities of applying computers to improve many human activities. The influence of computers outside of science is large, and at the same time modern science itself is unthinkable without them. NASA's projects Voyagers 1 and 2, the spaceprobes to the planets Jupiter and Saturn, and the space shuttle "Columbia" would have been impossible without computers. An example of their application in fluid dynamics is weather forecasting. With the aid of computers the temporal development of atmospheric flows all over the earth can be computed—for the time being, of course, with very simplifying assumptions. With this tool, meteorologists have the capability to predict weather theoretically, and perhaps eventually even to predict earth's climate. Details on these possibilities will be presented in Chapter 9.

Computers can also be used to calculate time-dependent flow phenomena and to create detailed pictures of the action taking place. Everyone who has seen movies whose individual frames have been calculated and drawn by computer-controlled equipment is impressed with these "numerical experiments." The computations are based on mathematical models, and the results are the solutions of complicated equations. The computer is given a certain initial state of the flow as input, and, then, following the programmed algorithms, the computer calculates and "pictures" the time development of the evolving motion. The output can be as unexpected as in a physical experiment in a laboratory, thus the term "numerical experiment" [3]. The great similarity between abstract calculations and real events is often amazing. For instance, laminar flow patterns computed by solving the Navier-Stokes equations[4] with sufficient accuracy are so realistic that they match photographs of actual fluid motions. The accuracies of the computed force and pressure coefficients are within the limits of experimental error in the wind tunnel (see Figs. 6.16, 6.25, 6.26, and 6.45). Therefore, in many cases, numerical experiments can replace tests in wind tunnels or at least can reduce the number of difficult observations required to answer specific questions in meteorology and oceanography. In numerical experiments it is possible to control or even avoid the undesirable side effects of physical limitations built into wind tunnels, and the complex initial conditions for weather prediction can be changed or adjusted systematically in order to study their various influences. Computers can also calculate fields of physical quantities, which are difficult to measure with instruments. For example, the vorticity field can be calculated and plotted.

The revival of interest in the vortex theory since the 1950s can also be attributed to the power of computers. Indeed, all this progress in fluid dynamics is closely connected with the new knowledge in vortex theory. In order to understand the relation between computer and vorticity dynamics, some basic concepts of numerical analysis are necessary.

In fluid dynamics physical processes are described mathematically by differential equations that represent the conservation laws for mass, momentum, and energy at a point in the overall fluid continuum. The solution of these differential equations, that is, their "integration," over an extended fluid region yields the desired flow field. Since there are no simple mathematical solutions of these differential equations, except under grossly simplifying assumptions, they must be solved numerically point by point for a whole network of points superimposed on the flow region. This is, of course, possible only for a finite number of points, although a continuum consists of infinitely many points. Therefore, the larger the number of points in such a network, the better will be the approximation of the flow field. On the other hand, as the number of points increases, so do the computer costs. To appreciate the extent of the computations for such a network, picture the plane flow around a simple two-dimensional body. Figure 4.4 shows a section of a net with $60 \cdot 19 = 1140$ points in the vicinity of a thin elliptic cylinder. Such a network or mesh was used for the computation of the flow pictures in Chapters 5 and 6. To increase the accuracy of the computation, it is often advantageous to select for the network a coordinate system in which the body surface is one of the coordinate lines. In Fig. 4.4 the lines form an elliptic–hyperbolic system. For the numerical solution of the differential equations, finite-difference, finite-element, panel, or spectral methods may be applied [Reference 51, Chapter 1; 4–7], and each method has its advantages and drawbacks, depending on the problem to be solved.

To compute a plane flow with the accuracy of a few percent for the velocity, a mesh of at least $100 \cdot 100$ or 10,000 points is necessary. With this mesh, however, only flows of relatively small Reynolds number can be computed (Chapters 5 and 6), and this restriction will apply throughout the present discussion. For flows with a three-dimensional structure the third space coordinate must be included. Then, the number of points may jump to at least $100 \cdot 100 \cdot 100 = 10^6$, that is, to 1 million points or more. So far, however, only flows at a certain instant have been considered. If the time development of a flow field is computed, at least 5000 time steps are needed. For a plane flow problem, then, $10,000 \cdot 5000 = 5 \times 10^7$ space–time points are required, and for a spatial problem $10^6 \cdot 5000 = 5 \times 10^9$, or 5 billion space–time points. Then consider that for each point an entire system of equations must be solved. The fact that the computer can handle so many calculations serves to convey an idea of the power of modern computers.

Why has the computer gained such importance in vorticity dynamics? To answer this question, once more the experiment is imagined, to set a body abruptly into motion. The spreading of the pressure disturbance with the speed of sound is not crucial for the time development of the flow. Rather this development is dictated by the transfer of vorticity, which takes place much more slowly [2]. An extreme case will make this clearer: If one assumes an incompressible fluid, the speed of sound is infinite, and every pressure wave travels infinitely fast. However, the vorticity keeps its finite spreading velocity, and because the vorticity determines the flow development, the model of an incompressible fluid can be used in computer calculations, since it simplifies the differential equations considerably. Payne in 1958 was the first to compute the laminar time-dependent flow past a circular cylinder on this basis [8].

Physical flow phenomena can also be described more simply by the concept of vorticity than with the usual quantities of velocity and pressure. This will be demonstrated in the following sections with the aid of the Reynolds number and later with the phenomena of flow separation and vortex formation (Chapter 5).

4.4. FLUID MOTION AT VERY SMALL REYNOLDS NUMBERS

The Reynolds number has been defined as the ratio of inertial force to viscous force. When viscosity dominates, as in very slow motions, or for very small objects, or in very viscous fluids, the Reynolds number is small with respect to unity. An example of this is the flow past a plate, as shown in Fig. 4.2. For small Reynolds numbers the dipolelike distribution of vorticity is typical. In the limiting case, when Re = 0, vorticity spreads only through diffusion. Although this situation cannot be realized in practice,

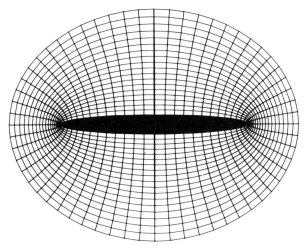

Figure 4.4. Section of an elliptic–hyperbolic network for the computation of plane flows around a plate of elliptic cross section (computer-generated picture).

it is of theoretical interest and is useful in applications as an approximation for very small Reynolds numbers.[5] The steady flow past a sphere will be used as an example. At Re = 0 the flow field (except for the direction) is the same in front and in the rear of the sphere. In Fig. 4.5a the flow can be reversed merely by changing the direction of the arrows. The lines of equal vorticity are dipolelike and completely symmetric with respect to the line A–B (Fig. 4.5b). In contrast to the lines of equal vorticity in Fig. 4.1, the rolls here do not have straight axes but are ringlike because of the axisymmetry of the flow. The symmetry before and after the body is surprising at first glance, since a certain asymmetry is expected because of the flow direction from left to right. However, this is exactly what happens when convection is neglected. One can see this immediately in the flow for Reynolds number 5 (Fig. 4.6). The small asymmetry indicates that now the inertial force has an influence, however small. Convection carries vorticity. In front of the sphere, therefore, convection works against diffusion, in the rear it supports diffusion. At higher Reynolds numbers, however, the interplay of convection and diffusion becomes quite involved, and, consequently, wake flows of various types form. Chapters 5 and 6 will deal with these phenomena. It may be mentioned that with increasing Reynolds number the vorticity in front of the body is pushed more strongly against the body and forms a "boundary layer" close to the surface. This concept, which is discussed more extensively in Section 4.5, is important in the study of very high Reynolds numbers, where viscosity can "almost" be neglected.

For the understanding of many phenomena in nature the description of flows around disks and rods is important. This is particularly true for the discussion of the free fall of these bodies in the earth's gravitational field. For Re = 0, disks and rods keep their initial orientation during fall, even when the bodies are dropped obliquely, that is asymmetrically, to gravity (Fig. 4.7). That means the bodies experience no torque during fall [10]. This statement, however, is valid only for Re = 0 and is based on the antisymmetry of the streamlines and lines of equal vorticity with regard to the straight line A–B in Fig. 4.8. However, it does not follow from the antisymmetry of the equivorticity lines that the lift forces of the upper and lower halves of the disk cancel each other; rather they add. Owing to this lift (or side force) the rod falls obliquely (Fig. 4.7c). The side force is important for the propulsion of microorganisms as will be described later. The side force is also responsible for the fact that a freely falling helical rigid filament rotates (Fig. 4.9).

In nature, tiny objects falling in air and water have very small Reynolds numbers (Re ≪ 1). Bacteria, mushroom spores, pollen, dust particles, and certain seeds are so small that their velocity of fall-

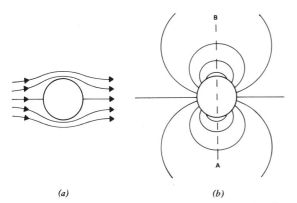

Figure 4.5. (a) Streamlines and (b) equivorticity lines around a sphere for Re = DV/ν = 0.

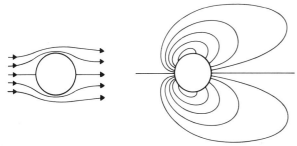

Figure 4.6. Streamlines and equivorticity lines around a sphere for Re = 5 (adapted from V. G. Jenson [9]).

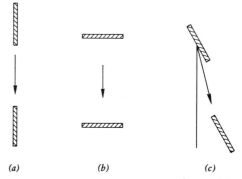

Figure 4.7. Free fall of a rod for Re = 0. The rod keeps its initial orientation during fall and does not rotate. Rod a moves about twice as fast as rod b. Rod c falls obliquely due to a side force.

50 Vorticity

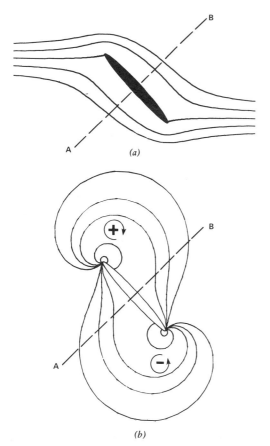

Figure 4.8. (a) Streamlines and (b) equivorticity lines around a plate obliquely placed in a parallel flow for Re ≪ 1 (computer-generated picture).

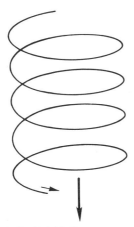

Figure 4.9. A helical rigid filament rotates during fall due to a side force, which acts on every rodlike element of the filament.

ing in air is very small. They are carried away by the smallest draft of air. The reason for the small velocity of fall is the high drag of these bodies. As is well known, the surface of a sphere increases with the square of the diameter; its volume, however, increases with the third power of the diameter. For small spheres, therefore, the surface is very large relative to the volume and so is the drag, which depends essentially on the size of the surface. For a constant velocity of fall (settling velocity), drag and gravity are at an equilibrium (Fig. 3.1a). The following examples give an idea of the order of magnitude of the Reynolds number for small objects [11].

Type of body	Diameter (mm)	Settling velocity (mm/s)	Reynolds number
Mushroom spores	0.005	4.5	0.0016
Pollen, birch	0.025	24	0.043
Pollen, beech	0.037	45	0.12

The Reynolds numbers are based on the kinematic viscosity of air, which is $\nu = 14.1$ mm^2/s at about 10°C.

The spherical form is not advantageous if a body is to fall as slowly as possible, because the sphere has the smallest surface area for a given volume. But this does not mean that its drag is at a minimum too. Admittedly, the drag of a sphere[6] is only 4.58% larger than the smallest possible drag of a body (Fig. 4.10).

The small drag of a sphere-type body is not detrimental for spores and pollen at very small Reynolds numbers. However, for aquatic organisms that

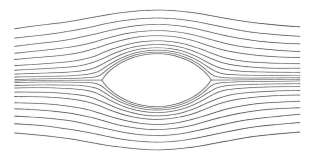

Figure 4.10. Body contour with the smallest drag for Re = 0 at a given volume. The body is axisymmetric (from J. M. Bourot, *Journal of Fluid Mechanics*, Cambridge University Press [12]).

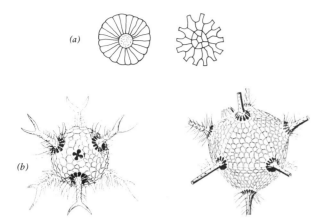

Figure 4.11. Microorganisms with small settling velocity in water: (a) disklike diatoms and (b) polygon-symmetric Radiolaria (from Haeckel's Challenger monograph [13]).

cannot propel themselves, the body form is often a matter of survival. Diatoms and radiolaria, which need the sunlight at the water surface for photosynthesis, must ride the smallest upward flow and must not sink to the dark bottom of the lake or sea, where they would perish. Figure 4.11 shows some organisms with large surfaces relative to their volumes. It is interesting that some radiolaria are able to control their upward and downward movements through generation of gas bubbles in their interior. Thus, they are not as helpless as diatoms.

The spherical and axial symmetry of many organisms, or to be more precise their polygonal symmetry about a point or an axis, is remarkable. Actually, these symmetries are not surprising as will be shown, and one should ask rather, why only lower organisms have such symmetries. When matter accumulates around a center without restricting conditions, for instance in crystals around a nucleus of condensation, the most probable arrangement is spherical symmetric (or, depending on the structure of the crystals, polygon symmetric around a point) [13]. If a force, like gravity, has a decisive influence on matter, the freedom of possible arrangements is restricted in the direction of this force. In this case one expects only axial symmetry around a line parallel to the acting force. The disk, which falls broadside down, is an example.

What influence does the flow have on the symmetry of microorganisms? At Re = 0 the flow around spherical, rodlike, and disklike bodies is symmetric in front and in the rear of the body (Fig. 4.5). For very small Reynolds numbers, therefore, the front part of the organism does not differ from the rear (Fig. 4.11a). This symmetry disappears at Reynolds numbers larger than zero (in practice larger than unity) (Fig. 4.6). The body must adjust to this new situation if it wants to resist the fluid in either a minimal or maximal way. The front–rear symmetry is then lost, and only the axisymmetry about an axis parallel to the flow direction remains (Fig. 4.12). Finally, even this symmetry is limited when organisms can propel themselves, for in addition to the direction of propulsion, the direction of gravity must be considered. Then only right–left or bilateral symmetry is present, which most creatures have, humans among them.[7]

Organisms that propel themselves exploit two different mechanisms at very small Reynolds numbers. One group, such as ciliates and certain bacteria, operates with cilia [11, 14]; the other group carries out wavelike motions, either with flagella, as do protozoa and spermatozoa [14–16], or through flexible change of the body contour, as do amoebas [17]. Figure 4.13a shows the various phases of the movement of the cilia. Between times 1 and 6 the cilia push against the water, generating a high resistance through lateral positioning. During the backstroke from 7 to 11 the cilia return to the initial position with a resistance as small as possible. In this way a thrust is produced during one period of the stroke. Figure 4.13b shows a ciliate as an example of an organism with cilia. The cilia are numerous and densely arranged, and their mutual influence must

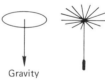

Figure 4.12. The seed of the dandelion is an example of the lost front–rear symmetry since the Reynolds number is larger than unity. The body still has axisymmetry in the direction of gravity.

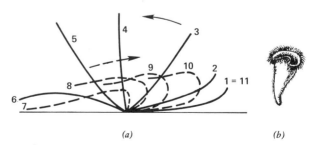

Figure 4.13. (a) The various phases of movement of cilia according to Jacobs [11]. (b) Ciliate; the cilia are about 0.005–0.01 mm long.

be considered [14]. Some ciliates (vorticella) are fixed by stems to a foreign body and can rotate through screwlike torsion of the stem.

In the evolution of marine creatures, locomotion with flagella is of much greater significance than that with cilia. Propulsions with flagella is not only faster but can be maintained at high Reynolds numbers. Thus during the evolution of larger sea animals, whose Reynolds number increases considerably with greater body length and speed, the propulsion mechanism did not need to be changed (see Section 5.4).

The transverse wave motion of a circular-cylindrical filament generates thrust. Figure 4.14 shows this way of functioning. During the transverse oscillation the elements of the filaments create a force component in the direction of locomotion when the wave travels from front to rear [15]. If V denotes the forward speed of the filament and U is the wave velocity, the approximate relation is $V = 0.4U$ for organisms with Re = 10^{-3}.

The wave motion is not restricted to a plane; it can also be helical motion or motion in the form of solid-body rotation. In both cases a torque is produced, which causes a rotation of the filament and an opposite rotation of the head. There are organisms without a head (spirochete) and with a head. The head can be very large in relation to the flagellum (Fig. 4.15a). In solid-body rotation, the flagellum rotates like a shaft in a ring. The head thus propels the flagellum like a rotary motor (Fig. 4.14b) [14, 18].

The optimal conditions for locomotion with fla-

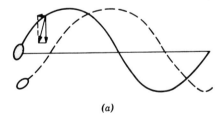

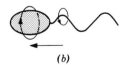

Figure 4.14. (a) Generation of thrust through the movement of flagella. The backward traveling transverse wave causes on an element of the filament a force in the direction of locomotion (see also Fig. 4.7c). (b) Solid-body rotation of bacteria with rigid filaments.

Figure 4.15. Organisms that propel themselves with flagella for Re ≪ 1: (a) Coccolithophoride and (b) sperm.

gella can be obtained theoretically [15]. The amplitude is about one-sixth of the wavelength, and the angle between an element of the filament and the direction of locomotion is a constant, about 42° (Fig. 4.14a). For plane wave motion these conditions produce a serrated curve, which can only be approximated in reality. This difficulty does not exist for spatial movements.

The following data give an indication of the sizes and motions of organisms with flagella.

1. **Plane waves.** (i) Sperm of bulls (Fig. 4.15b) [16]—Length $L = 0.07$ mm, diameter of filament 0.0002 mm, diameter of head 0.0005 mm, velocity of locomotion $V = 0.2$ mm/s, wave velocity $U = 0.46$ mm/s, frequency 10 Hz, amplitude of wave 0.013 mm, Reynolds number Re = $VL/\nu = 0.2 \cdot 0.07 / 10 = 1.4 \times 10^{-3}$. (ii) Free-living nematodes [15]—Length $L = 1$ mm, velocity of locomotion $V = 0.4$ mm/s, wave velocity $U = 1$ mm/s, Reynolds number Re = 0.04.

2. **Helical waves.** *Euglena viridis* [15]—Length $L = 0.1$ mm, length of head 0.06 mm, diameter of head 0.01 mm, wavelength 0.035 mm, amplitude 0.006 mm, frequency 12 Hz, body rotation 1 Hz.

3. **Solid-body rotation.** *Chromatium okenii* [18]—Length of filament 0.025 mm, diameter of head 0.008–0.015 mm, body rotation 6–8 Hz, rotation of filament 60 Hz, velocity of locomotion $V = 0.1$ mm/s.

4.5. BOUNDARY LAYER AND SHEAR FLOW

In 1904 L. Prandtl published a paper that has become of fundamental importance in fluid mechanics. He divided the flow field around a body at very high Reynolds numbers into two regions: a thin layer near the surface in which vorticity exists, and the region outside, which can be considered as essentially frictionless [19]. Prandtl called the thin friction layer a "boundary layer" (Fig. 4.16). This model permits large mathematical simplifications because potential flows and boundary-layer flows can be computed separately much more easily than can the whole field without these simplifications. In engineering the boundary-layer concept has been of far-reaching consequence. It can explain many phenomena in a simple way and has contributed much to the progress of technology, especially in the aircraft and turbine industries. The idea of the boundary layer will be applied extensively in this book.

Far away from the body surface, fluid layers can also exist in which the velocity across the layer changes rapidly and in which vorticity is present in concentrated form. These regions are generated when boundary layers detach from the wall and are carried into the fluid. This process will be described in Section 5.1. Such "free boundary layers" are usually called "shear layers" or "shear flows."[8] Figure 4.17 shows two examples of fluid motions which have the same vorticity distribution. Flow field (b) is generated from (a) simply by superposition of a constant parallel stream, which itself has no vorticity.

This behavior illuminates an important property of the vorticity field: Unlike the velocity field, the vorticity field is invariant with regard to changes of the inertial frame. As described in Section 2.3, the

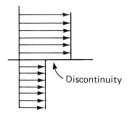

Figure 4.18. Discontinuity in a velocity profile.

dependence of streamlines and pathlines on the reference frame causes considerable difficulties in the study of fluid flows, particularly in defining a vortex (Fig. 2.13). Because of its invariance property, the vorticity field is, therefore, quite suitable for analyzing time-dependent flows, as will be demonstrated in Chapters 5 and 6. The question then arises as to whether the vorticity field can also be used to define a vortex. The answer is positive only for freely moving vortices in a homogeneous fluid (Section 5.3 and Chapter 8).

As the shear layer becomes thinner, eventually an infinitely thin layer will result. This is called a "discontinuity surface," since the corresponding velocity profile has a jump—a discontinuity. In plane flows in which the discontinuity surface is normal to the drawing plane and is represented by a line, one simply talks about a "discontinuity line" (Fig. 4.18). On this line the vorticity has an infinitely large value. The discontinuity line is an abstraction that does not exist in reality. However, as in many a theory, it has the advantage of simplifying and illustrating certain events. For a short time period or locally, discontinuities can be approximated in an experiment or can occur in nature. For instance, if two very thin plates slide along each other in water and are suddenly removed from the water, a discontinuity surface is generated, which decays quickly. In the earth's atmosphere a cold front is a discontinuity line: the temperature shows a sharp jump across the line of the front.

A discontinuity line of the velocity decays by

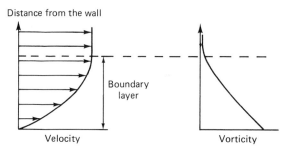

Figure 4.16. The concept of boundary layer.

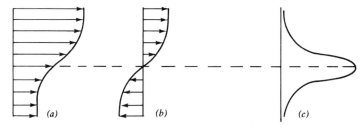

Figure 4.17. Two shear flows (a) and (b) with equal vorticity distribution (c).

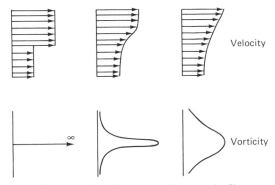

Figure 4.19. Decay of a discontinuity line.

spreading out vorticity (Fig. 4.19), which for high Reynolds numbers is associated with the occurrence of instability (Fig. 3.29).

The spreading of a discontinuity can be imagined not only in time but in space. The sequence of sketches in Fig. 4.19 then represents not profiles for one place at different times but profiles for a succession of locations. An example of this is the spatial spreading of a jet (Fig. 5.24).

Decay and spreading are viscous effects. The larger the Reynolds number, the slower is the diffusion process relative to convection. For very large Reynolds numbers, diffusion can often be neglected, and discontinuities can be assumed in a flow. This model will be often applied in this book.

4.6. THE CLASSICAL THEORY OF VORTICITY

Before Helmholtz published his famous work "On integrals of the hydrodynamic equations which correspond to vortical motions" in 1858 [20], the opinion was generally held that flow without vorticity was a well-founded theoretical assumption. However, large discrepancies with reality had been recognized, and Euler had already pointed out that the assumption of potential flow is not always justified. Helmholtz rejected this assumption and, with his paper on vortical motions, opened the door to important new discoveries. This was the beginning of the classical theory of vorticity.

Helmholtz did retain the idea that viscous forces may be neglected in a fluid with vorticity. These subtle but extremely important differences among the concepts of potential flow, fluid with vorticity, and viscous flow are expressed in precise terms as follows for the understanding of the text in the ensuing sections.

Assumptions	Terminology
Vorticity is zero in the whole fluid (except for singular points), and viscous forces are neglected	Potential or irrotational flow[9]
Vorticity is present but viscous forces are still neglected (no diffusion of vorticity)	Ideal or inviscid or frictionless flow
No restricting assumptions at all on vorticity and viscous forces	Real or viscous flow

In his paper Helmholtz derived two theorems on vorticity. The first theorem is general and valid for real flows. It says that the strength of vorticity in a vorticity tube is the same in all cross-sections. In addition, the theorem states that a vorticity tube must be closed or must end at a boundary. To understand this more clearly, one may think of the analogous case of a stream tube, like the flow of water through a pipeline. If the cross-section diminishes, the velocity of water must increase, since the flux through all cross-sections is the same according to the conservation law of matter (Fig. 4.20a). Vorticity behaves in the same way. The total strength of the vorticity remains constant over a cross-section. If the vorticity tube narrows, vorticity (or to be exact the vorticity component perpendicular to the cross-section) must increase. Such vorticity components are drawn in Fig. 4.20b. Arrows on the circumference indicate the corresponding tangential velocities. Also, the statement that a vorticity tube cannot end inside a fluid but must be closed or end at a boundary is analogous to the stream tube. The same is true for the branching of a tube (Fig. 4.20c), for which an example is the propeller vortex (Fig. 4.34).

Helmholtz's second theorem is valid only for ideal flows of incompressible fluids.[10] It says that vorticity in such a flow can neither be generated nor destroyed, since during the movement of vorticity the fluid particles cannot leave the vorticity line on which they are positioned. Leaving this line would be possible only through diffusion. For plane flows Helmholtz's second theorem becomes considerably simpler and more illustrative: The fluid particles carry vorticity without either losing it or acquiring new vorticity. An example for Helmholtz's second

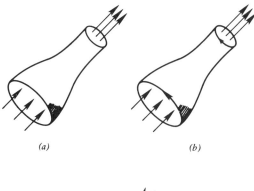

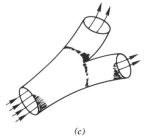

Figure 4.20. (a) Velocity component perpendicular to the cross section in a constriction, (b) vorticity component, and (c) branching of a stream or vorticity tube.

theorem is the smoke ring. Although this example demonstrates the usefulness of the ideal-flow model, it reveals that it is impossible to neglect viscous forces in general. If a particle can neither acquire rotation nor give it away, how can it rotate in the first place? The fact that these thoughts are not as obvious as they appear can be seen in the classical controversy between Hadamard [21] and Klein [22], in which Hadamard rejected the possibility of vorticity production in a inviscid fluid, but in which Klein argued that Helmholtz's vorticity theorem must be interpreted in such a way that it includes vorticity production through discontinuities. Subsequent discussions in the literature were either in support of Klein (Betz [23, 24]) or against him (Ackeret [25]), but none offered compelling proof. Prandtl, although in general supporting Klein (see Prandtl [26]), nevertheless offered a solution to the problem in his ingenious, practical way by cutting the Gordian knot as follows: He argued that viscosity cannot be neglected. One therefore must look at the limiting process Re → ∞ rather than analyze the existence properties of the equations for inviscid fluids at Re = ∞. (This limiting process is the basis for his boundary-layer concept in Section 4.5.) If one has to start with an inviscid-flow solution, because the Navier–Stokes equations are too difficult to solve, one must introduce the discontinuity line artificially, that is, in an "axiomatic" way [27].

Helmholtz's second theorem has an interesting consequence. Since a vorticity field is invariant with regard to changes of the inertial frames, it follows that a streak line, whose particles carry vorticity, cannot alter when the inertial frame is changed. For instance, a discontinuity line, which rolls up because of instability, is invariant to changes in inertial frames (Fig. 3.29).

The application of Helmholtz's two theorems is illustrated in a vorticity tube perpendicular to the water surface of a river and extending to the river bed (Fig. 4.21). This whirlpool moves with the current in which it was generated. If one neglects the decay of the vortex over a short distance, then the total strength of the vorticity is constant in time according to Helmholtz's second theorem. When the vorticity tube reaches a region of deeper water, the tangential velocity of the vortex increases through stretching of the tube according to Helmholtz's first theorem. The increase in velocity is visible in the deeper dimple in the water surface, an effect which has already been explained in Section 3.3.

A whirlpool can also occur without being in contact with the river bed. In this case both ends of the

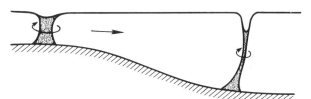

Figure 4.21. The stretching of a vorticity tube.

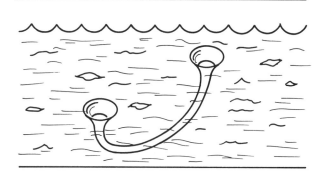

Figure 4.22. Both ends of a vorticity tube are terminated at the surface of the water.

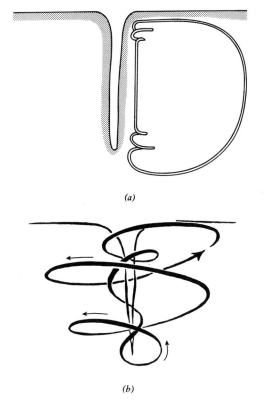

Figure 4.23. (a) Closed loops of vorticity tubes near a free surface. (b) Corresponding pathline.

vorticity tube must be at the free surface (Fig. 4.22), or they must form closed loops (Fig. 4.23). These tubes spread out at the upper and lower parts of the whirlpool in an axisymmetric fashion (only one is sketched in Fig. 4.23). The corresponding streamline (or particle path) looks like the one drawn in Fig. 4.23b. An explanation of the various types will be given in Chapter 7.

4.7. THE LANCHESTER-PRANDTL HYPOTHESIS OF FLYING

Shouldn't human beings also be born with the ability to fly?

C. F. Meerwein, 1784

Shortly before the end of the last century, time was ripe for the development of airplanes. The gliding of large birds proved that it ought to be possible, in principle, to build a heavier than air flying apparatus with which a human could fly. Since Icarus, the legendary Greek flier, there have always been humans, among them no less a person than Leonardo, who wanted to realize this dream of mankind despite the scorn and laughter of detractors. Perhaps Meerwein from Emmendingen, Germany, was the first man to actually glide. He is said to have glided down a hill in 1784 with a self-built flying machine and to have landed softly on a dunghill [28]. Ironically, the fact that the Right Honorable "Landbaumeister" had chosen a landing place so below his rank seemed to have been of more hilarious importance to the spectators than the significance of the event itself. Some historians attribute the first glider to Jean-Marie Le Bris (1808–1872).

About the middle of the 19th century the number of people who seriously accepted the challenge of solving the problem of flying increased significantly. Among them were Stringfellow, Lilienthal, Chanute, Penaud, and Langley [29]. They experimented and looked for theories, but the prospects for a theoretical explanation of flying remained slim. Although Cayley (1773–1857) argued that flying with an apparatus heavier than air should be feasible, it was "proven" at the beginning of the 19th century, on the basis of a mistaken application of Newton's sine-square law (which is approximately valid only for very rarefied gases), that flying "is impossible" [Reference 45, Chapter 1]. In addition, potential flow theory, which at that time was used as the basis for most fluid flow studies, stated that no force acts on a body moving with constant speed; that is, such a body has neither drag nor lift (d'Alembert paradox), which obviously contradicts reality. Then, in 1858, Helmholtz published his theorems on vorticity, but they were not directly applicable to the problem of flying. This was the situation in 1894 when Lanchester (1868–1946) found a remedy [Reference 45, Chapter 1]. He expounded the hypothesis that a flat plate has lift if it is placed at an oblique angle in a potential flow and a rotation is superimposed (Fig. 4.24). This rotation is called "circulation" in aerodynamics. Lanchester had grasped the essence of lift.

Lanchester's hypothesis was formulated mathematically by Kutta and Joukowsky. The amount of lift was then determined by Kutta by the condition that the rear stagnation point shifts to the trailing edge of the plate (Fig. 4.24b). The fruitfulness of this working hypothesis for airfoil theory was immense. A comparison with real fluid flows shows (Fig. 5.29) that friction forces do in fact cause exactly what the circulation concept does in potential flow: the shifting of the rear stagnation point to the trailing edge in such a way that the resulting pressure distribution around the plate generates lift.

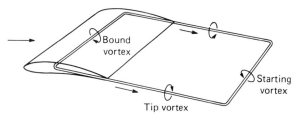

Figure 4.25. Bound vortex, tip vortices, and starting vortex form a closed vortex ring.

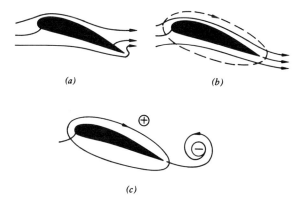

Figure 4.24. Generation of lift of an airfoil through superposition of circulation in a potential flow. (a) Potential flow without circulation, no lift; (b) potential flow with circulation, lift; (c) the sum of starting vortex $(-)$ and circulation $(+)$ (bound vortex) is zero.

Lanchester's hypothesis contradicts Helmholtz's second theorem that vorticity adheres to the fluid particle and not to the body [30]. However, the hypothesis worked. One could now talk about "bound vortices" in contradistinction to free vortices. But although lift is obtained with this model, the body still does not experience drag. Lanchester, therefore, improved his theory by assuming a wing of finite (not infinite) length. At the end of the plates two vortices must occur because the pressure above the plate is smaller than below. The exact description of these "tip vortices" is given in Section 5.3. The tip vortices at the ends of the plate cause a drag, the so-called "induced drag." With the bound vortex they form a "horseshoe vortex." This vortex must be closed, according to Helmholtz's first theorem, since a vortex filament cannot end inside a fluid. The missing vortex is the "starting vortex," which is generated when the plate begins to move (Figs. 4.24c and 4.25).

It is of historical interest that Lanchester was considered an outsider and that his theory found little understanding among his countrymen in England. His first paper on circulation around a plate with nonzero angle of attack was even rejected for publication.

Meanwhile people on the continent were also working on the theory of flight. Independently from Lanchester (although later) Prandtl developed the hypothesis of plate circulation as well as the model of the tip vortex [27]. These ideas became the basis for airfoil theory. From these beginnings together with the previously mentioned boundary-layer theory, the famous Göttingen School around Prandtl evolved. This school exerted great influence on the development of aerodynamics between the two world wars.

The flat plate with a nonzero angle of attack does not have the highest lift. The optimal shape of a wing profile depends on many factors, and their explanation exceeds simple potential flow theory. Flow separation is one factor which influences lift considerably. For Reynolds numbers up to 10^5, thin, slightly cambered plates are advantageous for obtaining a large lift with a drag as small as possible (Fig. 4.26a). Insects, birds, and model planes fall in this range of Reynolds number. At Reynolds numbers larger than 10^5, streamline profiles are advantageous in subsonic flow (Fig. 4.26b) [31].

In Fig. 4.27 lift coefficients c_L are given as a function of the angle of attack α for some airfoils.[11] The lift coefficients of thin, flat plates calculated according to circulation theory (a) agree well with experimental data (b) for Reynolds numbers in the turbulent range (ca. Re $= 2 \times 10^6$) and up to $\alpha = 7°$. At higher angles of attack the flow separates behind the leading edge, and the lift coefficient drops. Streamline shapes have about the same lift curve in this Reynolds number range, but the flow separates at larger angles of attack. For smaller Reynolds numbers in the laminar flow range ($< 7 \times 10^4$) the curves are lower. Thin plates (c,d) are better suited for lift generation in this range than streamline profiles (d). Cambered plates have particularly good lift coefficients (e). By 1874 the Lilienthal brothers had recognized the advantage of cambered plates.

There is a distinction between passive and active flight, depending on whether the body is merely car-

Figure 4.26. (a) Slightly cambered plates and (b) streamline profiles generate larger lift than flat, straight plates at nonzero angle of attack. The optimal wing profile depends not only on the Reynolds number but also on other factors.

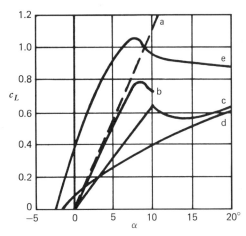

Figure 4.27. Lift coefficients as a function of the angle of attack (*a*) for a thin, flat plate according to the circulation theorem, (*b*) for a thin, flat plate at Re = 2×10^6, (*c*) at Re = 4×10^4, (*d*) streamline profile at Re = 4×10^4, (*e*) cambered plate at Re = 4×10^4 (curves *c*, *d*, and *e* from Hertel [16]).

ried by the air or can propel itself. In passive flight there is also a distinction between "gliding" and "soaring." Gliding is simply falling along an inclined path, although the body must have a certain lift (Fig. 4.28). Soaring is considered gliding in an upwind whose velocity can even exceed the downward velocity of the body. Then, the body ascends while soaring. Seagulls, vultures, condors, and gliders benefit from the upwinds in heat convections ahead of thunderstorms and fronts, in leewaves of mountains, and in the upwinds of mountain slopes (Chapter 8).

The glide angle in Fig. 4.28 is a measure for the quality of a gliding body. The tangent of this glide angle is called the "glide number" and is the ratio of drag to lift. The glide number depends not only on the angle of attack and the wing profile but also on the ratio of span to chord. For a ratio of 6:1 glide numbers of 1:20 and more can be obtained with favorable profiles. This means that a body travels a horizontal distance of 20 m as it drops 1 m. In recent years gliders made of carbon-fiber-supported plastics have reached glide numbers of more than 1:50. For example, the "SB 10" of "Adademische Fliegergruppe" in Braunschweig, Germany, has a span of 29 m and an aspect ratio of 20.

Gliding and soaring are exploited not only in the feathered world and in the sport of gliding, there are winged fruits and seeds, flying squirrels, lizards, frogs, and fish [11]. The technique of ski jumping also involves an aerodynamic lift problem.

Active flying is done by birds and insects, and fish that swim by flapping their tail fins use the same principle. The tail fin is an oscillating flat, elastic plate, which not only flaps to and fro in a translational way but also rotates and can be deformed. Through suitable change of the angle of attack during a flap period, thrust is produced (Fig. 4.29). Active flying rests on the same principle, but, in addition, lift must be provided to counteract gravity. When Fig. 4.29 is turned by 90° and a certain angle of attack for lift is added, the typical flapping of bird and insect wings is obtained (Fig. 4.30). In addition to lift circulation (and the corresponding induced drag) flapping wings and fins generate complicated vortices, which are shed from the body and which cannot be described with simple airfoil theory (Chapters 5 and 6).

Lift can also be generated by the alternate separating and folding of two plates. Although the total circulation is zero for ideal flows, according to Helmholtz's second theorem, the two plates have after separation the same amount of circulation but

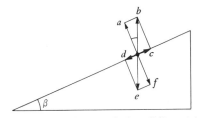

Figure 4.28. Balance of forces during gliding with constant speed: *a*, lift; *b*, aerodynamic force; *c*, drag; *d*, thrust; *e*, gravity; *f*, downward force; β, glide angle.

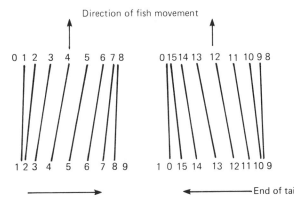

Figure 4.29. Angles of attack of a fin during a flap period for the production of thrust. The cycle is divided into 16 time intervals. Reprinted (with modification) with permission from *Mathematical Biofluiddynamics* by Sir James Lighthill, p. 22, 1975. Copyright 1975 by Society for Industrial and Applied Mathematics.

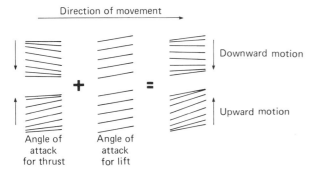

Figure 4.30. The principle of flight of birds and insects. Reprinted with permission from *Mathematical Biofluiddynamics* by Sir James Lighthill, p. 29, 1975. Copyright 1975 by Society for Industrial and Applied Mathematics.

with opposite sign (Fig. 4.31). This "Weis–Fogh mechanism" explains the hovering of certain wasps *(Encarsia formosa)*. Figure 4.31*d* shows schematically the process during the opening of the wings.

4.8. WHEEL, PROPELLER, AND BOOMERANG

The wheel is the symbol for endless periodicity. It is believed that it was invented 5000 years ago in Mesopotamia, perhaps earlier in southern Russia. From the first cart wheel and pottery disk, the application of the wheel has been of enormous significance in technology [Reference 64, Chapter 1].

The wheel is also a symbol of the human intellect, for the principle of the wheel could not be applied in the evolution of locomotion in higher animals. There are, of course, monocellular organisms, which move by rotating in water; there are tree seeds, which are carried away by the wind on helical wings; and there are ball-turning scarabs.[12] Some organisms of microscopic dimensions probably have a mechanism that enables a flagellum to rotate like a shaft in a ring (Fig. 4.14*b*). But in higher animals blood vessels and nerve fibers would have to be severed and bridged by gliding connections to make wheellike motion possible. The circular motion of the human arm and leg, therefore, can be only translational (Fig. 4.32). For instance, the pedals of a bicycle must transmit the translational circular motion of the feet into the rotation of the wheel. Nature has circumvented its inability to develop a wheel with a trick. Through the translational periodicity of swinging and whirling it has created effective mechanisms for propulsion: the flight of birds and insects and the swimming of fish.

As a transport element in fluids, the wheel can generate thrust in three directions, corresponding to the three spatial coordinates: radially to the axis, parallel to it, and in the azimuthal direction.

The paddlewheel generates an azimuthal thrust by using the drag of a plate normal to the flow (Fig. 4.33*a*). In nature the azimuthal thrust of swimming aquatic birds and sea turtles is generated with their

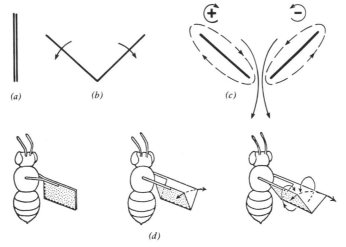

Figure 4.31. Weis–Fogh mechanism for the explanation of hovering of *Encarsia formosa*. Two adjacent plates (*a*) produce two circulations of equal strengths but opposite signs during unfolding (*b*) and separation (*c*). (*d*) Sketch of the arrangement of the wings according to Weis–Fogh. The effect of vortices at the ends and the sides of the wing is neglected. These vortices will be discussed in Chapter 6. Reprinted (with modification) with permission from *Mathematical Biofluiddynamics* by Sir James Lighthill, p. 177, 1975. Copyright 1975 by Society for Industrial and Applied Mathematics.

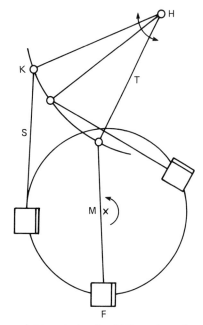

Figure 4.32. Translational whirling through two joints. Schematic representation of a rotating foot when bicycling: F, foot; K, knee; M, center of the circle of rotation; T, thigh; H, hip-joint; and S, shank (adapted from Hertel [16]).

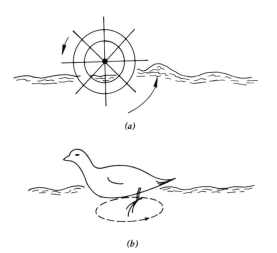

Figure 4.33. Azimuthal thrust in nature and technology.

feet. The circular movement, of course, is here translational (Fig. 4.33b). During propulsion the area of the spread foot is used to oppose the water with a drag as large as possible. When the foot is drawn forward, its area facing the flow is kept as small as possible. This process is reminiscent of the movement of cilia in Fig. 4.13a. In the locomotion of aquatic birds, of course, the Reynolds number is much larger and the flow behind the feet is whirled.

Axial thrust is of much greater importance in technology than are azimuthal and radial thrusts, because it is more efficient. The most important devices for producing axial thrust are propellers, turbines, water screws, and wind mills. Propellers and water screws were proposed by Leonardo (the water screw actually by Archimedes) but were not utilized until the 19th century. Their mode of operation can be explained by airfoil theory (Figs. 4.34 and 4.35). The cross section of a blade is simply an airfoil profile. When wind blows at an airfoil in an axial direction, a force component is produced in the azimuthal direction, and the propeller rotates. If, on the other hand, the propeller is driven by a motor, a suction force is generated in the axial direction, which corresponds to a thrust. For an airplane propeller the forward thrust is horizontally directed; for a helicopter the thrust is mainly vertical to counteract gravity [33]. But suppose the motor stalls for some reason. If the propeller can be disconnected from the motor quickly enough to allow the blades to rotate freely, the braking effect of the rotating blades can enable the helicopter to descend slowly. This phenomenon is the same as blowing air against a freely rotating propeller. The axial thrust caused by the fall is transformed to a large extent into rotary energy of the blades.

The braking effect of the rotary motion during fall is used in nature in various ways. For very small Reynolds numbers (Re $\ll$ 1) helical filaments have already been mentioned (Fig. 4.9). In this case, of course, only viscous forces are acting. Fruits and seeds often have the form of propellers. Figure 4.36 shows the path of a maple seed.

In nature the hovering flight of a humming bird is very similar in principle to that of a helicopter.

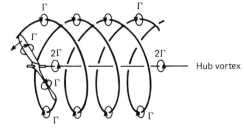

Figure 4.34. Vortex system of a propeller. The vortex tube of the axis with circulation 2Γ branches to the bound vortices of the blades. These again pass over in the helical tip vortices with circulation Γ. The lift force on the blades has a component in the axial direction that generates a thrust (see also Fig. 4.35) [32].

Wheel, Propeller, and Boomerang 61

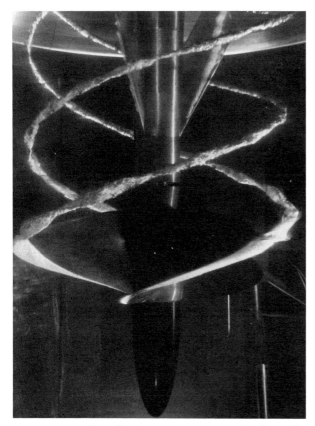

Figure 4.35. Propeller of a ship with three blades under water. The tip vortices, which start at the ends of the blades and wind around the axis helically, are clearly visible through the hollow cavitation cores (photograph, David Taylor Naval Ship Research and Development Center).

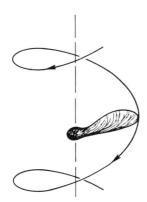

Figure 4.36. Falling maple seed.

Although the whirling of the wings of a humming bird is translational, the overall effect is similar to that of the rotating blades of a helicopter as can be seen when the "actuator-disk theory" is applied (Fig. 4.37). In this model thrust is uniformly distrib-

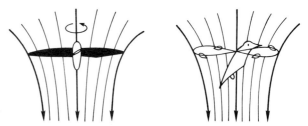

Figure 4.37. The hovering of a humming bird can be compared approximately to the functioning of a helicopter.

uted over the rotor disk across which a sudden jump in pressure occurs. Air moves through the rotor, but beyond it the air is undisturbed. At the border of these air masses slip is assumed. A more realistic model for the helicopter requires a helical sheet (Fig. 4.38) and for hovering birds a sequence of detached vortex rings [33, 34] (see Section 5.7).

The freely rotating propeller blades are formed in such a way that they start rotating from rest whenever they are blown at. A flat board would not start to spin (Fig. 4.39). However, if it receives an initial impulse, it may under certain conditions continuously rotate by itself in an airstream. This phenomenon is called "autorotation" [35]. The "Lanchester propeller" is explained in the following way

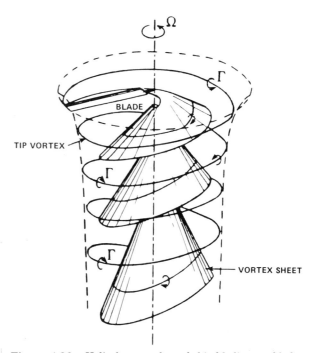

Figure 4.38. Helical vortex sheets behind helicopter blades (from A. R. S. Bramwell [33], Edward Arnold Ltd.).

62 Vorticity

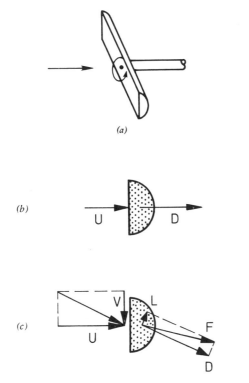

Figure 4.39. Lanchester propeller. The explanation is given in the text.

(Fig. 4.39): (a) The propeller is set into an airstream parallel to its axis of rotation. (b) The velocity of the airstream is U but there is no initial spin, and the resulting force F is equal to the drag D. (c) With small initial spin $V = \Omega R$ (where Ω is the angular velocity) the flow velocity relative to the propeller is composed of velocities U and V. In addition to drag D, lift L is generated. The resulting force F now has a component opposite to the direction of rotation so that the propeller slows down until it comes to rest. (d) If the initial spin is large enough, however, the resultant F may have a component in the direction of rotation, and the propeller autorotates [Reference 47, Chapter 1; 36]. The propeller blade need not have a profile but can consist of a thin plate. Glauert [37] derived an approximate criterion

for autorotation: the slope of the lift curve $dL/d\alpha$ (with α the angle of attack) must be negative and larger than the drag of a blade element—Glauert's criterion: $dL/d\alpha + D < 0$.

Autorotation of airfoils as shown in Fig. 4.40 is of significance in aerodynamics. With small initial rotation the lift of the downward moving wing half is greater than that of the upward moving part because of the larger angle of attack. The resulting torque acts against rotation and returns the wing to the horizontal position. However, with sufficiently large initial rotation the flow behind the downward-moving wing half stalls from a certain angle of attack on, and the lift on this part will be smaller than on the upward moving part. According to Fig. 4.27 the drop of the lift corresponds to a negative lift slope and Glauert's criterion can be met (provided that $dL/d\alpha$ is larger than D). The resulting torque then acts in the direction of rotation, and the wing autorotates. Glauert, by the way, found the criterion for autorotation when he studied the spin of airplanes in 1919 [37].

Autorotation is an essential element of "spin," which is the spiral descent of an airplane (Fig. 4.41). To explain steady spinning in a very simplified way, the fuselage is assumed to be a rectangular thick plate and the angle of attack to be 90°. Then, a yawing motion around the spin axis is similar to the roll of the wing. In reality, spin motion is much more complicated and, in a free fall at an angle of attack less than 90°, consists of both rolling and yawing motions [38]. Spin can be very dangerous to airplanes and has caused many crashes, especially of light aircraft. During World War I (until 1916) spin was almost always fatal. Then, spin was mastered and became part of air maneuvering. Modern jet aircraft with large nose sections and swept wings are also prone to spinning [39].

Radial thrust is advantageous for thrust generation from rest or for forward movement that is small

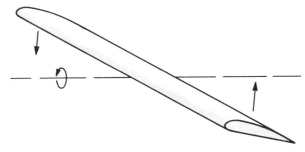

Figure 4.40. Autorotation of wings.

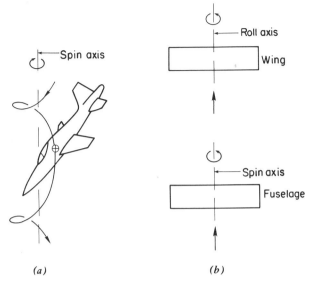

Figure 4.41. (a) Spin of an aircraft which (b) is composed of autorotation of wing as well as fuselage.

relative to the tangential velocity of the blade. To obtain radial thrust, the angle of attack of the blade must be changed during a single revolution in such a way that the lift force always has a component in the direction of motion. Figure 4.42 shows the blade positions of a Voith–Schneider propeller during one revolution.[13] The blades do not rotate around their axes; they turn in a translational way. The ratio of forward speed to translational velocity is unity in Fig. 4.42. With a smaller ratio, more thrust is generated. In this case the blades do rotate about their axes.

A curiosity in the application of airfoil theory is the boomerang. The aborigines of Australia use it for hunting and warfare as a ballistic projectile, which returns to the sender if the target has been missed. The theory of this legendary throwing stick was studied some years ago by Hess of the University of Groningen [40]. The various types and applications of returning and nonreturning boomerangs will not be discussed here, but the principle of the returning type will be investigated more closely.

The boomerang is a banana-shaped piece of wood. It is held with the hand at one end and thrown forward in a horizontal direction so that the plane of the boomerang is about vertical. As it is released, the boomerang is made to rotate rapidly. Whether the ends of it are directed forward or backward is not essential; its flying performance depends on its profile (see next paragraph). After the boomerang has been released, it first flies in the direction of the initial momentum, turns slowly to one side and upward, and returns in an arc.

The flight principle is a combination of airfoil theory and the theory of gyros. The curved form of the boomerang is of only secondary importance and serves merely to stabilize its flight. Cross- or star-shaped types can also be used. The important feature is the cambered profile of the arms (Figs. 4.43 and 4.44). Because one side of the profile is more convex than the other,[14] a lift force perpendicular to the plane of the boomerang is generated, and this force is larger for the forward moving arm than for the following one (Fig. 4.45). The translational ve-

Figure 4.43. Profile of a boomerang of the Australian aborigines.

Figure 4.42. Simplified representation of the translational movement of the Voith–Schneider propeller. The engineering design of the propeller is quite complicated.

Figure 4.44. The airfoil properties of a boomerang.

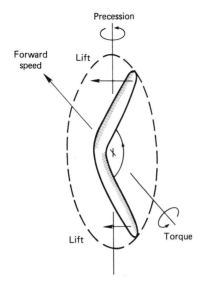

Figure 4.45. Forces and torques, which act on a boomerang (adapted from F. Hess, "The Aerodynamics of Boomerangs," *Scientific American*, Nov. 1968, p. 124. Copyright © 1968 by Scientific American, Inc. All rights reserved).

locity of the leading arm is added to the rotational velocity; for the following arm it is subtracted. A torque results, which tries to turn the boomerang, but because of its rotation, it behaves like a gyro. This means that the boomerang does not yield to the torque but changes its direction around an axis, the axis that is normal both to the axis of the rotating boomerang and to the axis of the torque. The boomerang in Fig. 4.45 carries out a precession to the left.

Three other flying devices may be mentioned: the discus, the "frisbee" (trademark), and the cylindrical wing (Fig. 4.46). The first two are circular disks; the frisbee has a downward curved edge and the discus is flat. Both are thrown with a small angle of attack and are simultaneously given a spin (Fig. 4.46a). This rotation stabilizes the disk's position during flight. If the discus were thrown without rotation about its axis of axial symmetry, it would autorotate about an axis of lateral symmetry (Section 6.5), and the flight distance would be much shorter [41]. The plastic disk of the frisbee is lighter than the wooden disk of the discus, and the flight velocity is, therefore, smaller. To obtain the necessary lift for a long gliding path, the edge is bent downward (Fig. 4.46b). The disk can also be replaced by a ring (Figure 4.46c) [42]. The elegant flight is aesthetically beautiful to watch.[15]

The third device is the annular wing (Fig. 4.46d). Although this lifting body has attracted most atten-

tion in supersonic aerodynamics [43], it also displays flying properties at low speed. A toy model of it can easily be made by gluing a rectangular piece of paper to a cylinder, with the leading-edge end made a little heavier than the rear. Stability is again increased by throwing it with a spin.

Remarks on Chapter 4

1. The statement that vorticity is generated at the boundary area of a fluid is valid only under the assumption that the fluid is homogeneous and barotropic, and that nonconservative forces are excluded [Reference 39, Chapter 1]. For instance, vorticity can be created inside a fluid in stratified flows (Section 8.4).

2. The adherence of fluid to a wall is a phenomenon that cannot be explained by the theory of continuum but only by molecular theory. In gases an interaction with the surface takes place, in which the bouncing gas molecules transmit a certain percentage of their energy and momentum to the wall. If the total tangential momentum is given to the wall (inelastic molecular collisions), macroscopically, an adherence of the fluid to the wall results. However, if the tangential momentum is totally reflected (elastic collisions), one talks about "perfect slip." In this case the tangential shear stress at the wall is zero. Between these two extremal cases slip flow occurs. The adherence of liquids to a wall is ascribed to molecular adsorption and is, thus, different from that of gases. Under normal conditions liquids and gases adhere to solid walls. However, rarefied gases slip if the mean free path lengths of the molecules are of the magnitude of the body dimensions [44].

3. The concept of a dipole is taken from the theory of electricity. A positive and a negative point charge generate field lines, which are shown in Fig. 4.47. When both charges come closely together, one talks about a dipole charge. In fluid dynamics the positive and negative charges are replaced by a source and a sink. If the source produces fluid which is sucked into the sink, then the dipole is a source–sink flow. In a vorticity field the dipole is a source of positive and negative vorticity.

4. The equations of motion, which include friction terms to describe the diffusion of vorticity, are called "Navier–Stokes equations" (see also Section 1.5). They model the flow of Newtonian fluids like water, air, and oil remarkably well.

5. In the limiting theoretical case Re = 0 there is no flow around an infinitely long cylinder

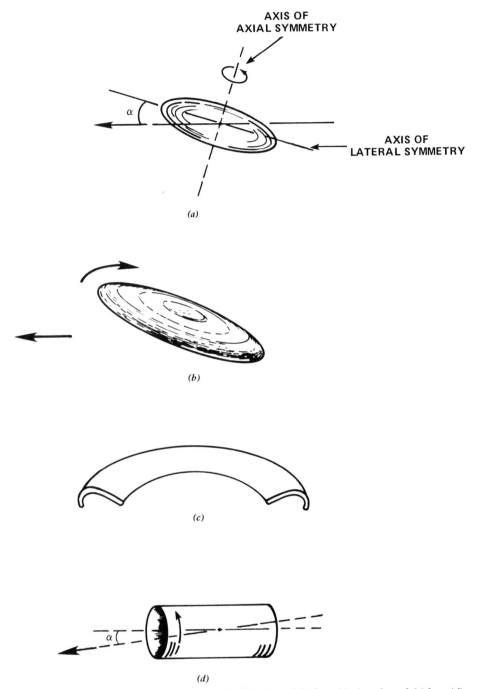

Figure 4.46. Flying devices: (a) discus, (b) disk-shaped frisbee, (c) ring-shaped frisbee, (d) cylindrical wing.

(Stokes' paradox), but a flow around a body with a finite surface like a sphere is possible.

6. For very small Reynolds numbers (Re = $VD/\nu \ll 1$) the drag of a sphere is, according to Stokes' law, $D = 3\pi\mu DV$, where μ is the dynamic viscosity, D is the diameter of the sphere, and V is the velocity. If one introduces the drag coefficient c_D, which is related to the drag D by $D = c_D A V^2 \rho/2$, one obtains for c_D the simple, one-parametric formula $c_D = 24/\text{Re}$. ρ is the density of the fluid, and A is the cross-section of the sphere.

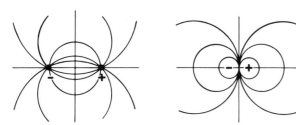

Figure 4.47. The concept of a dipole.

7. Bilateral symmetry is valid for human beings only with certain restrictions. The asymmetry of organs like the heart and intestines is, however, only of secondary siginifance made necessary by the length of the intestines [13].

8. Shear flows without vorticity exist. An example is the potential vortex.

9. In potential-flow theory, which is defined by vorticity $\omega = \text{curl } \mathbf{v} \equiv 0$, two cases must be distinguished: dynamic viscosity (1) $\mu > 0$ and (2) $\mu = 0$. In the first case, potential flow is a solution of the Navier–Stokes equations for viscous fluids. The frictional forces cancel each other, but dissipation takes place that must be balanced in a steady flow by work done at the boundary of the flow domain. An example is the potential vortex (or Rankine vortex for nonzero core) maintained by a rotating rod (Section 3.2). In general, however, nonrealistic boundary conditions must be imposed on a body to enforce curl $\mathbf{v} \equiv 0$ everywhere in the fluid. For instance, to obtain a potential flow past a circular cylinder, a certain tangential velocity distribution must be generated on the body surface. This requires work which is then dissipated into the fluid. The drag is zero. In this way, d'Alembert's paradox that no drag force exists in a steady potential flow is resolved. In case 2, any friction and dissipation effects are neglected by the assumption $\mu = 0$. In most practical applications, case 2 is meant when potential flow is assumed.

10. The conditions for the validity of Helmholtz's second theorem are the same as those given in Item 1 for the production of vorticity at a wall.

11. The lift coefficient c_L is defined in a way similar to the drag coefficient in Item 6 by $L = c_L A V^2 \rho /2$.

12. The Dutch graphic artist Maurits Escher has presented with his "Curl-ups" a hypothetical organism that is able to roll. In the Namib desert in Southwest Africa a spider actually lives, which is a bit like the "curl-up": It can contract itself to a sphere and roll down sand dunes.

13. Sculling a boat over the stern is based on the same principle. With a single oar, fixed in a rowlock at the stern, the boat can be propelled by rotating the oar obliquely. The oriental "Ro" is based on a similar principle [45].

14. In toy and sport shops boomerangs may sometimes be found. The cheaper versions often have no cambered profile; they cannot return.

15. With a certain skill frisbees can be thrown to perform a "boomerang flight." They return, however, only in the presence of a headwind.

5. *Separation*

5.1. FLOW SEPARATION AND VORTEX FORMATION

Vorticity is produced at a body surface and spreads from there into the fluid. Obviously, vortices exist that swim far away from a solid wall in a flow and have a higher vorticity than the surrounding fluid. How can vorticity in concentrated form occur in a fluid?

To answer this question the concept of "flow separation" in a plane and steady fluid flow must be explained.[1] In a steady flow the body surface is a streamline as well as a pathline. Although the velocity is zero at the surface (if the body is at rest relative to the reference frame), it has a nonzero value at an arbitrarily small distance from the wall. When particles approach each other on the surface streamline from opposite directions, they meet at a point and depart then from the wall. This phenomenon is called "flow separation," and the surface point of departure is the "separation point" (Fig. 5.1a). If the fluid moves toward the surface, the flow direction is reversed, an "attachment point" is obtained (Fig. 5.1b). A common name for separation and attachment points is "stagnation point," since the fluid near these points is stagnant.

The existence of such branching points follows from the law of conservation of matter, that is, from the fact that streamlines cannot end in the interior of a fluid. The streamline around a body or a closed fluid region must be a closed curve, or at least one attachment and one separation point at the surface must exist.[2] See Figs. 2.9 and 2.13c. If more than one attachment point exists, the same number of separation points must occur, and this in alternating sequence (Fig. 5.2).

Two types of branching points may be distinguished, depending on the way in which they are generated. The first kind is kinematic and follows from the fact that a body divides the parallel flow on its upstream side (points A_1 and S_2 in Fig. 5.2). Vorticity is here of only secondary significance in the sense that it does not determine the occurrence of the branching points but only their exact positions. The second type of branching point depends on the vorticity of the flow. This branching occurs only under certain conditions, but in fluid dynamics separation means the second type (points S_1 and A_2). How can vorticity help to explain flow separation? At the separation point (and at the attachment point) the vorticity is zero. Since the vorticity produced at the wall depends on the increase in velocity perpendicular to the wall, vorticity at the wall decreases if the streamlines near the wall diverge (Fig. 5.3a). If the vorticity profile has its highest value at the wall, a field of diverging streamlines is not sufficient to decrease the vorticity at the wall to zero. However, if the highest value of vorticity is within the flow (which can happen if convection of vorticity dominates over diffusion, see Section 4.1), vorticity induces an opposite flow at the wall. This means that separation occurs[3] (Fig. 5.3b). The single phases of the vorticity distribution perpendicular to the wall at consecutive locations are shown in Fig. 5.4. The corresponding streamlines and equivorticity lines near the separation point are shown in Figs. 5.5 and 5.6.

At the separation point vorticity of the boundary layer is carried into the fluid in the form of a shear layer. The occurrence of flow separation is, therefore, a prerequisite for the generation of vortices at solid walls. In general, it is difficult to determine the exact location of separation or attachement points. They must be obtained for each flow from observations or computations. An exception is the sharp edge, at which separation always occurs in an oblique flow.

The results of the considerations so far can be

68 Separation

Figure 5.1. Streamlines (a) near a separation point and (b) an attachment point.

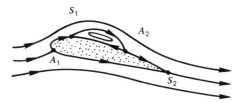

Figure 5.2. Surface of a body with two attachment points A_1 and A_2 and two separation points S_1 and S_2.

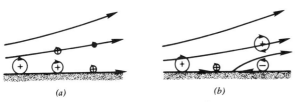

Figure 5.3. Diverging streamlines near the wall and additional vorticity in the flow are the reasons for flow separation.

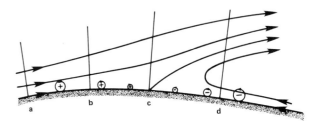

Figure 5.5. Streamlines near the separation point c.

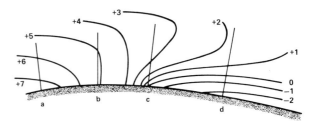

Figure 5.6. Equivorticity lines.

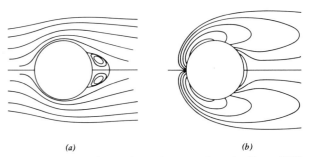

Figure 5.7. (a) Streamlines about a sphere for Re = DV/ν = 40 with D as the sphere's diameter; (b) equivorticity lines (from S. C. R. Dennis and J. D. A. Walker, *Journal of Fluid Mechanics*, Cambridge University Press [1]).

made clearer with the example of the flow past a sphere. Figure 4.6 illustrates the flow for Re = 5. Here diffusion dominates over convection, but the small influence of convection on the flow can be recognized. Flow separation has not yet occurred, but it is expected in the rear of the sphere where the streamlines diverge. If the Reynolds number is increased beyond 20, convection becomes so strong that the situation shown in Fig. 5.3b develops: The flow separates, and a vortex ring forms. Figure 5.7 shows the flow for Re = 40.

The point at which the flow separates from the wall depends on the Reynolds number. The higher the Reynolds number, the sooner the flow separates, and the larger is the vortex ring (Fig. 5.8). Axisymmetric flows around spheres with attached vortex rings occur in the Reynolds number range 20 ≤ Re < 400.

For Reynold numbers greater than 400 the flow becomes unstable and unsteady; the axisymmetry disappears and the vortex ring detaches from the body (Section 6.3).

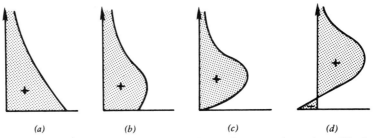

Figure 5.4. Vorticity distribution across the wall near the separation point c. The locations of the cross sections a through d are given in Fig. 5.5.

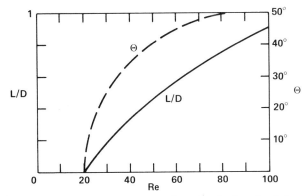
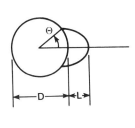

Figure 5.8. Approximate data for the length of the separation region and for the separation point as a function of the Reynolds number (adapted from H. R. Pruppacher, B. P. Le Clair, and A. E. Hamielec [2]).

If one compares the streamlines of the vortex ring in Fig. 5.7a with the corresponding equivorticity lines, one recognizes that the axis of the vortex ring does not coincide with the location of extremal vorticity but that the highest value of vorticity still occurs at the body surface where vorticity is produced. This result is surprising because intuitively the vortex axis is identified with the location of extremal vorticity. The equivorticity lines in Fig. 5.7b admittedly reveal the tendency to constriction, but in reality there is no closed equivorticity line (which would indicate an extremum in the vorticity field).[4] The question asked in the beginning of this section on the occurrence of concentrated vorticity within the fluid remains unanswered.

Examples of flows past spherical bodies in nature with Reynolds numbers smaller than unity are described in Section 4.4. An example of flows past spheres with Reynolds numbers up to instability is the raindrop. To indicate orders of magnitude for the diameter of the sphere, its terminal velocity, and the Reynolds number, some data are given for spherical raindrops [3]. The kinematic viscosity of air is assumed here at about 20°C, that is $\nu = 15$ mm²/s:

Diameter (mm)	Terminal velocity (mm/s)	Reynolds number
0.1	220	1.5
0.2	700	9.3
1.0	4000	267

Raindrops develop in saturated air through condensation of water vapor around impurities. However, they do not behave like solid spheres, since shear forces at the surface cause the water to move and form a vortex ring (Fig. 5.9). Streamlines and equivorticity lines for Re = 30 are shown in Fig. 5.10. Here, too, as in Fig. 5.7b, no extremum of vorticity can be found.

A raindrop has a spherical form only for small Reynolds numbers. With larger Reynolds numbers shear forces and surface tension deform the sphere to a spheroid, which finally bursts at still higher Reynolds numbers. In Fig. 5.11 a raindrop is sketched with an attached vortex ring of air. The separation point is located farther back than for a solid sphere because of the movement of water at the surface.[5] Impurities in water and air, which are caught within the vortex and fall with the raindrops to the ground, also influence the surface tension, which decides in an essential way the strength of the vortex within the raindrop. The second vortex inside the drop as sketched in Fig. 5.11 is so weak that

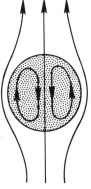

Figure 5.9. Flow about and inside a small raindrop for Re < 5. The observer moves with the drop.

70 Separation

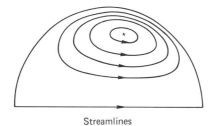
Streamlines

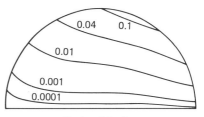
Equivorticity lines

Figure 5.10. Streamlines and equivorticity lines (in dimensionless units) in a raindrop (adapted from B. P. Le Clair, A. E. Hamielec, H. R. Pruppacher, and W. D. Hall [4]).

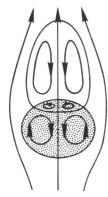

Figure 5.11. Flow around and inside a deformed raindrop for Re ∼ 200.

the water in the rear of the drop is practically motionless.

Dust particles are trapped not only inside the raindrop but also in the wake vortex. Hence, they fall to the ground faster than they would if they fell by their own weight. This knowledge is important in the study of air pollution [5].

The flow around an infinitely long circular cylinder behaves like that past a sphere, but here a vortex pair corresponds to the vortex ring behind the sphere. Its axes are parallel to that of the circular cylinder (Fig. 5.12). Since cylindrical bodies cause larger disturbances in a flow than spherical ones, the vortices behind cylindrical bodies are larger. Figure 5.12 shows the flow past a circular cylinder for Re $= DV/\nu = 40$; Fig. 5.13 gives the length of the separated flow region as a function of Reynolds number. The range of steady flow, in which a vortex pair behind the circular cylinder occurs, is $5 < \text{Re} < 45$.

Flow separations are also found in pipes and channels if the conditions given in Fig. 5.3 are met. Such a situation can occur in pipe constrictions. Figure 5.14 displays streamlines and equivorticity lines in a circular tube for Re = 50 (Re = DV/ν with V being the velocity at the axis far away from the constriction and D being the diameter of the

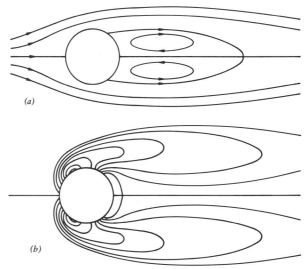

Figure 5.12. (a) Steady streamlines around a circular cylinder for Re = 40; (b) equivorticity lines (adapted from H. Takami and H. B. Keller [6]).

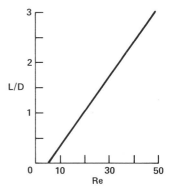

Figure 5.13. Length of the separation region behind a circular cylinder as a function of the Reynolds number (from S. C. R. Dennis and G. Chang [7]).

tube). Pipe flows with such small Reynolds numbers are found in blood vessels of higher animals. In arteriosclerosis the arteries are constricted by fatty wall cells and sedimentation of cholesterol. The thickening not only increases the pressure loss through constriction but can also detach and cause

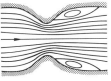 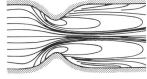

Streamlines   Equivorticity lines

Figure 5.14. Steady flow through a constriction of a tube for Re = 50. The production of vorticity at the wall is highest at the smallest cross section (adapted from J. Lee and Y. Fung [8], The American Society of Mechanical Engineers).

embolisms. Moreover, the walls in the separated flow regions behind constrictions are not supplied with sufficient oxygen and nutrients, so they harden and again favor the formation of thrombi. Also platelets concentrate in the separation regions and stick together [9].

The constriction as a place of highest vorticity production is also the location of greatest shear flow. Because a strong shear flow damages the red blood corpuscles and favors the permeability of cell membranes to lipoids, constrictions must be avoided. Since blood flow is not steady but pulsates, Fig. 5.14 is, thus, only an example of a steady or quasisteady state.

Blood flow is laminar in general. Exceptions in the form of temporary turbulence in certain phases of pulsation occur at heart valves without damage to the blood. However, if the aortic valves are pathologically constricted (aorta stenosis), the jet leaving the valve can become turbulent, a situation to be avoided.

The weakening of arterial walls due to high shear flow can lead to arterial dilation (aneurism). Particularly susceptible are branching points of arteries in the abdomen and brain (Fig. 5.15). Despite the small Reynolds numbers in the branches the flow inside the dilation can become turbulent [Reference 15, Chapter 4].

In compressible fluids flow separation can also be triggered by a shock wave, a phenomenon that is

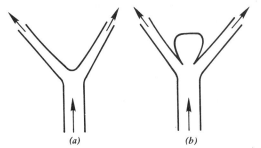

Figure 5.15. Branching of an artery (a) without and (b) with aneurism.

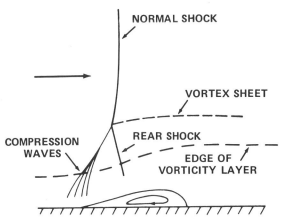

Figure 5.16. Shock-induced flow separation in transonic flow.

important in the aerodynamic design of aircraft in the transonic range. Figure 5.16 shows the interaction of a laminar boundary layer with a normal shock [10, 11].

Flow separation (in the narrow sense) does not necessarily require a convection process of vorticity as depicted in Fig. 5.3. Flows around bodies with concave walls can separate and form vortices even at Re = 0, that is, with no convection at all. Figure 5.17a illustrates such a situation [12]. A steady vortex at Re = 0, which is called a "Stokes vortex," can occur even in front of a body, as shown in Fig. 5.17b for slow motion past a convex–concave lens.

Stokes vortices can also be present in flows between two plates that form a dihedral and in shear flows around rotating cylinders [13].

5.2. VORTICES BEHIND AN EDGE

... It thus follows that every geometrically perfect sharp edge, at which fluid flows, even with the smallest velocity of the remaining fluid, must tear up and form a separation sheet.

Helmholtz, 1868 [14]

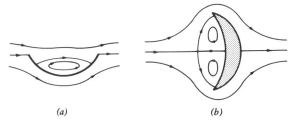

Figure 5.17. (a) Steady flow separation with Stokes vortex at a circular arc for Re = 0. (b) Stokes vortices in front of a convex–concave lens in a flow with Re = 0 (from J. M. Dorrepaal [12]).

Flows always separate at the sharp edge of a body (except for the theoretical limit Re = 0 and at an edge parallel to the flow, see Fig. 5.17). This behavior is caused by the large curvature of the edge. Upstream of the edge, streamlines converge so much that near the edge a large amount of vorticity is produced. This vorticity is not carried around the edge but moves away through convection, resulting in the situation of Fig. 5.3b and enforcing separation from the edge.

Figure 5.18 displays streamlines and equivorticity lines around a thin circular disk perpendicular to the flow. The motion is axisymmetric for Re = 100, and after a sufficient length of time a steady state is reached. Figure 5.18b reveals clearly that vorticity is generated mainly at the edge.

The temporal development of these steady flows may be described with the aid of computer calculations in the following way: At a certain time a body is abruptly put into steady motion. With a sphere separation begins in the center of the rear and spreads along the body (Fig. 5.19a); with a circular disk separation starts at the edge (Fig. 5.19b), spreads to the axis of symmetry, and then forms the vortex ring of Fig. 5.18.

Streamlines and equivorticity lines of a flow normal to an infinitely long plate are similar to those past a circular disk. However, the flow past an oblique circular disk or an oblique plate is more complicated because of the asymmetry of the flow. For Re = 15 the steady flow past a plate at an angle of attack of 45° produces only one vortex (Fig. 5.20), so somewhere between 90° and 45° a transition takes place from two vortices to one. After the sudden start of the oblique plate the steady state develops without oscillation as in the case of the asymmetric flow past a plate at 0° or 90°. At higher Reynolds numbers (ca. Re = 30) the transient

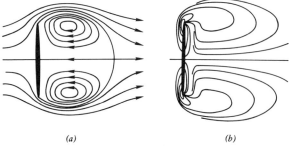

Figure 5.18. (a) Streamlines around a circular disk perpendicular to the flow for Re = 100. (b) Equivorticity lines (adapted from R. L. Pitter, H. R. Pruppacher, and A. E. Hamielec [15]).

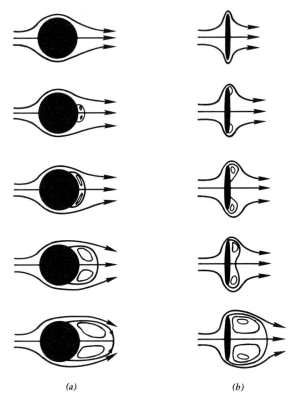

(a) (b)

Figure 5.19. Temporal development of a vortex ring (a) behind a sphere and (b) behind a disk for Re = 100 (adapted from Y. Rimon [16]).

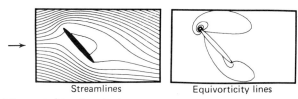

Streamlines   Equivorticity lines

Figure 5.20. Steady flow past a plate at an angle of attack of 45° for Re = 15. In the rear only one vortex forms (computer-generated picture by H. J. Lugt and H. J. Haussling [17]).

phase to steady state is oscillatory. Beyond Re = 45, a periodic vortex separation occurs, and a steady state is never reached.

Additional examples of flow separation at edges in steady motions are displayed in Figs. 5.21 through 5.26. Figure 5.21 shows the flow in a cavity, which is driven by the outside stream. One or several vortices form depending on the depth of the cavity.

If the cross section of a channel increases abruptly, the flow separates (Fig. 5.22). The streamlines drawn can be interpreted as plane motion (in a slot with abrupt widening) or as axisymmetric motion (in a circular pipe with abrupt widening). A

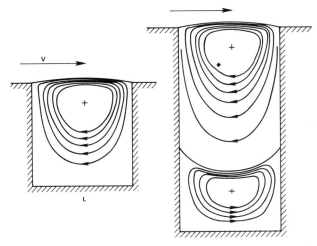

Figure 5.21. Vortices in rectangular cavity for Re = LV/ν = 100 (adapted from U. B. Mehta and Z. Lavan [18], The American Society of Mechanical Engineers).

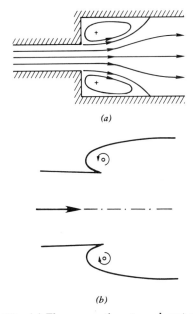

Figure 5.22. (a) Flow separation at an abrupt enlargement of a channel; (b) cusp diffusor for controlling separation by trapping vortices as proposed by F. O. Ringleb.

particular case of this kind of flow separation is the jet, which is formed when fluid is poured out of an opening [19] (Fig. 5.23).

At a hole in a wall or an orifice in a pipe the flow constricts behind the opening (Figs. 5.24 and 5.25). This constriction is called a "vena contracta" and diminishes the flux through the pipe. Figure 5.25 shows that flow separation can also occur in front of the orifice, caused by diverging streamlines along the walls.

Figure 5.23. Flow at an opening.

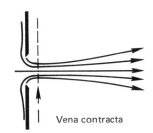

Figure 5.24. Vena contracta in a jet.

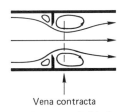

Figure 5.25. Vena contracta behind an orifice in a pipe.

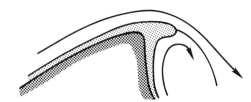

Figure 5.26. Snow cornice.

In nature snow cornices form through flow separation at a ridge (Fig. 5.26). The wind carries snow from both sides of a ridge to the tip and in time builds a dangerous overhang of snow, which may break off and cause an avalanche [20].

Before the separation of a vortex from a wall is described, the spatial distribution of vorticity in a steady vortex will be illustrated once more. Figure 5.27 shows the steady flow in an open box, a flow maintained by moving a flat plate over it. The box is assumed to be very long in the direction perpendicular to the page. This flow is very similiar to the motion in a cavity (Fig. 5.21), but with the differ-

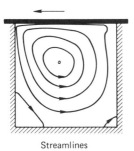
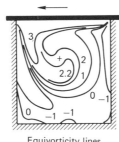

Streamlines   Equivorticity lines

Figure 5.27. Streamlines and equivorticity lines in an open box, over which a flat plate is pushed—Re = 400 (adapted from O. R. Burggraf, *Journal of Fluid Mechanics,* Cambridge University Press [21]).

ence that in this case the vortex is driven by the plate and not by the passing flow outside the cavity. The corresponding equivorticity lines are also shown. Again, one recognizes that the vortex center does not coincide with the location of the maximum value of vorticity but that the highest values appear at the walls. However, a tongue with a high value of vorticity surrounds in part the center of the vortex. This tendency to encircle the vortex center increases with higher Reynolds number and will finally lead to vortex separation via an instability [22]. It may be emphasized once more in connection with the explanations of the following section that the curves in Fig. 5.27 are time independent, that is, the tongue does not change with time.

5.3. VORTEX SEPARATION

Vortices remain attached to the body after the start from rest only for small Reynolds numbers, and in this case they form part of the steady flow field. At higher Reynolds numbers the flow becomes unstable, and the vortices are shed from the body. This flow instability will be discussed in Chapter 6. Instability initiates the transition to an asymmetric, time-dependent flow, but accelerated or decelerated flows may also trigger separation of symmetric vortices from the wall.

The concept of "vortex separation" must be distinguished from that of flow separation. Vortex separation is always a time-dependent process, in which vorticity assumes extremal values inside the fluid. By contrast, flow separation occurs in time-dependent as well as in steady motion.

The development of a vortex ring behind the sphere and disk in Fig. 5.19 did not result in vortex separation but in a steady state, since the Reynolds number was below the critical value at which the flow becomes unstable. In a stable flow a vortex separates through strong acceleration or braking only if a vortex already exists behind the body.[6] For instance, if the disk in Fig. 5.18 is suddenly slowed, the vortex ring will separate from the body. Beyond the critical Reynolds number, however, a vortex ring develops behind the sphere and disk after an abrupt start, and the vortex ring will separate from the body. This vortex is called a "starting vortex." Taneda has also observed "secondary vortices" behind circular cylinders (but not behind flat plates) after the dimensionless time $Vt/D \sim 2$ from the start for Reynolds number Re = 550, which is beyond the critical value (Fig. 5.28). More details have been given in References 24 and 25. For airplanes the starting vortex at the trailing edge of a plate or an airfoil at a nonzero angle of attack is of great importance. The existence of this vortex was postulated in Section 4.7 (Fig. 4.24) for the generation of circulation in a potential flow. The strength of the starting vortex and of the circulation is determined by the "Kutta condition," which says that the separation point must be located at the trailing edge of the airfoil (which makes the streamline through the separation point tangential to the lower side of the rear wedge, Fig. 4.24b). This condition is superfluous in a real flow. Pictures of real flows, however, indicate whether the hypothesis of the "bound vortex" with its center inside the airfoil is useful or not.[7] They also give information on the time it takes the separation point to migrate to the trailing edge, and the streamline through this point to be tangential to the lower side of the rear edge. Figure 5.29 demonstrates this development for a blunt edge at Re = 200 and at an angle of attack of 45° after the abrupt start. In this case the streamline through the

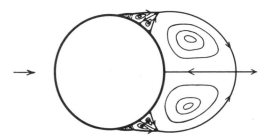

Figure 5.28. Starting vortex pair with secondary vortices behind a suddenly moved cylinder at the dimensionless time $Vt/D = 2$ for Re = 550 (from S. Taneda [23]. Excerpt from *Recent Research on Unsteady Boundary Layers,* IUTAM Symposium 1971, Vol. 2, 1972, E. A. Eichelbrenner (ed.), Les Presses de l' Université Laval, Québec).

Figure 5.29. Development of the Kutta condition for Re = 200 (from H. J. Lugt and H. J. Haussling [17]).

Figure 5.31 Vortex separation behind an edge perpendicular to the flow. Equivorticity lines at two consecutive times.

Figure 5.32. Roll up of a thin vorticity layer. The equivorticity lines are sketched.

separation point is parallel to the center line of the elliptic trailing edge at the time $Vt/D = 0.2$. D is here the chord of the elliptic wing. Figure 5.30 shows the flow field past a wing for Re = 1000 at a small angle of attack for the time at which the Kutta condition is "fulfilled." Vorticity has not yet had time to spread into the fluid, and it forms a thin boundary layer around the body. However, at the trailing edge the equivorticity lines have formed a "tongue" that constricts at about the same time to a ring. This starting vortex separates from the body and is carried away (Fig. 5.31) [26].

At higher Reynolds numbers the starting vortex develops in the following way: Because of the boundary-layer characteristics of the vorticity in front of the edge, the tongue of high vorticity shrinks to a thin layer (Fig. 5.32). This layer rolls about the vortex center as was indicated in Fig. 5.27 for the relatively small Reynolds number of 400 in the steady flow inside a box. At the center of the vortex, diffusion dominates convection. At this point a locally restricted solid-body rotation develops with an extremum of vorticity. The appearance of this extremum signals vortex separation.

At very high Reynolds numbers (Re > 10^5, laminar or turbulent) the vorticity layer is so thin that it can be considered a discontinuity sheet (or in the drawing a discontinuity line, see Section 3.5 [27, 28]. This line is also a streak line and can, therefore, be made visible through injection of dye at the edge (Fig. 5.36 below). It may also be recalled that the discontinuity line is invariant with respect to the change of the inertial system (Section 4.5). In Figs. 5.33 and 5.34 the roll up of a discontinuity line at a sharp edge is sketched. Figure 5.35 shows the vortex layer of Fig. 5.33 in perspective.

Although mathematically the spiral discontinuity line consists of infinitely many windings at the center (Fig. 5.34), even at extremely high Reynolds numbers the core of the vortex rotates like a solid body through viscous diffusion.[8] According to Moore and Saffman [29] the vortex outside the core is structured in the following way: Adjacent to the core an annular region exists in which the spiral is

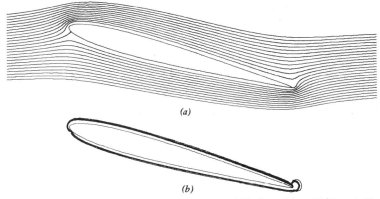

Figure 5.30. For Re = 1000 the Kutta condition is fulfilled at about $Vt/D = 0.08$ after the abrupt start. (a) Streamlines and (b) equivorticity lines at that instant (computer-generated picture from U. B. Mehta and Z. Lavan [Reference 9, Chapter 2].)

Figure 5.33. Roll up of a discontinuity line according to Prandtl [28]. The streamlines are drawn as dashed lines. Since the process is unsteady, the streamline at the edge does not coincide with the discontinuity line (which is a streak line).

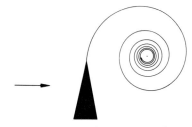

Figure 5.34. Discontinuity line from an edge with infinitely many windings.

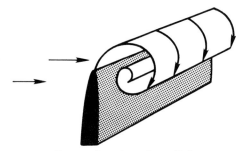

Figure 5.35. Perspective view of a rolled-up vortex layer (discontinuity sheet).

thickened by diffusion. The spiral becomes thinner, and it follows a region in which the tightly wound spiral can be identified. In the external region the spiral is connected with the body, which is the origin of the discontinuity line. When this feeding line weakens and then ceases, the vortex becomes detached. In the annular region (whether detached or not) in which the spiral thickens, laminar and turbulent flows must be distinguished (Section 6.7).

Shape and strength of the developing vortex depend on the wedge angle and the way the flow against the wedge is accelerated. In Fig. 5.36 photographs of streak lines reveal the growth of a starting vortex behind a 90° wedge in an accelerated flow (a,b). The bubble in the shear layer may indicate a secondary vortex. Between the instants (b) and (c) the flow suddenly stops. A vortex of opposite sign develops at the other side of the wedge (d) demonstrating the conservation law of vorticity [30].

The starting vortex behind a plate at a nonzero angle of attack is chosen as an example of the roll up of a vorticity layer at small Reynolds numbers and of a vortex layer at very high Reynolds numbers. After the abrupt start vortices develop at the leading and trailing edges for large angles of attack. In Fig. 5.37a the vortex layer for an infinitely large Reynolds number (discontinuity line) has been approximated by point vortices (Section 3.5). For comparison Fig. 5.37b displays the vorticity field around a thin elliptic cylinder at Re = 200 (flow field of Fig. 5.29 at later times). At first glance the two sequences of pictures have a certain similarity. The vortices at the edges develop simultaneously, and their extensions are about the same. However, the vorticity distribution in the vortex at the trailing edge for Re = 200 differs considerably from that of the point-vortex model. For Re = 200 the local extremum of vorticity has its highest value at the instant the extremum appears. After that the vortex decays and the extremal value diminishes. For Re = ∞ vorticity concentrates in the vortex during the entire process of roll up.

Vortex layers can also roll up in corners and at concave walls. If one holds a vertical plate fixed and pushes a horizontal plate beneath it, a vortex layer rolls up in front of the fixed vertical plate. Behind this plate the fluid moves in the way shown in Fig. 5.38. The roll up of a vortex layer at a concave wall can be demonstrated easily with a cup of coffee. If one sprays a jet of milk against the rim of the cup just above the free surface of the coffee, the roll up of two vortex layers can be clearly seen (Fig. 5.39).

In compressible fluids vortices can develop when shock waves encounter a corner or a plate asymmetrically [32, 33]. In Fig. 5.40 a shock wave from left hits a rectangular block where the shock is reflected and diffracted in such a way that a pressure drop around the edges occur. This causes a flow that separates behind the edges and forms vortices.

Vortex separation through acceleration and deceleration of a body produces a row of vortices. When a plate is abruptly moved back and forth, starting and stopping vortices are generated alternately. Of much greater importance, however, is the generation of alternating vortices by oscillating plates and disks. The motions involved include pure translational and pure rotary oscillations and a combination of the two (Fig. 5.41). In addition, a parallel flow can be superposed. Here, three categories of oscillating bodies may be distinguished: (1) "active" or "forced" oscillation of a body, that is, os-

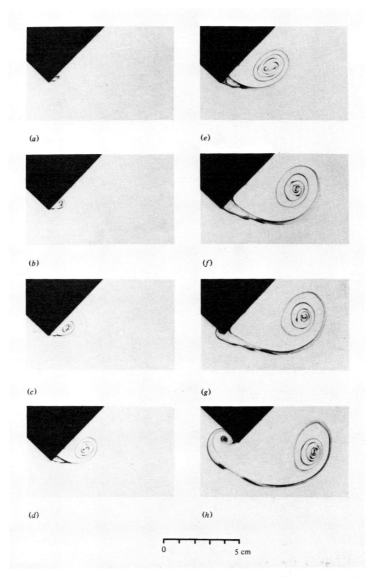

Figure 5.36. Sequence of photographs showing streak lines from the apex of a 90° wedge. The flow is accelerated and stopped between (c) and (d). (Courtesy of D. I. Pullin and A. E. Perry, University of Melbourne, Australia.)

cillation of a body by forces other than those of the fluid flow; (2) "passive" or "flow-induced" oscillation of a body, that is, the movement of a body which is free to yield to the fluid forces; and (3) "neutral" oscillation, which is a special kind of forced oscillation. No net energy is exchanged between body and fluid (Fig. 5.42).

Examples of forced oscillation are the flights of birds and insects and the swimming of fish. Flying and swimming were described in Section 4.7 in the context of airfoil theory. Swimming of fish will be taken up again in the next section.

Flow-induced oscillation will be discussed in Section 6.4 in the larger framework of "flow instability." Neutral oscillation is approximated by a swimming snake in calm water [Reference 16, Chapter 4].

5.4. NOTE ON THE EVOLUTION OF FAST-SWIMMING FISH

Organisms have existed in water for a billion years. During this time evolution has taken many roads to adapt organisms to their environment and to give

78 Separation

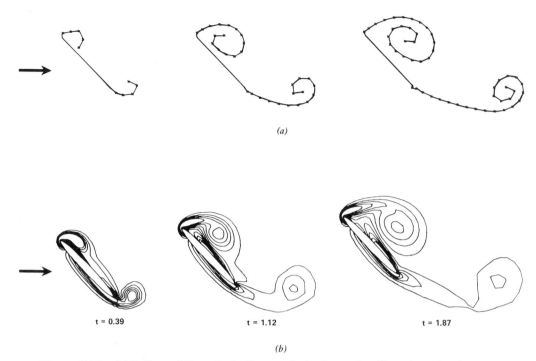

Figure 5.37. (a) Roll up of discontinuity lines at the leading and trailing edges of a plate at an angle of attack of 45° shortly after the abrupt start. The discontinuity lines are approximated by point vortices and represent vortex layers for Re = ∞. (b) Vorticity field around a thin elliptic cylinder for Re = 200. The time is dimensionless as in Fig. 5.29. Closed lines indicate extrema of vorticity in the fluid.

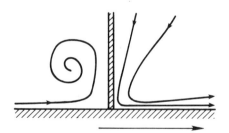

Figure 5.38. Vortex and sink flow in a corner. The horizontal wall moves (adapted from R. J. Tabaczynski *et al.*, *Journal of Fluid Mechanics*, Cambridge University Press [31]).

Figure 5.39. A jet of milk sprayed against the rim of a cup of coffee rolls up into two vortices.

Figure 5.40. Vortex generation at the edges of a block after a shock wave from the left hits the block (Mach–Zehnder interferogram by H. U. Hassenpflug, Aerodynamisches Institut, Aachen).

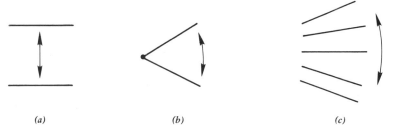

Figure 5.41. (a) Translational oscillation, (b) rotary oscillation, and (c) combination of both types.

them a chance for survival. The selection process has created various forms of adaptation: protective armour, mimicry, and fast propulsion during flight or hunt. The following survey is quite brief and will indicate only the general trend of evolution for fast-swimming fish.

The Reynolds number assigned to an aquatic animal is determined by the animal's length and velocity. The magnitude of the Reynolds number reveals immediately whether viscous or inertial forces are decisive for propulsion. The following table gives the magnitude of the Reynolds number for a variety of organisms:

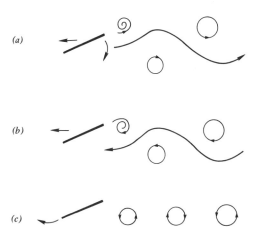

Figure 5.42. Vortex separation during (a) active or forced oscillation of a plate, (b) passive or flow-induced oscillation, and (c) neutral oscillation. Note the rotary direction of the vortices in these special cases and compare them with those of an oscillating circular cylinder in Section 6.4.

Organism	Reynolds number
Bacteria	10^{-6}
Protozoa with flagella	10^{-3}
Nematodes	10^{0}
Tadpoles	10^{2}
Leeches	10^{3}
Eels, carps	10^{5}
Dolphins, sharks	10^{7}
Blue whales	10^{8}
for comparison:	
Humans	10^{6}
Submarines	10^{9}

Organisms with Reynolds numbers smaller than unity use friction forces to propel themselves, whereas animals with higher Reynolds numbers apply inertial forces.

In Section 4.4 propulsion mechanisms at very small Reynolds numbers were described and two groups were identified: Thrust by means of cilia and thrust through flagella or wavelike body movement. For Reynolds numbers larger than unity cilia are ineffective for propulsion but wavelike body movement remains efficient. This fact has been of tremendous importance in the evolution of aquatic animals because the propulsion mechanism did not have to change during the evolution to higher animals. In addition, nature has developed still another propulsion mechanism, which is effective only at large Reynolds numbers: jet propulsion.

When wavy body movement is used as a means of propulsion, no essential change in the body form or in the kind of movement is observed in the Reynolds number range from 1 to 1000. The entire circular-cylindrical body participates in wave motion. With increasing Reynolds number, however, vorticity produced at the body is pushed farther and farther downstream, and the oscillation at the rear of the body gains in importance. There, the wave amplitude increases with Reynolds number.

The first modification of the basic circular-cylindrical structure is found in animals with Reynolds numbers larger than 1000. To increase the efficiency of the oscillation at the rear of the body, this part became flat in the direction perpendicular to the wave motion. A typical example is the eel. Vortex separation, which generates thrust behind an oscillating plate (Fig. 5.42), became here for the first time an efficient means for propulsion. The vortices shed by the eel are, of course, not two dimensional but ring shaped.

With the emphasis on the end of the body, oscillation of the front part lost its significance. It can even be shown that oscillation of the front part becomes a source of unfavorable vorticity. The eel-type structure is thus improved by suppressing these oscillations, and this was achieved through decrease of the link between the front and rear parts and through further development of the platelike body end. The tail of the fish was thus evolved [Reference 15, Chapter 4].

With the development of fish to higher Reynolds numbers another problem in addition to the optimization of thrust appeared: the deceleration of the body. At Re < 1 the body stops immediately when the force of the propulsion ceases to act, but at higher Reynolds numbers the body continues to move a certain distance owing to inertia until it comes to rest through friction. Hence the fish needs a means to control braking. For this reason nature has developed the remaining fins, which, in addition to braking, have the important task of stabilizing the body.

The typical form of the fish body, such as that of perch and goldfish, is by no means the best for achieving high speed. Nature has approached the ideal form, independently, from which initial form evolution started, whether from fish or mammal. A high-speed swimmer has a streamlined body that faces the water with minimum drag and a lunate tail fin to obtain maximum propulsion. The reason for the lunate tail form is not yet completely clear. It is conjectured that the vortex rings shed from this tail fin have optimal efficiency [Reference 15, Chapter 4]. Among fish, tuna and swordfish come closest to the ideal form (Fig. 5.43a), and among mammals, dolphins and whales (Fig. 5.43b). The following table gives some data for the speed of these and other animals:

	m/s
Swordfish	above 20?
Dolphin, whale	10
Salmon	5
Trout	3.5
Goldfish	1
Human, olympic record	2

Mammals like dolphins and whales differ in their body form from fish in the horizontal position of the tail fin. Since the dolphins and whales do not have swim bladders to balance buoyancy and sinking, they obtain balance with their horizontal fins. There are also fish that do not have swim bladders. An example of a high-speed swimmer without a swim bladder is the shark, which counteracts sinking through its asymmetric tail fin (Fig. 5.44) [Reference 15, Chapter 4].

Jet propulsion of fish actually is characteristic of the oldest creatures in the animal kingdom, although the name sounds so modern. The principle of propulsion is indeed the same as that of a rocket. Besides the jellyfish, which generates propulsion by contracting its parachutelike body, many cephalopods have excelled in jet propulsion. By a onetime or periodic expulsion of water from their mouths, they can reach velocities up to 4 m/s. However, their endurance is very limited. The animals usually remain motionless in water and suddenly shoot away like an arrow when in danger (Fig. 5.45).

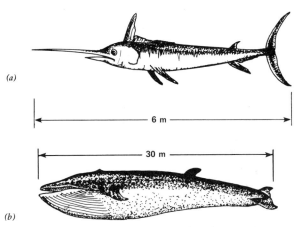

Figure 5.43. (a) Swordfish and (b) blue whale.

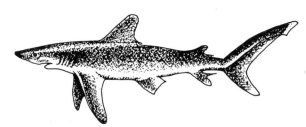

Figure 5.44. Shark.

Figure 5.45. Squid (M = mouth).

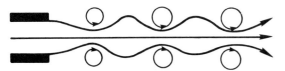

Figure 5.46. Generation of a jet through periodic expulsion of water.

When water is expelled, a vortex ring forms. During periodic expulsion, which is comparable to active oscillation, a jet is generated between the vortex rings (Fig. 5.46). The vortices perform here a kind of "support" [34]. A similar principle is used in schooling. For fish (or birds) that swim in the wake of others, the expenditure of energy in locomotion is lessened [Reference 14, Chapter 4].

5.5. THREE-DIMENSIONAL VORTICES BEHIND BODIES

The roll up of a vortex layer can be delayed and even stopped if a flow perpendicular to the drawing page of Fig. 5.33 is superposed. This means that the fluid moves against the flat side of the plate at an angle other than 90° (Fig. 5.47). The plate can no longer be assumed as infinitely long (or two dimensional) but must be limited by a side edge. From the tip of the plate a conical vortex layer, called a "tip vortex," rolls up. (In contradistinction, the vortex in Fig. 5.35, which does not have a velocity component parallel to the tip, is called an "edge vortex"). The tip vortex can reach a steady state if the angle between the plate area and the flow is not larger than about 35°. In this case the streamlines lie in the vortex layer (Fig. 5.47) [35].

Tip vortices are observed at the side edges of a rectangular plate at a nonzero angle of attack (Fig. 5.48). However, they are unstable and form with the vortex of the front edge complicated, short-lived horseshoe vortices. In contrast, the tip vortices at a delta wing are usually stable because of the "conical similarity" between axial vortex spreading and increase in the wing area (Fig. 5.49). These vortices stabilize the airplane in both subsonic and supersonic flow and induce a high velocity field that results in an additional lift called "vortex lift." Tip vortices have been carefully investigated by the aircraft industry [36, 37]. The vortex lift, which with the potential-flow lift forms the total lift (Fig. 5.50), increases with the angle of attack and with decreasing slenderness ratio. However, this lift is limited when vortex breakdown (see farther along) occurs at a certain critical angle of attack above the wing.[9]

In supersonic flows the roll up is influenced by shock waves and expansions caused by the body. Figure 5.51 shows the situation around a delta wing at an angle of attack. The lower side and its boundary layer are enveloped by the bow shock. The flow remains attached at the leading edge where an expansion fan originates but separates a little away from the leading edge. Toward the centerline but outside the viscous flow regime a "realignment" or "embedded" shock forms. This "separation with shock" is not the only type of separation in supersonic flows. At higher Mach numbers "shock-induced" separation, and at higher angles of attack "leading-edge separation with shock" occur. All three types are sketched in Fig. 5.52 [38–40].

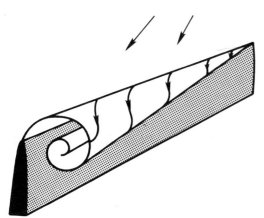

Figure 5.47. Steady tip vortex.

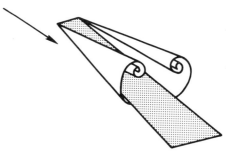

Figure 5.48. Tip vortices at a rectangular plate in an oblique flow.

82 Separation

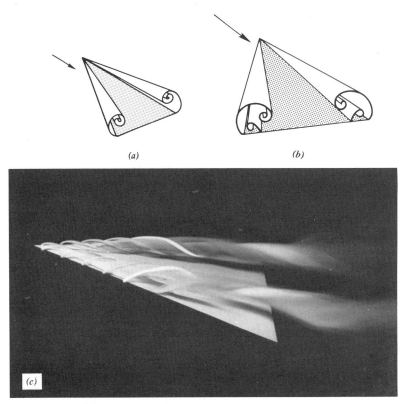

Figure 5.49. Tip vortices on the lee side of a delta wing in subsonic flow. (a) Primary vortices and (b) secondary vortices induced by the primary vortices through separation of the boundary-layer flow of the wing surface. (c) Photograph of tip vortices above a delta wing by H. Werlé, ONERA.

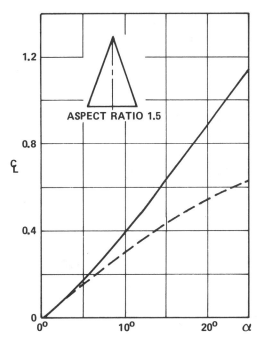

Figure 5.50. The total lift of a delta wing (solid line) is the sum of potential-flow lift (dashed line) and vortex lift (adapted from E. C. Polhamus [36]).

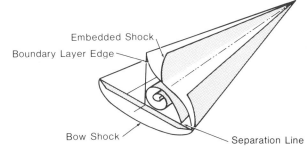

Figure 5.51. Supersonic flow past a delta wing with shock waves and separated vortex layer (adapted from D. J. Peake and M. Tobak [38]).

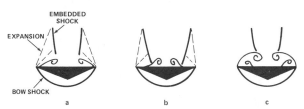

Figure 5.52. Various types of leading-edge separation with shock. (a) Separation with shock, (b) shock-induced separation, and (c) leading-edge separation with shock (adapted from D. J. Peake and M. Tobak [38]).

The tip vortices do not end at the rear edge of the wing but can extend many chord lengths behind. They can persist for several hundred meters, depending on the size of the aircraft, and endanger smaller airplanes which follow (Fig. 3.26) [Reference 22, Chapter 3; 41].

Tip vortices form around every vehicle with sharp edges when wing or water passes by at an angle. They can be observed around cars if exhaust gases make them visible[10] (Fig. 5.53) [42, 43]. A ship with sharp or strongly curved bilges generates "bilge vortices." Since water flows downward behind the bow, a vortex develops under the bilge, and its direction of rotation can be seen in Fig. 5.54. At the stern water flows upward, and the generated bilge vortex has the opposite direction of rotation [44, 45]. Bilge vortices contribute to the ship's drag and affect the propeller performance. To be distinguished from these steady bilge vortices are unsteady bilge vortices that can form if the ship rolls, sways, or heaves, that is, if it oscillates or swings about its longitudinal axis. The periodically generated vortices are similar to those behind a falling plate (Section 6.4). If in addition the ship has a forward speed, complicated unsteady tip vortices form.

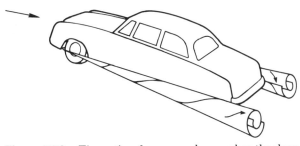

Figure 5.53. Tip vortices form around a car when the shear flow underneath the car escapes sideways. The vortices become visible through exhaust gases. Vortices from the upper part are omitted.

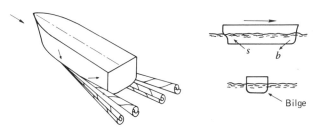

Figure 5.54. Bilge vortices underneath a ship's hull. For full ships the bilge is a rounded or sharp edge at which vortices form if the flow has a component normal to it. At the bow (b) the water flows downward, at the stern (s) upward. The tip vortices formed thus have opposite directions of rotation.

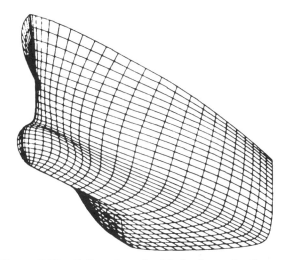

Figure 5.55. Bulbous bow of a ship for decreasing the wave drag and the strengths of the bilge vortices (computer-generated picture, David Taylor Naval Ship Research and Development Center).

The drag of a ship caused by bilge vortices (as well as the wave drag) can be diminished by means of a bulbous bow beneath the waterline (Fig. 5.55).

As long as the flow separates at a sharp edge, the three-dimensional vortex layer is roughly conical and quite simple. However, general three-dimensional separation phenomena are complicated because the separation line can be curved on a body surface and its position is not known in most cases. The vorticity vector has two components at the surface. Points on the separation line, at which the two vorticity components are zero, are called "singular"; other points on the separation line are "regular" (Fig. 5.56). A "closed" separation line prevents fluid from entering the region behind the separation line, and the separation area forms a bubble (Fig. 5.56a), while an "open" separation line pushes fluid only away from the surface by forming a vortex layer (Fig. 5.57) [46, 47].

Separation points in plane and axisymmetric flows are always singular. One may consider the examples of Figs. 5.7 and 5.12 in perspective. In plane flows the separation line is always a straight line

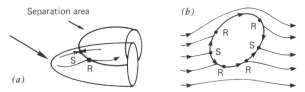

Figure 5.56. (S) Singular and (R) regular separation points on a body surface. They form a "closed" separation line: (a) in perspective and (b) seen from above.

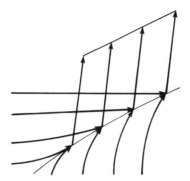

Figure 5.57. "Open" separation line.

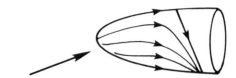

Figure 5.59. The separation line is a streamline.

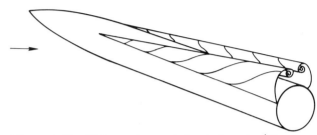

Figure 5.60. Wake vortices on the back of a rocket. The separation lines are open.

normal to the page surface of Fig. 5.12; in axisymmetric flows it is always a circle, whose plane is also normal to the page. Since for these types of flows only one nonzero vorticity component exists and it vanishes at separation points, the separation points are singular. In a general three-dimensional flow three types of singular separation points are distinguished, as shown in Fig. 5.58 [Reference 2, Chapter 4].

The properties of regular separation points have not yet been fully clarified [48]. It appears, however, that the interpretation of Eichelbrenner and Oudart [49], that the separation line formed by regular points is the envelope of streamlines at the body surface, must be abandoned. Instead, Lighthill's argument [Reference 2, Chapter 4] holds that the separation line is a streamline (Fig. 5.59).

The distinction between open and closed separation lines depends on whether the separation line starts on the body surface or is endless. The tip vortices in Figs. 5.47 through 5.49 are generated at sharp edges, which are open separation lines. Figure 5.60 shows an example of the open separation line on a round body surface. All axisymmetric bodies in an axisymmetric flow, however, have closed circular separation lines if the flow is steady. The general case of noncircular closed separation lines (Fig. 5.56b) and their transition to open separation lines is shown in Fig. 5.61 for the steady flow past a cigar-shaped body (prolate spheroid). At an angle of attack of 0° (Fig. 5.61a) the flow is axisymmetric and the separation line thus circular. At a small angle of attack (Fig. 5.61b) this separation line deforms and changes at higher angles of attack (Fig. 5.61c) to an open separation line. With further increase in the angle of attack the beginning of the separation line migrates upstream (Fig. 5.61d). Here, a "nose vortex" may form, and Werlé [50] has made photographs of such nose vortices. At still higher angles of attack the separation line closes again (Fig. 5.61e). The purely qualitative description of the sequence of sketches in Fig. 5.61 holds for both laminar and turbulent flows. More information is found in [References 51 and 52].

The type of vortices generated at a body surface thus depends on the form of the separation line, and

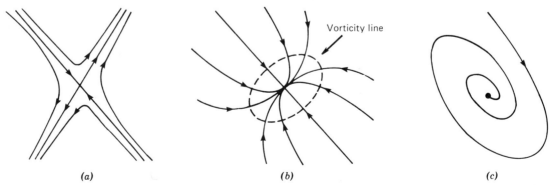

Figure 5.58. Singular separation points: (a) nodal point, (b) saddle point, and (c) spiral point.

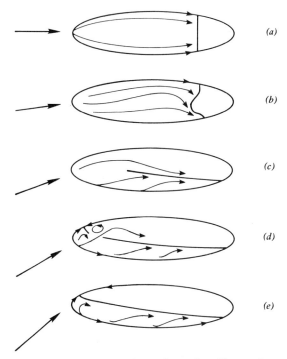

Figure 5.61. Flow around a prolate spheroid at various angles of attack. In cases (a), (b), and (e) the separation lines are closed; in (c) and (d) they are open (adapted from K. C. Wang [48]).

this again on the shape of the body surface and the surrounding flow. From the multitude of vortices generated on three-dimensional bodies several typical examples may be mentioned. On a swept wing, nose vortices can develop between fuselage and wing tip at large angles of attack in subsonic flow (Fig. 5.62). Airplanes with strakes form tip vortices called "strake vortices" (Fig. 5.63) [38, 53]. Tip vortices can also be generated in corners, and one may imagine the corner vortex of Fig. 2.9 on which a ve-

Figure 5.63. Strake vortices (photograph by G. E. Erickson, Northrop Corp.).

locity component perpendicular to the page is superposed. A similar type of vortex is the one that develops along concave walls, as for instance toward the stern along ship hulls (Fig. 5.64). Such a vortex is called a "stern vortex" and must be distinguished from the bilge vortex of Fig. 5.54 [54].

A cylindrical obstacle, placed on a flat surface forms a corner in the plane of symmetry (Fig. 5.65). The boundary layer on the flat surface separates in that corner and is swept around the obstacle to produce a "necklace vortex" (sometimes called "horse-

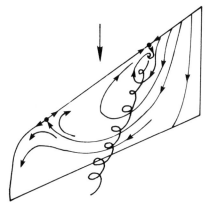

Figure 5.62. Nose vortex on a swept wing near the fuselage (from K. C. Wang).

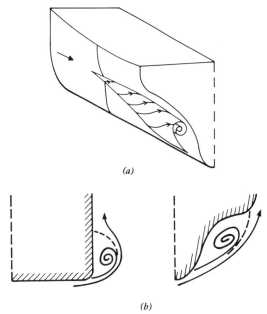

Figure 5.64. (a) Stern vortex on concave ship hull. (b) Comparison between bilge vortex and stern vortex (from H. J. Lugt [54]).

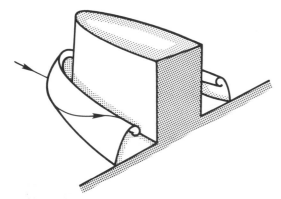

Figure 5.65. Roll up of the vorticity layer at a surface around an obstacle placed on it (from E. C. Maskell [55]).

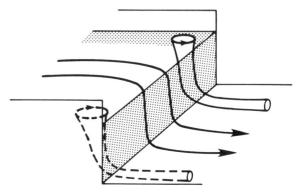

Figure 5.67. Vortices inside a weir.

shoe vortex"). It can occur in strands that are steady, oscillating, or irregular, depending on the Reynolds number. Necklace vortices can be laminar or turbulent [56, 57]. They prevent snow in drift winds from accumulating at the base of tree trunks, and the same kind of vortices form indentations in the sand behind stones and bridge piers in a river bed. The flow past an obstacle of short height is quite complicated (Fig. 5.66) [58]. In the structural analysis of buildings it is important to know the wind force caused by these vortices and by the tip vortices on the roof (Section 6.4).

Vortices form not only at solid surfaces but also at liquid ones. Figure 5.67 shows two vortices inside a weir, and Fig. 5.68 shows the bow vortex around the front part of a ship, which may be considered a type of necklace vortex (see Leonardo's drawing in Fig. 1.8). For bow vortices, of course, the bow wave determines the shape of the free surface, whereas boundary-layer separation causes the necklace vortex. A particular kind of necklace vortex is formed at the stern of surface ships as shown in Fig. 5.69 for the flow behind a transom stern. Waves break behind the ship by generating vortex layers in the flow direction. At high speed, that is for large Froude numbers (Section 8.4), the two layers shown in Fig. 5.69a extend far downstream; at lower speed a breaking wave perpendicular to the flow also occurs (Fig. 5.69b), forming a necklace vortex in a way opposite to that previously described.

The multitude of vortex flows formed around bodies may be classified into the following basic types:

Edge vortex	See Fig. 5.35
Tip vortex	Edge vortex with axial velocity component (Fig. 5.47)
Wake vortex	In the narrow sense, see Fig. 5.7
Corner vortex	With and without axial velocity component, see Fig. 2.9
Necklace vortex	See Fig. 5.65
Nose vortex	See Fig. 5.61d
Hub vortex	As the extension of the axis of a rotating body; see Fig. 4.34

To illustrate the flow around a body that displays almost all types of vortices, a moving submarine in a slight turn is sketched in Fig. 5.70.

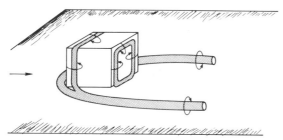

Figure 5.66. Topology of vortex filaments around a cube placed on a horizontal surface.

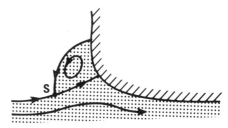

Figure 5.68. Bow vortex of a ship.

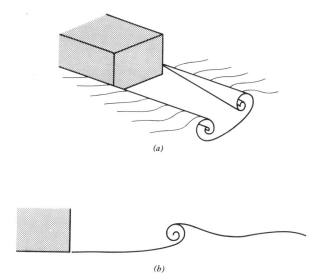

Figure 5.69. (a) Vortex layers caused by a wave breaking behind a transom stern. Perspective view for high speed, (b) side view for low speed (from H. J. Lugt [54]).

5.6. SWIRLING FLOWS

Vortices with an axial velocity component, also called "swirling flows" (particularly in a pipe), deserve a more thorough description because of their importance. A simple example of a plane vortex with a superposed axial flow is solid-body rotation with constant axial motion (Fig. 3.7). The vorticity lines are straight and parallel to the axis as for solid-body rotation, since the superposition of a parallel constant flow does not change the vorticity field. How do vorticity lines behave if the cross section of a vorticity tube or a pipeline changes? Because Helmholtz's second theorem (Fig. 4.20) gives information only on the vorticity component normal to the entrance and exit areas (the vorticity flux remains constant in a vorticity tube), the behavior of a vorticity line must be determined with the aid of the velocity profile. In the presence of a radial velocity component and when the azimuthal velocity component is changed by narrowing the cross section, the vorticity lines are bent to spirals (Fig. 5.71). Here, the axial velocity near the axis is increased more than it is at the outer edge. When the cross section becomes wider, the axial velocity decreases near the axis more than at the wall. This behavior is of particular interest if a vortex is hindered by a fixed wall perpendicular to the vortex axis. Because of the nonslip condition the vorticity lines (except for the axis itself) cannot end at the wall. They are bent away from the axis toward the wall in a spiral way and form a boundary layer, which is called the "Ekman layer" (Section 7.2).

Nonaxisymmetric disturbances in a circular tube and nonaxisymmetric flows also have spiralling vorticity lines. An important example is the flow through a bent circular pipe. No axial vorticity component exists in the flow through a straight circular pipe, but in a curved pipe the centrifugal force acts in such a way that the fluid is pushed against the outer wall and a secondary circulation develops

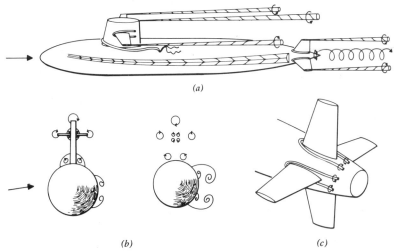

Figure 5.70. (a) The major vortex configurations around a submarine at a small angle of attack in a slight turn. (b) Two cross sections of the hull at the sail and behind. In the latter case the local angle of attack (drift angle) is larger. (c) Necklace vortices around the control surfaces at the stern (from H. J. Lugt [54]).

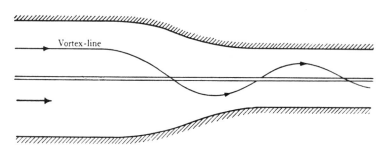

Figure 5.71. Vorticity line in a vorticity tube of changing cross section (from G. K. Batchelor, *An Introduction to Fluid Dynamics*, Cambridge University Press, 1967).

(Fig. 5.72), which has an axial vorticity component. The streamlines become spirals and two vortices form (see also Fig. 3.20) [59]. This phenomenon can also be observed in river bends where the secondary flow causes erosion of the outer bank and a sharpening of the bend [60].

In an oscillating motion with zero mean flow four circulation cells form as depicted in Fig. 5.73 [61]. Oscillating pipe flow with nonzero mean flux may cause resonance between the axial flow and the secondary motion [62, 63]. The resonance plays an important role in the development of atherosclerosis in curved or branching large arteries (see Fig. 5.15) [64].

Tip vortices and swirling flows in general, which do not receive rotational energy downstream, do not always decay by spreading vorticity in the way the decaying potential vortex does. The tangential velocity decreases downstream, and the flow is stable as long as the pressure also decreases in the downstream direction [65, 66]. However, the pressure has a tendency to increase because the tangential velocity decreases. If this tendency prevails at the axis, a separation point may occur at the axis, which indicates zero axial flow. The vorticity tube abruptly expands and forms a closed egg-shaped region or a spiral. This phenomenon is called "vortex breakdown" (or sometimes "vortex burst"). In addition to the pressure increase along the axis, the swirl must be sufficiently large for the occurrence of vortex breakdown, and the streamlines must diverge near the axis (Fig. 5.74).

Vortex breakdown was first observed by Peckham and Atkinson in 1957 [67] in tip vortices of delta airplanes and independently by the author in 1959 [68] in circular pipes behind orifices. Vortex breakdown can also occur in straight pipes without constrictions (Harvey [69]), on wings in nose vortices (Fig. 5.62), in necklace vortices (Fig. 5.70), in rotating fluids inside a container (Section 7.5), and in tornadoes (Section 10.3). Discussion of the explanation of this phenomenon was quite lively in the 1960s, and even today, no theory is unanimously accepted. From the theories offered, three groups are mentioned [70]: (1) vortex breakdown is a kind of flow separation within a rotating fluid (Gartshore, 1962; Hall, 1967; Grabowski and Berger 1976); (2) vortex breakdown is a consequence of instability (Ludwieg, 1962); (3) vortex breakdown depends on the existence of a critical state (Squire, 1960; Ben-

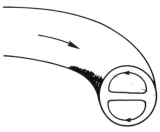

Figure 5.72. Steady laminar flow in a curved pipe with two secondary circulation cells.

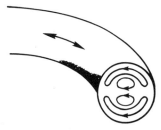

Figure 5.73. Oscillating laminar pipe flow with zero mean flux. The secondary flow consists of four cells.

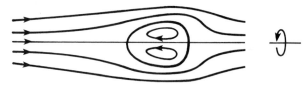

Figure 5.74. Vortex breakdown. The flow is sketched in the meridional plane. The streamlines drawn do not contain the azimuthal velocity component ("meridional flow").

jamin, 1962; Bossel, 1967; Escudier, 1980). In addition to the experimental work by Peckham, Atkinson, Lugt, and Harvey, research has also been done by Kirkpatrick (1964), Hummel (1965), Sarpkaya (1971), Leibovich and Randall (1973), Tsai and Widnall (1980), and Escudier, Bornstein, and Zehnder (1980). Survey articles are found in References 70–72.

The following explanation for vortex breakdown may be offered in which all three groups mentioned above have their place: the stronger the swirl, the lower the pressure minimum at the axis. Downstream along the axis, vorticity spreads away from the axis through diffusion and convection. Thus, the pressure increases along the axis. This increase may be augmented by diverging streamlines in the case of a diffusor. The pressure increase causes a deceleration of the axial flow which, if sufficiently strong, can lead to a stagnation point on the axis with flow reversal (Fig. 5.74). Near the stagnation point the streamlines diverge considerably, and the tangential velocity diminishes as a result of radial convection and angular momentum conservation (ignoring viscosity). This process goes beyond the state of equilibrium between centrifugal acceleration and pressure gradient, and is, in effect, an "overshoot" caused by the wave characteristics of rotating fluids (Section 7.4) [73]. A repulsive force causes the fluid particles to converge. Converging streamlines again increase the tangential velocity and decrease the axial velocity near the axis, and the separation region closes. This process may repeat itself because of the wavy character of the rotating fluid. In most cases, however, the flow becomes unstable and turbulent behind the first "egg." Photographs of flows with vortex breakdown reveal that the separation region is not necessarily completely axisymmetric and impenetrable but that fluid may enter the separation region from behind. Figure 5.75 shows that a dye filament from upstream has colored the "egg" during the experiment.

At higher Reynolds numbers instability causes nonaxisymmetric vortex breakdown that manifests itself in spiral vorticity tubes, either in form of a simple spiral or of a double spiral [72]. Figure 5.76a shows the spiral vortex breakdown of a turbulent flow behind a circular orifice, made visible by hollow-core cavitation. A segmental orifice (Fig. 5.76b) not only constricts the pipe but causes a strong nonaxisymmetric disturbance. The hollow core winds helically through pipe and orifice.

Some more information is given on the conditions for vortex breakdown in an axisymmetric swirl. Consider first a straight tube of circular cross section with given boundary conditions. The flow field is then completely determined by the Reynolds

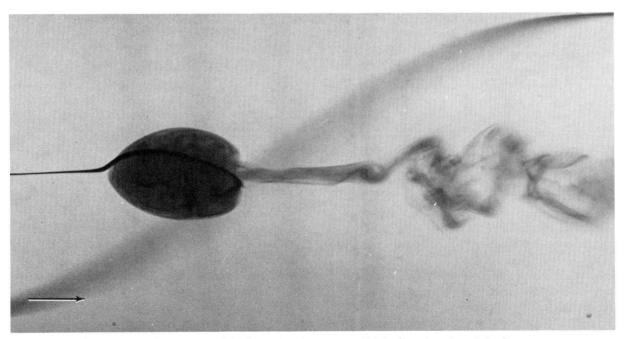

Figure 5.75. Vortex breakdown of a swirling flow in a slightly diverging pipe of circular cross section. The flow is from left to right; the Reynolds number is of the order of 5000. (Photograph by T. Sarpkaya, Naval Postgraduate School, Monterey, California.)

90 Separation

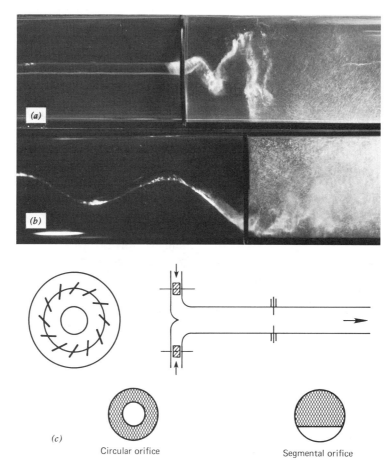

Figure 5.76. (a) Turbulent swirling flow in a pipe with a circular orifice. Vortex breakdown of the spiral type, made visible by a hollow core cavitation of gas along the axis. Re = 10^5 defined by the average axial velocity and the inner diameter of the pipe. (b) The circular orifice is here replaced by a segmental orifice. (c) Sketch of the apparatus for the flow pictures of Figs. 5.75 and 5.76. The diameter of the pipe in Fig. 5.75 increases slightly downstream (Photographs by the author [68]).

number (based on the axial velocity) and by the velocity profiles at the entrance and the exit of the tube. The exit profile is assumed to be so far downstream that it has no appreciable effect on the vortex breakdown. Then the entrance velocity profile and the Reynolds number determine the flow field.

In experiments and engineering applications it is cumbersome and not practical to use the entrance velocity profile (consisting of the profiles of the three velocity components) as input. Instead, it is approximated by a number of more convenient flow parameters or by a set of constants that constitute a family of polynomials representing the entrance profile. The latter approach is used in numerical computations [73, 74]. Typical flow parameters, in addition to the Reynolds number and the geometry of the tube, are the swirl parameter

$$\tan \sigma = \frac{v_\varphi}{v_z}$$

upstream of vortex breakdown somewhere on the profile and v_r/v_φ; the latter parameter is for ring-shaped entrance chambers [72] (v_r, v_φ, and v_z are the velocity components in cylindrical polar coordinates r, φ, z).

Because of the approximate nature of the flow parameters chosen, the experimental data are scattered. For instance, vortex breakdown occurs from a minimum swirl angle σ on, which lies between 30°

and 50°. The location of vortex breakdown moves upstream with increasing swirl angle and with smaller ratio of the entrance axial velocity at the axis to that outside the core [75]. For constant Reynolds number, vortex breakdown of double-helix form occurs at relatively small swirl; at higher swirl a helix forms, which then changes at still higher swirl to the "egg" form.

This general behavior is also observed for vortex breakdown in other swirling flows. For instance, for incompressible fluids the flow field around an airfoil is completely determined (without approximations) by the Reynolds number and the angle of attack. The larger these flow parameters are, the stronger is the swirl and the farther upstream the vortex breakdown occurs. In aerodynamics it is important to know the angle of attack (the Reynolds number is of less influence) at which vortex breakdown reaches the trailing edge of the wing, since at small angle of attack vortex breakdown may occur far downstream of the wing and moves toward the wing with increasing angle of attack. Vortex breakdown over the wing can cause sudden changes in lift, pitch, and roll. For flat delta wings the location of vortex breakdown is plotted in Fig. 5.77 as a function of α.

Vortex breakdown for swirling flows in a container will be discussed in Section 7.5.

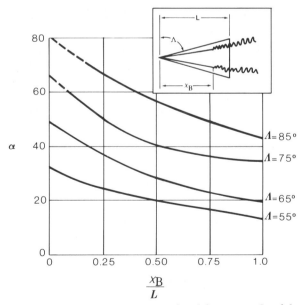

Figure 5.77. Location of vortex breakdown on a flat delta wing as a function of the angle of attack (adapted from A. M. Skow and A. Titiriga [76]).

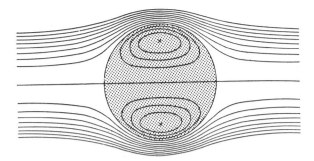

Figure 5.78. Hill's spherical vortex from 1894.

5.7. FREE VORTEX RINGS

Within a finite and closed stream area surrounded by a parallel flow, the vortex ring is the simplest motion (Fig. 5.78). Special examples have been cited in connection with flows past spheres and disks, with jet flows, and with Bénard cells. These vortices are time independent and receive their energy and vorticity from the boundary. The following text describes vortex rings that have separated from the body through acceleration or instability and swim freely in the fluid. Such "free" vortex rings were described in Section 3.6 for vortex tubes with thin cores by a model that neglected friction forces. This model can be improved if the thin ringlike vortex tube is replaced by a vortex ring with a thick core of vorticity. The core can even occupy the total space of the vortex ring. An example is Hill's spherical vortex [Reference 21, Chapter 3]. The shaded area in Fig. 5.78 designates the spherical vorticity field. The fluid in this sphere cannot escape and is carried away with the vortex ring. The ability of a vortex ring to transport fluid over a certain distance has made the vortex ring well known among scientists and laymen. The vortex ring of a cigarette smoker illustrates this clearly. Also the swiftness and seeming lightness with which vortex rings move contribute to their popularity. Reynolds once remarked with respect to vortex rings that nature prefers rolling over gliding. The reason for the eigenmotion has already been given in Section 3.6: The individual ring elements mutually induce a velocity component perpendicular to the ring's plane.

Ideal vortex models, however, cannot describe the generation and decay of vortices, nor can they explain phenomena due to the influence of friction. The following arguments show that vortex rings are in reality more-complicated structures.

Free vortex rings can develop through expulsion

of fluid from openings, through abrupt or oscillating motion of bodies, or through differences of temperature and density in a fluid.

Vortex rings can be produced simply with a box that has a circular opening on one side and a membrane on the other. If one fills the box with smoke and then taps on the membrane, smoke is pushed out of the opening for an instant. The smoke forms a vortex ring that moves quickly away from the opening and then decays. Smoke rings produced in this way have become famous in the advertisements for "Camel" cigarettes on Times Square in New York. There are also toy pistols based on this principle. Candles can be extinguished with them and light curtains can be made to move. Of more serious nature are attempts to influence the weather with vortex rings. Around 1900 antihail cannons were used in northern Italy to prevent hail by means of vortex rings of hot air. When this method was abandoned, it is said that about 2500 antihail cannons had been produced. During World War II vortex rings are said to have been tried as antiaircraft weapons [77]. Also at rocket launching, exhaust pipes, locomotives, and cannons vortex rings can be observed.

In nature vortex rings are found of all sizes. Vortex rings produced during the locomotion of fish have been mentioned already, but in general they cannot be observed because they are invisible. An exception is the vortex ring of the cuttlefish *(Loligo marmorae)*, which belongs to the family of cephalopods and propels itself by means of a jet. When in danger, it mixes an inklike fluid in the jet which forms a vortex ring during ejection. This ring keeps its form for a while and serves either as mimicry or as a diversion to enemies until the cuttlefish has escaped. Flying and hovering birds and insects, men-

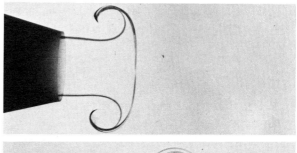
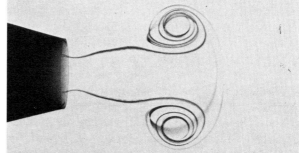

Figure 5.79. Development of a vortex ring in water at a nozzle, made visible by dyed streak lines; Re = 1350 (Photograph by M. Gühler, Fachhochschule Mannheim).

tioned in Section 4.7, also produce vortex rings that are realistic compared to the simple model of Fig. 4.37 [78].

Vortex rings also form during the eruption of volcanoes and consist of steam, ashes, and hot gases.

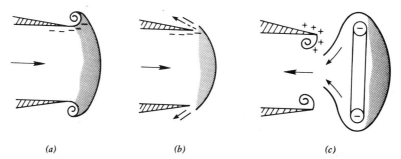

Figure 5.80. (a) The vorticity of the boundary layer (−) generates a discontinuity surface, which rolls up to a vortex ring. (b) If the boundary layer is sucked off at the opening, no vortex ring can develop. (c) The detached vortex ring induces a secondary vortex with vorticity of opposite sign (+) [sketches (a) and (c) according to T. Maxworthy, *Journal of Fluid Mechanics*, Cambridge University Press [79], sketch (b) after D. Sallet, University of Maryland].

Figure 5.79 shows the development of a vortex ring in water at a nozzle. The water is expelled from a tube with the aid of a piston. Simultaneously, dye is injected into the water near the edge of the nozzle. The vorticity of the vortex ring originates in the boundary layer formed at the opening during the expulsion of the fluid. This boundary layer separates from the edge of the pipe or nozzle and rolls up in the form of a vorticity layer to a vortex ring. The generated streak lines coincide at the beginning of the roll up with the thin, separated vorticity layer (Fig. 5.80). Later, when vorticity has been distributed over the vortex ring, streak lines are no longer lines of concentrated vorticity. If the boundary layer is sucked away during the expulsion of the fluid, no vortex ring can form (Fig. 5.80b).

Figure 5.81 shows the structure of a vortex ring farther away from the place of origin. The Reynolds number is defined by the constant velocity of the piston and the inner diameter of the edge of the nozzle.

As soon as the vortex ring is detached, it induces a flow toward the tube's axis, which produces vorticity of opposite sign at the edge of the tube or nozzle. This vorticity forms a secondary vortex ring with rotation opposite to that of the primary vortex

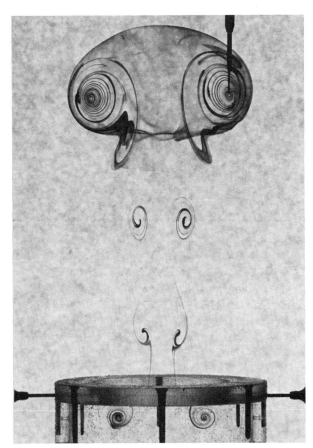

Figure 5.82. Vortex ring in water, made visible by dyed streak lines. The secondary vortex inside the tube is clearly recognizable; Re ≈ 2000 (Photograph by D. Sallet, University of Maryland).

ring (Fig. 5.80c). Figure 5.82 reveals clearly the secondary vortex at the end of the tube, where an orifice is located.

An example of the development of a free vortex ring behind an axisymmetric body is displayed in Fig. 5.28 (when the body is interpreted as a sphere instead of a circular cylinder). In Reference 68, Chapter 6, photographs can be seen in which the generation of a vortex ring is visible when a falling disk hits the bottom of a tank.

The generation of free vortex rings does not depend on the presence of solid walls. When a drop hits the surface of a liquid, a vortex ring develops (Fig. 5.83). The density of the drop and thus the vortex ring is heavier, therefore, the ring travels longer before it decays. This vortex ring obtains additional kinetic energy from gravity in the same way as the rain drop does.[11] In air, vortex rings that are hotter than their surroundings and that are produced at the ground ascend. It has been observed

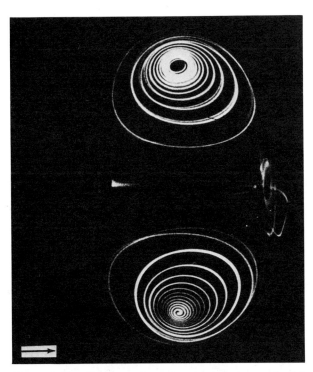

Figure 5.81. Structure of a vortex ring in water, made visible by dyed streak lines; Re = 1500 (Photograph by M. Gühler, Fachhochschule Mannheim).

Figure 5.83. Drops generate vortex rings when they hit the free surface of a liquid.

during explosions that 10–15s after the explosion a smoke ring formed which persisted for about 10 min and reached a height of almost 2 km during this time [80]. These vortex rings will be described in more detail in Section 8.2.

In contrast to inviscid vortex rings, which move with constant velocity without decay, real vortex rings lose speed. They decelerate not only through loss of energy due to friction but also through entrainment of fluid from the surroundings (Fig. 5.84). This fluid, which is entrained over the whole surface of the ring, must be set into rotation and increases the size of the vortex ring. The entrainment of fluid was first described in 1939 by Krutzsch [81]. The velocity of laminar vortex rings decreases inversely with time, that is, with t^{-1}. Real vortex rings also eject matter (and vorticity) in the wake, as first observed also by Krutzsch and later more closely studied by Maxworthy [82]. Additional information on the formation and decay of real vortex rings is contained in References 83 and 84.

The fact that a vortex ring is the simplest flow structure, an "element," suggests that it be considered a building block for more-complicated flow configurations. For instance, the concept of a vortex ring may be useful in studying turbulence (Section 6.6). It is, therefore, of interest to know how a vortex ring reacts with other rings and with solid walls.

Real vortex rings behave differently from ideal vortex rings near solid walls. When a vortex ring approaches a wall, vorticity of opposite sign is created

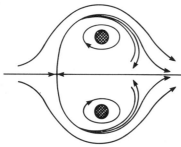

Figure 5.84. Real vortex ring with entrainment and wake (from S. E. Widnall [Reference 26, Chapter 3]).

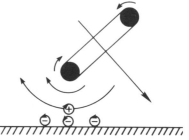

Figure 5.85. Vortex rings vanish near a solid wall through production of vorticity of opposite sign.

Figure 5.86. Development of a secondary vortex when a vortex filament approaches a wall (from J. K. Harvey and F. J. Perry [86]).

at the wall. The two vorticity fields then cancel each other (Fig. 5.85). During this process secondary vortices may develop at the wall where the streamlines diverge, and separation can occur. For a straight vortex filaments this development is sketched in Fig. 5.86. If a vortex ring approaches the wall at 90°, several vortex rings, depending on the strength of the primary vortex, can be induced at the wall sequentially. The first induced vortex tries to push the primary vortex away from the wall. During this time the influence of the induced vortex on the primary one diminishes, and the primary vortex approaches the wall again. If it is still strong enough, it induces the second secondary vortex, and so on [85].

A similar, although more-complicated situation, occurs when a vortex ring hits the edge of a plate. In Fig. 5.87a (side view) a vortex ring approaches the edge. In Fig. 5.87b a secondary vortex ring of opposite rotation has formed, which has a smaller diameter than the primary one [86]. (The secondary vortex ring is lighter in the photographs since the dye that makes it visible is obtained from the primary one.) In Fig. 5.87c and Fig. 5.87d both rings start to disintegrate. The downstream half seems to coalesce in this process.

Finally, two real vortex rings, whose paths cross at a small angle, combine to form an elliptical vortex ring. However, if the angle is larger than 32°, the vortex rings separate again [87]. For straight vortex filaments the merger during decay is described in Reference 88. Other vortex interactions and the be-

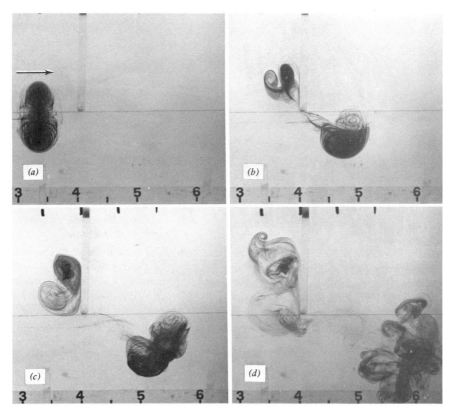

Figure 5.87. A vortex ring approaches the lower edge of a vertical plate from left to right. The sequence of pictures shows the formation of a secondary vortex ring and the disintegration of both vortex rings (Photograph by P. E. M. Schneider, Max-Planck-Institut für Strömungsforschung, Göttingen).

havior of noncircular vortex rings are discussed in References 89 and 90.

Remarks on Chapter 5

1. The description of steady flow separation does not apply to unsteady motions without further explanation. For example, the flow in Fig. 5.1 may be considered time dependent in such a way that the separation point migrates along the surface from right to left with velocity V. Now, the steady-state criterion of zero vorticity (or zero shear stress) on the surface for flow separation does not apply any more. This is easily seen if the unsteady flow is made time independent by fixing the reference frame not to the wall but to the separation point; the wall will move then with velocity $-V$ from left to right (Fig. 5.88). Hence, no separation point can occur at the wall. It has migrated into the fluid (a). But the vorticity field has not changed, that is, it is invariant with respect to changes of the in-

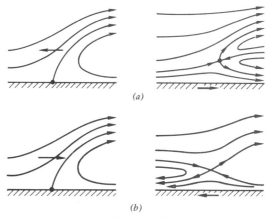

Figure 5.88. An unsteady flow field, with a separation point moving along a solid wall, can be made steady (at least locally around the separation point) by superposing a velocity in such a way that the separation point becomes fixed to the reference frame. (a) The separation point moves right to left (from D. P. Telionis and M. J. Werle [92], The American Society of Mechanical Engineers), (b) from left to right (from D. T. Tsahalis [93]).

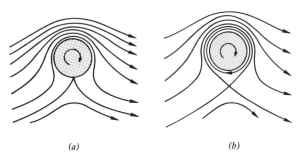

Figure 5.89. Potential flow past a rotating circular cylinder. (*a*) The stagnation points coincide on the body surface. (*b*) The stagnation points are inside the fluid.

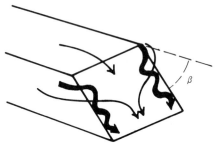

Figure 5.90. Tip vortices in the rear of a body with a slanted base (from R. Sedney [98]).

ertial system. In a similar way an upstream-moving separation point can be made steady by superposing the velocity V. Now the wall moves with velocity V from right to left. Here too, the separation point is shifted into the fluid (*b*), and again is the vorticity field invariant. The conditions for separation are thus more complicated [91–93]. Unsteady three-dimensional flow separation is discussed in Reference 94.

2. The points of separation and attachment can also coincide as in the flow past a rotating cylinder under certain conditions (Fig. 5.89).

3. Flow separation may also be explained, as is done in textbooks [Reference 35, Chapter 1], by the pressure increase in the flow direction. Bernoulli's theorem for potential flows shows that diverging streamlines in an incompressible fluid cause a pressure increase since the velocity decreases:

$$\frac{p}{\rho} + \frac{V^2}{2} = \text{const.}$$

In a viscous fluid particles lose kinetic energy because of their adherence to the surface, and they cannot overcome the pressure increase from a certain point on. The particles separate from the surface.

4. Sometimes the opinion is expressed [Reference 9, Chapter 2] that vortices are defined by the existence of an extremum of vorticity, a criterion that is used in this book for the definition of vortex separation. Such a definition appears to be too restricting to the author [Reference 1, Chapter 2] since rotating fluids near a wall (Fig. 5.7) are no longer considered vortices, and the vortex definition is not then unique for inhomogeneous fluids (Chapter 8).

5. In flows around liquid spheres instability occurs at lower Reynolds numbers, according to Reference 95 at about Re = 270.

6. The reason for vortex separation cannot always be clearly divided into acceleration effect and instability. This is particularly true for symmetric flows past bodies. For this reason certain problems like those of oscillating bodies are treated in both Chapters 5 and 6.

7. It has not yet been proven mathematically that the flow model of Lanchester–Prandtl together with the Kutta condition corresponds to real flow in the limit Re = ∞ [96].

8. A theorem by Prandtl [Reference 19, Chapter 4] says that vorticity inside closed streamlines is constant in steady, inviscid flows. If diffusion of vorticity into the core thus diminishes with time, the core must assume solid-body rotation.

9. An edge vortex can also be trapped on the upper surface of a straight wing by specially designed flaps ("Kasper wing"), and additional lift can be obtained. See the trapped edge vortices in Figs. 5.22*b* and 5.26 [97].

10. Vortices, which influence the drag of a car, develop also on the rear window. Depending on the angle β, a closed separation bubble for approximately $\beta > 30°$ occurs, and open separation for $\beta < 30°$ (Fig. 5.90) [98].

11. Raindrops are vortex rings according to the vortex definition 1 in Section 2.1. However, in contradistinction to free vortex rings, they have no extremum of vorticity because they receive their total rotational energy from the outer boundary through shear forces. The free vortex ring, on the other hand, receives its rotational energy during its creation. However, when the free vortex ring falls in the gravitational field, it receives additional energy if the density of the fluid in the vortex is larger than that of its surroundings.

6. Instability and Turbulence

6.1. WHAT IS INSTABILITY?

No phenomenon in fluid dynamics has such far-reaching significance as that of flow instability. This condition was first recognized by Helmholtz [Reference 14, Chapter 5], Lord Rayleigh [1], and Lord Kelvin [2] at the end of the last century. Instability is a mechanism by which a fluid accommodates to strong forces, and new flow patterns are created. For instance, in a steady flow in which only viscous and inertial forces are acting, diffusion cannot counteract convection above a certain Reynolds number. In most cases instability leads to turbulence, and almost all flows in nature and technology are turbulent. Despite the overwhelming importance of turbulence there is still no satisfying theory in the sense that turbulent motion can be derived from first principles. This problem will be discussed in Section 6.6. In the transition range between the first appearance of instability and the development of turbulence beautiful vortex configurations can form. They are observed everywhere in nature.[1]

When a flow is disturbed and the disturbance decreases with time so that after a while the original state of flow is regained, the flow is said to be "stable." When a stone is thrown into water, the disturbance spreads in rings and then vanishes. The original state of rest is restored; the flow is stable. However, when the flow is disturbed and the disturbance increases and completely changes the original flow, this flow is labeled "unstable"; instability leads to a new kind of flow. A well-known example from mechanics will give a better understanding: A ball is put into a mould (Fig. 6.1a). When the ball is displaced, it moves back into its previous position. This state is thus stable. However, when the ball is placed on an elevation and displaced a little, it will move away from the original position. This state is unstable (Fig. 6.1b). Fluids behave in a similar way, although the situation is much more complicated. The interplay of the participating forces determines the occurrence of instability, which can be of several types:

Instability due to inertial forces
Instability due to centrifugal and Coriolis forces
Thermal instability (buoyancy)
Instability due to gravitation, electromagnetic forces, and surface tension

The occurrence of instability is described quantitatively by characteristic parameters. For instance, it is the Reynolds number that determines the onset of instability due to inertial forces in the presence of viscous forces. Below a critical Reynolds number the flow is stable; above it, unstable. Above such a critical characteristic parameter, however, the original flow can exist, provided that any disturbance is avoided that would unbalance the unstable state. An example is the flow through a pipe. The critical Reynolds number is about Re = 2300. Below this number the flow is laminar; above it, turbulent. By carefully avoiding any disturbance, the flow could be kept laminar up to Re = 40,000 [3]. In practice, of course, disturbances are always present, and the flow is turbulent at that Reynolds number.

The new flow pattern obtained through instability can become steady after a sufficiently long transient time or it can remain unsteady (periodic). An example of the first case is the Bénard cell in Fig. 2.10, which will be described in more detail in Section 8.3. Examples of the unsteady case are periodic vortex streets and turbulent flows.

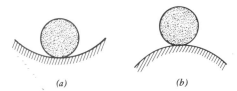

Figure 6.1. The concept of instability: (a) stable state and (b) unstable state.

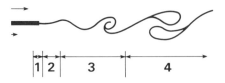

Figure 6.2. The four zones of an unstable shear layer (see text).

6.2. INSTABILITY OF SHEAR FLOWS AND BOUNDARY LAYERS

When inertial forces prevail over viscous forces, shear flows and boundary layers can become unstable above a certain critical Reynolds number.

For simplicity, the instability of an infinitely long and infinitesimally thin shear layer, called "Helmholtz instability," is considered first. A small disturbance in the form of a wavy motion is superposed on a parallel shear flow [4]. If the shear flow is stable, the disturbance will vanish with time. If it is unstable, discrete vortices will develop. This process has already been demonstrated in Section 3.5 by means of the model of a discontinuity line (Fig. 3.29) or a band of point vortices (Fig. 3.30) which has extremely high vorticity (Fig. 4.19). This line or band is always unstable in an ideal fluid, but in real fluids viscosity keeps the shear layer stable up to a certain Reynolds number.

In reality a shear layer is neither infinitely long nor infinitesimally thin but originates somewhere, say, behind a plate or at the exit of a nozzle when the boundary layer separates and becomes a free shear layer having a finite thickness.

According to Sato [5] and Freymuth [6] four zones may be distinguished when a free shear layer becomes unstable (Fig. 6.2):

1. The place of origin of the shear layer, that is, the location at which the boundary layer separates from the surface.
2. The zone in which small disturbances of the shear layer grow (range of "linear" growth). The disturbance can occur accidentally through environmental effects or it can be produced artificially by the sound from a loudspeaker or by a vibrating ribbon with the frequency f.
3. The zone in which the growing disturbances cause the shear layer to roll up into discrete vortices. Through nonlinearity higher harmonics $2f$, $3f$, etc., with smaller amplitudes than the basic harmonic occur [7].
4. The zone in which these vortices merge through "pairing"; in turn, there can be pairing of those vortices that have already "paired." This successive pairing process continues until an apparent disintegration to turbulence sets in. Pairing is a process in which two adjacent vortices of the same rotary direction spin around each other (Fig. 3.27), come closer, and then merge. Such "coalescence" results in the appearance of a subharmonic $f/2$ [6, 8].

Unstable shear layers may interact with each other if the distance between them is small. Figure 6.3 shows the development of a staggered row of vortices from two bands of point vortices representing two shear layers (Section 3.5). They can originate behind a flat plate parallel to the flow (Figs. 6.4 and 6.5). The boundary layers above and below the plate (Fig. 6.4a) join in the wake to a single shear

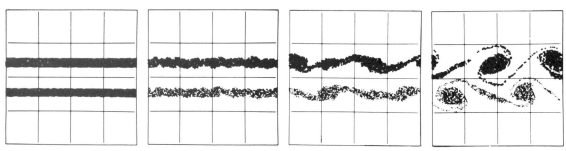

Figure 6.3. Roll up of two vortex bands, representing two shear layers, to a vortex street. The bands consist of hundreds of point vortices (computer-generated picture by K. V. Roberts and J. P. Christiansen [Reference 25, Chapter 3], North Holland Publishing Co.).

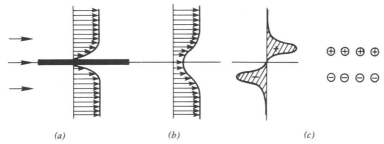

Figure 6.4. Wake behind a flat plate parallel to the flow. The shear layer can be described by two rows or bands of point vortices for large Reynolds numbers.

layer (Fig. 6.4b), which consists of vorticity with positive as well as negative sign. This shear layer can be approximated for large Reynolds numbers by two rows or bands of point vortices (Fig. 6.4c).

Shear layers can also form the boundary of cylindrical and planar jets (Fig. 6.6). Interaction of the two shear layers results in vortex arrays that are either symmetric or antisymmetric, depending on

Figure 6.6. The boundary of an axisymmetric jet dissolves into individual vortices at sufficiently high Reynolds number. The shear layer is made visible by streak lines. The Reynolds number is 1800 (adapted from H. Schade and A. Michalke [9]).

the upstream conditions and unsteadiness generated by downstream regions of the flow [10]. Also, the frequency of vortex shedding depends on the kind of shear layer in the opening. With thin boundary layers the Strouhal number (Section 6.3) is St = 0.012 $\sqrt{Re}$ for plane and axisymmetric jets, where the constant 0.012 varies somewhat with design of the nozzle contour. With fully developed velocity profiles (flow out of a long pipe or between two plane plates) the Strouhal number is a constant. With background noise always present, plane jets oscillate at a "preferred frequency." However, if one strong frequency dominates in that background noise, the frequency of the jet, within limits of those frequencies amplified by the jet, will adjust to that frequency. This process is called synchronization or "lock-in" (see Section 6.4) [11]. The range of Reynolds number for periodic vortex development in a jet is not known exactly. Experiments so far have revealed that vortices occur in the order of magnitude from 10 up to and beyond 10^4.

More-complicated types of shear-layer instability occur in flows past blunt bodies. Details will be presented in Sections 6.3 and 6.4, where three-dimensional unstable shear layers will also be mentioned.

An unstable shear layer that has rolled up to dis-

Figure 6.5. Vortices behind a flat plate for Re = 230,000 (photograph by F. N. M. Brown, University of Notre Dame, Indiana).

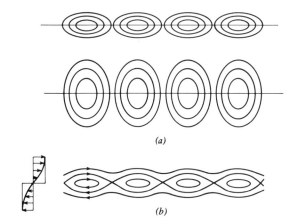

Figure 6.7. Decaying row of vortices (a) in a fluid at rest and (b) in a shear flow (Kelvin's cat eyes).

crete vortices decays if it no longer obtains energy from the flow. A simple row of vortices decays by the spreading of each vortex essentially normal to the row (Fig. 6.7a). In this case streak lines are not always suitable for illustration and are often difficult to interpret [12]. In the ensuing text, therefore, streamlines at certain time intervals will also be used. Streamlines of vortices depend on the reference frame and the superposed flow (Fig. 2.14). In Fig. 6.7b a shear flow free of disturbances is superposed on a row of vortices. The streamlines form "Kelvin's cat eyes," which vanish after a certain decaying time.

The flow pictures in Fig. 6.7 are valid only for a certain disturbance frequency (or wavelength) in the final stage of decay when nonlinear inertial effects are absent. In general, however, a periodic disturbance is composed of many frequencies, which may be determined mathematically by a "Fourier analysis." Vortices with high frequency, that is, with many small vortices per length unit, decay faster than those with smaller frequency. On the other hand, the strengths of the vortices with high frequency can be larger at the beginning of the decay than those with smaller frequency. Consequently, small vortices dominate in the beginning and are visible on flow pictures. They decay quickly, and larger vortices with smaller frequency appear. Figures 6.8 and 6.9 show the decay of an unstable circular discontinuity line. Such circular shear layers occur when vortices develop. They are sketched for stable shear layers in Fig. 5.34.

The unstable shear layer of a vortex is drawn in Fig. 6.10. The number of developing secondary vortices at the border of the core of the primary vortex depends on the kind of initial disturbance. In Fig.

Figure 6.8. Decaying circular discontinuity line without inertial effects. The sequence of streamline pictures starts with eight vortices. They decay, and six vortices of smaller frequency appear. Finally, the whole fluid is at rest (computer-generated pictures from the author [13]).

6.8 the number of secondary vortices in the beginning was arbitrarily chosen to be eight (without superposed shear flow). The sequence of pictures shows the fast decay of these eight vortices and the appearance of six new vortices (although they have always been there and have become visible only through weakening of the first eight vortices). Figure 6.9 consists of two photographs of a real experiment in a water tank. The discontinuity line between the rotating outer flow and the core at rest has been produced here with the aid of two circular cylinders, which are separated like a telescope [15].

Boundary layers at the surface of a body can also become unstable above a certain Reynolds number. This process of instability and subsequent transition to turbulence may be divided into the following steps [16, 17]:

1. Initial disturbances in form of two-dimensional waves travel downstream and grow. An observer riding with the waves would notice "cat eyes" (Tollmien–Schlichting waves).

2. Next, three-dimensional disturbances (in the downstream direction as well as perpendicular to it) occur and grow with local high-frequency fluctuations.

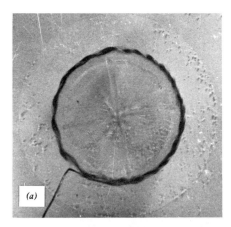

Figure 6.9. A circular discontinuity line begins to decay. Streak lines in a water tank: (a) 24 vortices, (b) 9 vortices a little later (photographs by J. R. Weske, University of Maryland, College Park, Maryland).

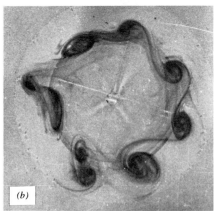

Figure 6.10. Curved discontinuity line behind a wedge. During roll up the discontinuity line itself becomes unstable and decays into discrete vortices (sketch drawn from a photograph by D. Pierce [14]).

3. Local eruptions form "turbulent spots" (Emmons [18], which may be envisioned as rolled up vorticity sheets composed of loops of vortex filaments (Leonard [19]. The formation of such loops on a surface was predicted by Theodorsen in 1955 [20]).

4. The final step is the transition of the total boundary layer to turbulence (Section 6.6).

6.3. PERIODIC VORTEX SHEDDING

The streamlines for the flow past a circular cylinder for Re = 40 are pictured in Fig. 5.12. In this range of Reynolds number the wake far away from the body can become unstable. According to Taneda [21] such instability can start at about Re = 30 (Fig. 6.11). With increasing Reynolds number the instability moves upstream, and at Re = 45 (or perhaps a little sooner) the total wake becomes unstable. At the smallest disturbance the two vortices behind the cylinder separate one after the other and swim away with the flow. New vortices form in an alternating way, and they also detach. In distinction to the steady flow of Fig. 5.12 a new, periodic, and asymmetric flow is generated, which is called a "Kármán vortex street" after the fluid dynamicist von Kármán.

What does such a system of vortices look like? If one wants to observe streamlines at certain times, one must decide, according to the explanations in

Figure 6.11. Onset of instability in the far wake of a circular cylinder at Re = 30.

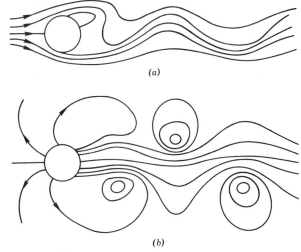

Figure 6.12. Kármán vortex street for an observer who moves (a) with the cylinder and (b) with the vortices (sketched from computer-generated streamline pictures by D. C. Thoman and A. A. Szewczyk [22]).

102 Instability and Turbulence

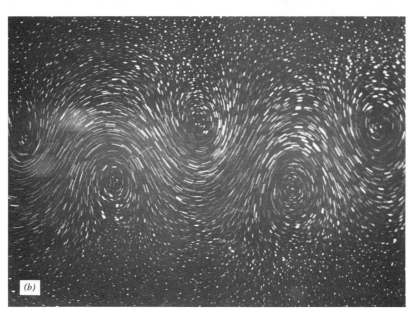

Figure 6.13. (a) Periodic vortex street for Re = 190, made visible by streak lines (photograph by O. M. Griffin, Naval Research Laboratory, Washington, D.C.). (b) Photograph of streamlines by A. Timme [23].

Section 2.3, whether to move with the cylinder or with the vortices. In the first case the streamline picture looks wavelike except in the immediate vicinity of the body (Fig. 6.12). In the second case the individual vortices of the street become visible through closed streamlines. Photographs usually reveal streak lines, which have been made visible by dye or smoke injection (Fig. 6.13).

Whether the vortices are arranged in a symmetric or alternating way and the distances among the vortices both depend on the stability of the new flow. A symmetric vortex street is unstable. The ratio of vertical to horizontal distance of the vortices must be about 0.28 in the stable state at the beginning of the vortex street, and this ratio becomes larger downstream during the decay of the vortices. The velocity distribution of a vortex corresponds quite well to that of a decaying potential vortex (Fig. 3.12) [23, 24].

Of interest are the forces that act on the cylinder and the frequency with which the vortices separate from the body. In the steady-flow situation Re < 45 the streamlines around the circular cylinder are symmetrical (Fig. 5.12) so that only one component of the force is present: the drag. In asymmetrical flows, however, other components appear: a force component normal to the flow (the lift) and a torque around the axis of the cylinder. The directions of the forces are sketched in Fig. 6.14.

When a vortex has shed from the body (for instance, the lower vortex in Fig. 6.14), the attachment point A migrates to the lower side of the cylinder. The fluid now needs a higher velocity to pass over the upper part of the cylinder. Because of the larger kinetic energy and the lower surface pressure, the separation S_1 is delayed while the slower flow on the lower side separates earlier due to the higher surface pressure, point S_2. However, the thinner boundary layer at the upper part produces more vorticity (the wall shear is larger), which then is assembled to a strong vortex that separates and swims away. Lift and torque reach an extremum about the time that the vortex separates. The directions of these forces are given in Fig. 6.14. After separation of the vortex a new cycle starts with the development of a new vortex on the lower side [26].

The frequency with which the vortices are shed from the body can be made dimensionless with the flow velocity and the diameter of the circular body. The new parameter, which is named in honor of the physicist Strouhal, depends only on the Reynolds number:

$$\text{Strouhal number} = \frac{\text{frequency} \cdot \text{diameter}}{\text{velocity}}$$

The Strouhal number for a circular cylinder is plotted versus the Reynolds number in Fig. 6.15.

Although the curve in Fig. 6.15 indicates that vortex streets exist up to Re = 1000, they actually occur into the turbulent range of Re = 10^7 and higher, with scattering of the experimental data for the Strouhal number St in the transition region 10^5

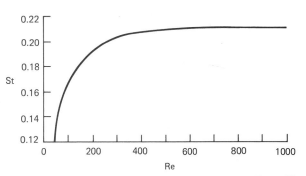

Figure 6.15. Strouhal number as a function of Reynolds number for a circular cylinder (from A. Roshko [27]).

$<$ Re $< 5 \times 10^6$. At Re = 10^7 the Strouhal number is about 0.28 [28]. However, not all vortex streets over the whole Reynolds number range have the same structure. Purely laminar, stable vortex streets behind a circular cylinder exist only in the range from Re = 45 to about Re = 200. Beyond these values, the structure of vortex streets is more complicated. The separating boundary layers can become unstable laminar shear layers or become turbulent. With higher Reynolds numbers vortex separation becomes more irregular, and three-dimensional disturbances emanating from the ends of the cylinder have greater influence on the formation of the vortex street [29].

Every cylindrical hindrance causes a periodic vortex street in the proper range of Reynolds number. For an asymmetrical flow or a flow around an asymmetrical body no special disturbance is necessary to initiate vortex shedding as is required in the case of symmetrical flow past a circular cylinder. The sequence of vortex separation after the start of the flow is uniquely determined. Figure 6.16 shows a sequence of flow pictures for an inclined plate at Re = VD/ν = 200 with D being the width of the plate. Lift is now always upward and the torque clockwise. Lift and torque oscillate during a cycle of vortex shedding, with a minimum about the instant (Fig. 6.16a) the lower vortex separates and a maximum when the upper vortex leaves (Fig. 6.16b).

For a flat plate parallel to the flow the fluid does not separate from the body. Rather, the vortex street develops farther downstream. It is not a result of flow separation from the body but of an unstable wake (Fig. 6.5). For arbitrarily formed cylindrical bodies the flow is naturally very complex. Figure 6.17 shows the results of an attempt to calculate by computer the flow past an irregularly formed body. Farther downstream, however, quite a regular vortex street may develop.

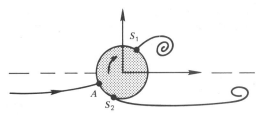

Figure 6.14. Schematic representation of the force components in an asymmetric flow around a circular cylinder. The directions of lift and torque depend on the position of the vortex close to the body (from S. K. Jordan and J. E. Fromm [25]).

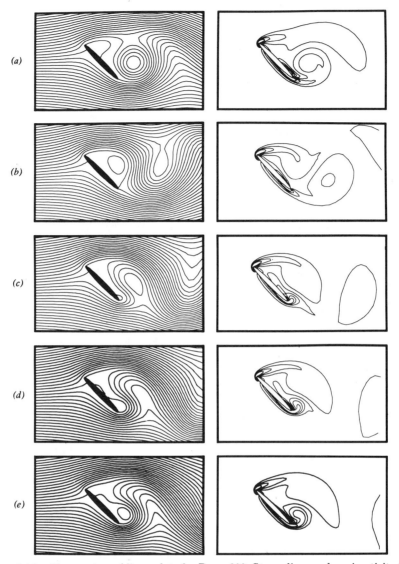

Figure 6.16. Flow past an oblique plate for Re = 200. Streamlines and equivorticity lines for one cycle of vortex shedding (computer-generated pictures from Reference 17, Chapter 5).

A group of cylinders arranged perpendicular to the flow or in a staggered pattern will also cause periodic vortex streets. Stability, vortex arrangement, and frequency depend on the cylinder positions [30–32].

Vortex shedding from bent, coned, and whirled cylinders has been studied by Ehrhardt [33], Gaster [34], and Taneda [21]. Cylinders placed in a shear flow produce vortex streets of greater stability and slightly higher Strouhal number [35, 36].

So far it has been assumed that the cylindrical bodies and the plane jets were infinitely long, that is, that the flow around them was two dimensional. What happens at the ends of cylinders of finite lengths? How do vortices separate when the length of the cylinder is shortened and the body finally becomes spherical or cubical? To answer these questions it is instructive to consider first the flow as ideal and the vorticity as confined to vortex tubes. Helmholtz's first theorem says that vortex tubes cannot end within a fluid. With photographs of flows behind a circular cylinder Taneda [37] has demonstrated that periodic vortex tubes are connected to each other at the ends of a cylinder in the way sketched in Fig. 6.18. The curved vortex tubes at the ends of the cylinder are, however, unstable (Section 3.6). Downstream they twist more and more and destroy the periodic order of the vortices.

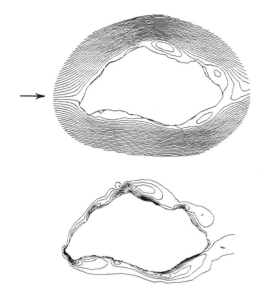

Figure 6.17. Streamlines and equivorticity lines shortly after the start of an irregularly formed cylinder for Re = 500 (computer-generated picture from F. C. Thames, Mississippi State University).

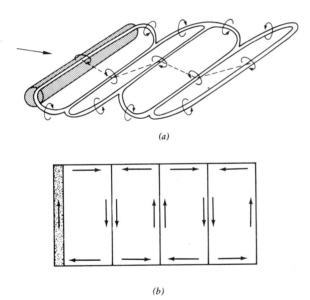

Figure 6.18. (a) Vortices of a periodic vortex street are connected to each other behind a cylinder of finite length; (b) the linkage seen from above.

Figure 6.19. Behind spheres periodic vortex streets change to two screwlike vortices (sketch from H. P. Pao and T. W. Kao [38]).

This is the reason for the disturbing influence of finite cylinders on two-dimensional effects in wind tunnel studies.

For spherical bodies the linked vortex tubes in Fig. 6.18 simplify to two helical vortices connected at the cross points (Fig. 6.19). The vortices detach behind spheres at Re = DV/ν = 400. Below Re = 400 the vortex rings attached to the spheres are axisymmetric (Fig. 5.7); however, from Re = 130 on they react to disturbances and oscillate. When vortices separate at Re = 400, the flow becomes asymmetric, and lift and torque occur in addition to drag. Moreover, according to Achenbach [39] the location of separation rotates about the rear half of the sphere. If a sphere falls freely, it does not follow a straight vertical line for Re > 400 but traces a helical path. The motion of the fall is, however, sensitive to small irregularities of the sphere. For instance, if the center of gravity and the center of the sphere do not exactly coincide, the sphere wobbles during fall [40]. Achenbach [39] has measured periodic vortex separation behind spheres for the range 400 < Re < 3.7×10^5. In the range 6000 < Re < 30,000 the Strouhal number changes from 0.125 to 0.18 with increasing Reynolds number.

Three-dimensional wakes behind spheres may be compared with three-dimensional vortices in a jet as shown in Reference 41.

Of historical interest is the fact that research on vortex streets started in the field of acoustics. In 1878 Strouhal [42] investigated sound-producing vibrations in wind. A year later Lord Rayleigh [43] discovered that wires that generate tones in an air stream need not themselves move. This understanding may have been the birth of the study of "aerodynamic noise." It was only in 1908 that Bénard [44] correlated the tones generated in a stream with the existence of a vortex street.

Vortices generated periodically cause the "singing" of wires. Acoustical tones are nothing more than periodic pressure and density fluctuations, and the pitch of a note depends on the frequency of the vibration. Von Kármán studied vortex streets theoretically by means of potential-flow theory. Bénard later complained about the name "Kármán vortex street," but this notation went into the literature anyway.

With the aid of the curve in Fig. 6.15 the order of magnitude of the pitch of a note or the diameter of a cylinder can be estimated. "Pure" vortex streets occur in the range 45 < Re < 200. This corresponds, for air with the kinematic viscosity 0.15

cm²/s, to the range DV = 6.75 to 30 cm²/s for the product of velocity and diameter. From the Strouhal numbers (Fig. 6.15) one obtains the frequency range $0.7/D^2$ to $5.9/D^2$. The audible frequency range is roughly between 16 and 20,000 Hz. Thus, tones can be heard for wire diameters from 0.06 to 6.1 mm.

The howling of a fast-moving thin stick in air is also caused by a periodic vortex street. Here, both the flow and the body oscillate. Romantic tales are connected with the aeolian harp, a string instrument, played by the wind and called after Aeolus, the Greek god of the winds. King David is said to have hung such an instrument over his bed at night so that the midnight breeze could play it. The explanation is less romantic: Here, too, the vibrations of the strings are generated by the vortex street.

The vortex street past a cylindrical body, however, changes if the body itself also vibrates. This condition will be examined more closely in the next section.

6.4. FLOW-INDUCED VIBRATION

The "rhomboi" which are whirled about in the mysteries produce a low note when whirled gently, but a high one when whirled virogously.
Archytas, 4th Century BC [Reference 9, Chapter 1]

On November 7, 1940 the Tacoma bridge in Washington State crashed during a storm with winds of 67 km/h. The bridge was more than 1.5 km long and had been opened to the public a few months before. Every phase of the collapse was filmed, since the disaster was anticipated hours before. This event drew wide attention for two reasons: Never before had a suspension bridge of that size collapsed, and the cause of the crash was unusual. The investigating committee, of which the fluid dynamicist von Kármán was a member, concluded that aerodynamic forces (that is, wind forces) came into resonance with the eigenfrequencies of the bridge. The oscillations were caused by periodic vortices behind the bridge. Although there were suspension bridges both smaller and larger than the Tacoma bridge, none had been built to be so flexible, a consequence essentially of the small width of the roadway. Even today, however, there is uncertainty whether the periodic vortices caused the oscillation of the bridge or whether oscillations of the bridge induced the vortices. Von Kármán himself believed in the latter interpretation [45].

Periodic generation of vortices by oscillating bodies was mentioned in Section 5.3, where vortex separation was explained as being caused by acceleration or deceleration of the fluid rather than by instability. Before vortex-induced vibration is more closely examined, the oscillation of a body or a fluid without parallel flow will be considered.

A circular cylinder may oscillate perpendicular to its axis (in Fig. 6.20 on a horizontal line indicated by the arrow). Frictional forces then cause an almost steady flow at a distance sufficiently far away from the cylinder. In the vicinity of the cylinder four vortices develop (Fig. 6.20) [Reference 35, Chapter 1; 46].

Kundt's dust figures, which are evaluated in acoustics for the determination of the velocity of sound, can be explained by the oscillating column of air in a pipe [47]. Standing sound waves in a horizontal pipe produce a flow near the wall in the direction of decreasing amplitude (Fig. 6.21). If dust happens to be on the bottom of the pipe, it moves to the nodal points and accumulates there. The distance of these nodal points from each other (l) is equal to one-half the wavelength of the sound. The velocity of sound is then equal to $2lf$ with f being the known frequency of the excited tone.

The properties of the vortex street behind a cylinder change when the cylinder also vibrates. There

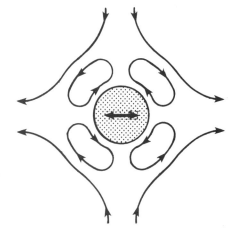

Figure 6.20. Almost steady flow field caused by the vibration of a cylinder in the direction indicated.

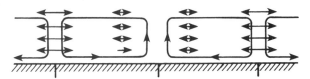

Figure 6.21. Kundt's dust figures.

is a difference between vortex separation behind an oscillating cylinder whose frequency f_c is both prescribed and independent of the frequency of vortex shedding f_v (forced oscillations), and vortex shedding which stimulates a freely moving cylinder to vibrate (induced oscillations).

In forced oscillations of the cylinder the frequency of vortex shedding f_v and the arrangement of the vortices can change, mainly because the position of the separation points on the body surface and the amount of vorticity shed from those locations vary. Near the natural frequency f_{vn} of the vortex street, that is, the frequency of vortex shedding from a stationary cylinder, synchronization of the frequencies (lock-in, Fig. 6.22) occurs: f_v adjusts to the enforced frequency f_c [48]. Figure 6.23 shows streak lines for the "lock-in" situation $f_v = f_c$. The width of the vortex street decreases at first with increasing amplitude of the cylinder oscillations (Fig. 6.23a) until a single row of vortices with alternating sign is formed (Fig. 6.23b). With further increase of the amplitude, vortices are generated that change the original alternating vortex configuration (Fig. 6.23c). The vortex street also deforms when the frequency f_c is varied. If, for instance, $f_c = 2f_{vn}$ and the cylinder oscillates parallel to the flow ("in-line" oscillation), the vortex configuration becomes symmetrical (Fig. 6.24) [49].

Forced oscillating cylinders increase the stability of the vortex street. For lock-in the upper limit of the "pure" vortex street moves from Re = 200 to Re = 350 [50]. The drag of the cylinder also increases up to 80%.

If the circular cylinder is replaced by a plate or a wing, the angle of attack (and its change with time) must be added as a third parameter in addition to frequency and amplitude. Figure 6.25 shows a sequence of computed instantaneous streamlines around an airfoil oscillating in a rotatory way in a parallel flow for Re = 5000 and St = 1 defined by the chord. Transverse oscillation of wings has been studied by Okajima et al. [52], torsional ones by Taneda [53], and longitudinal oscillations by Maresca et al. [54].

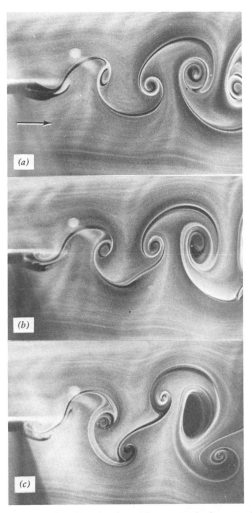

Figure 6.23. Synchronization of frequency f_v of vortex shedding with the enforced body frequency f_c. Streak lines behind a circular cylinder for Re = 190, $f_c/f_{vn} = 0.85$. The amplitudes of the oscillating cylinder normal to the flow are: (a) 0.5D, (b) 0.8D, and (c) 1.0D (photographs by O.M. Griffin, Naval Research Laboratory, Washington, D.C.).

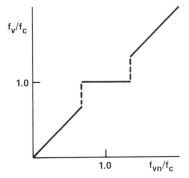

Figure 6.22. Concept of "lock-in" presented in a schematic diagram. Near the enforced frequency of the body f_c, the frequency of vortex shedding f_v adjusts to f_c; f_{vn} is the frequency of vortex shedding behind a stationary cylinder.

Figure 6.24. Change of vortex configuration in a vortex street through vibration of a cylinder. For $f_c = 2f_{vn}$ the vortex street becomes symmetric.

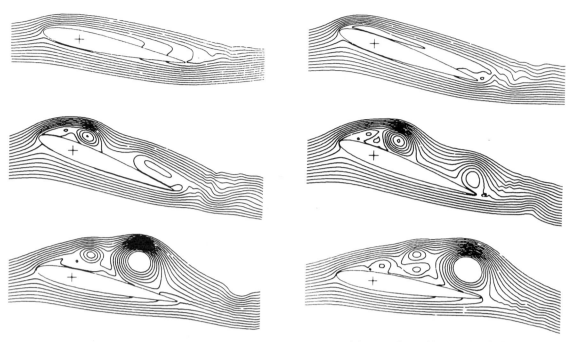

Figure 6.25. Flow past an oscillating wing for Re = 5000 and St = 1, defined by the chord. The maximum α is 20° (computer-generated pictures by U. B. Mehta [51]).

In nature swimming fish and flying birds and insects use the principle of oscillating plates to produce thrust and lift (see Section 5.4). Once more the Weis–Fogh mechanism for the generation of lift is mentioned as an example of the periodic generation of vortices with two bodies. As a complement to Fig. 4.31, Fig. 6.26 shows flow pictures that reveal frictional effects and that cannot be described with the model of potential-flow theory. As in the case of opening a book, air flows into the space generated by the opening of the two wings, and two vortices form at the upper edges inside the gap. The complete separation of the wings is not shown in Fig. 6.26, but the closing of the wings immediately after opening is shown. Air then escapes out of the gap, and two vortices of opposite sign are generated at the edges outside the gap. The vortices created during opening decay quickly [56, 57].

Flow-induced oscillation of bodies is caused essentially by four different mechanisms: (1) flutter, (2) galloping, (3) buffeting, and (4) vortex-induced oscillation [58–61].

1. If an elastically mounted body vibrates in a steady flow, it will induce the flow to vibrate also. The vibrating flow, in turn, exerts a force on the body that may dampen or increase the body vibration. The latter situation, which is triggered by instability, is possible for small flow disturbances only if the body has at least two degrees of freedom, for instance, a wing fixed at one point with an aileron fixed to the trailing edge of the wing [62]. This "linear" excitation mechanism is called "flutter."

2. An excitation with one degree of freedom is possible only for blunt bodies with a wake. The principle is the same as that of the Lanchester propeller (Fig. 4.39). Owing to stall a side force can act in such a way as to amplify the vibration of the body until a balance with frictional forces is reached. This kind of oscillation is called "galloping." The frequency of body oscillation is far below that of vortex shedding: to be exact, St < 0.1 [60]. Galloping can be demonstrated with a simple experiment. A long rectangular board, such as a common yardstick, is held at one end and the other end is dipped into water in a bathtub. It is difficult to pull the yardstick broadside through the water along a straight path. Instead, the yardstick vibrates perpendicular to the direction of pull (in the horizontal plane). The disaster of the Tacoma bridge incident, mentioned in the beginning of this section, was attributed by von Kármán to galloping.

3. Buffeting is turbulence-induced vibration [63].

4. Vortex-induced vibration has certain similarities to forced vibration near the natural frequency

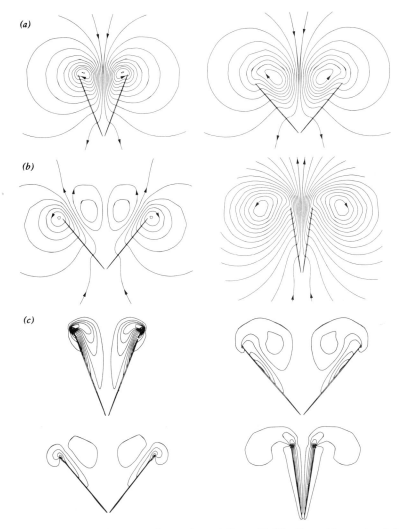

Figure 6.26. Opening and closing of two wings at Re = 30. The Reynolds number is based on the maximum velocity of the wing's edge and the width of the wing. The reference frame for the streamlines is the fluid at rest far away from the wings. (a) The wings open, (b) they close, and (c) the corresponding equivorticity lines (computer-generated pictures by H. J. Haussling [55]).

of the elastic structure. Lock-in occurs, and the diagram of Fig. 6.25 remains valid if f_c is replaced by the natural frequency. Moreover, the vortex arrangement in the wake differs from that behind a stationary cylinder. The freedom of the elastically mounted cylinder (for instance, one tied to a spring or fixed at one end) results in large amplitudes of body vibration in the lock-in region. This behavior is dictated by the action of the lift force on the body. The vortex strength is greater for the vibrating cylinder than for the stationary cylinder mainly because of the increased generation of vorticity in the boundary layer (Fig. 6.14). The drag force also increases with the amplitude of the oscillation.

For noncircular cylinders like plates the degree of freedom to react to fluid forces increases, especially if the body is not rigid. Then the elastic stiffness enters as an additional parameter. A vivid example is the wind-beaten flag. Examples of engineering interest are the vibration of buildings and other structures caused by wind forces and the vibration of towing cables [64, 65, 66].

Springs or other mechanical means can be used to restrict the movement of the freely oscillating body, but outer forces can function in the same way. As an example, freely falling plates and disks in the earth's gravitational field will be studied more closely.

110 Instability and Turbulence

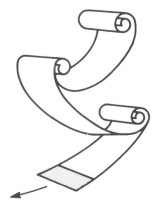

Figure 6.27. Roll up of vortex layers during the wobbling fall of a flat plate.

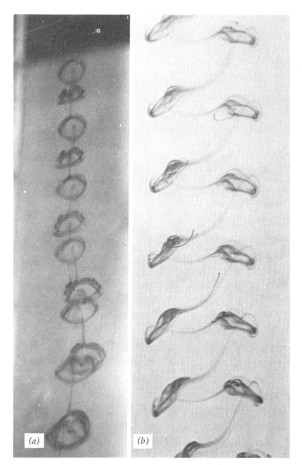

Figure 6.28. Alternating shedding of vortex rings behind a disk that wobbles as it falls. Re = DV/ν = 170, where V is the fall velocity. (a) View perpendicular to the plane of wobbling and (b) view parallel to the plane of wobbling (photographs by W. W. Willmarth, University of Michigan, Ann Arbor, Michigan).

A piece of paper wobbles during its fall, and the roll up of vortex layers behind the sheet is sketched in Fig. 6.27. Below Re = 70 an initial wobbling motion (caused by an initial oblique position) is damped during fall. Above this Reynolds number the amplitude of the oscillating motion increases up to a certain terminal value [67]. The wobbling, however, can also pass over to autorotation (Section 6.5). The movement of flat disks is similar, but instead of plane vortex layers, horseshoe vortices or vortex rings separate in alternating fashion (Fig. 6.28). The following list summarizes the various types of motion for a flat disk [68]. The Reynolds numbers are almost independent of the moment of inertia of the disk.

Re < 1   The falling disk maintains its initial position.

1 < Re < 100   The disk falls with its flat side in a horizontal position. If the initial position is oblique, the disk wobbles at first, but is damped into a horizontal position.

Re > 100   The disk wobbles without damping or passes over to autorotation. In this range the moment of inertia influences the type of motion.

Irregularly shaped bodies exhibit a similar behavior, and symmetric bodies whose center of gravity does not coincide with the geometric center have similar properties [40].

So far, periodic vortex shedding has been described for bodies in a fluid at rest and in a steady stream. What happens if vortices encounter a body periodically? These vortices may be generated in a jet or behind a body.

In Section 6.2 the occurrence of vortices in a plane or axisymmetric jet was discussed. By placing a body into the jet at a certain distance downstream from where the jet originates, causes, in general, an enhancement of the organization of the whole flow field. This means that amplification, resonance, or self-sustained oscillation occurs, essentially due to a feedback mechanism [69].

Figure 6.29 shows a planar jet impinging on a wedge. Vortices in the jet flow hit the wedge and cause a pressure minimum on the surface of the wedge closest to the vortex. The vortices themselves are distorted by the wedge and induce secondary

Figure 6.29. Planar jet impinging on a wedge.

vortices of opposite sign on the other side of the wedge at the apex. These secondary vortices interact with the following primary vortices, and the resulting complicated flow produces pressure fluctuations that influence the upstream flow at the nozzle. The amplitude of the disturbance at the nozzle is increased, and the original frequency is adjusted to a new one that essentially depends on the distance between nozzle and wedge. After a few cycles of this feedback mechanism a quasi-steady state of oscillation, self-sustained, is reached. In contrast to nonimpinging jets, the amplitude peaks are much higher and sharper; coalescence of primary vortices is suppressed [70].

Shear-layer vortices impinging on a corner are shown in Fig. 6.30. This corner may be part of a cavity (Fig. 5.21), and the way the vortices hit the corner can influence feedback and resonance of the cavity. In Fig. 6.30a a vortex approaches the corner; in Fig. 6.30b the vortex is "clipped" and in Fig. 6.30c it is partially clipped, and in Fig. 6.30d the vortex escapes [71]. Other edge geometries are discussed in Reference 69.

Oscillations of a body in the fluctuating wake of another body are reported in Reference 72. Practical applications in nature are the schooling of fish and certain flight formations of birds. Here, the lower drag in the wake of the animal in front is utilized (Section 5.4).

Self-sustained oscillations of impinging jets are of great importance in acoustics. As was mentioned in Section 6.3, vortex streets behind cylinders generate sound, and so do periodic vortices in a jet. However, the intensity and control of the sound are increased considerably by impingement. The independence of the tone from the Reynolds number is particularly crucial for musical instruments. Further amplification and stabilization of the tones can be achieved by cavities that act as "resonators" [69].

The generation of sound in vortical motion does not depend on the presence of a body on which the vortices impinge, as the example of a periodic vortex street in Section 6.3 demonstrates. Unsteady vorticity fields and vortex motions are sources of sound, and theories reveal that unsteadiness means the existence of second time derivatives of the vorticity [Reference 53, Chapter 1]. The highest values of these derivatives and thus the greatest sound occur in the development phase of a vortex and when vortices interact. Both are lower during decay.[2] It is interesting that in the study of vortical sound generation the fluid at the source of the sound can be approximated as incompressible, a seeming physical paradox for generating sound. However, this assumption for the source field far away from the

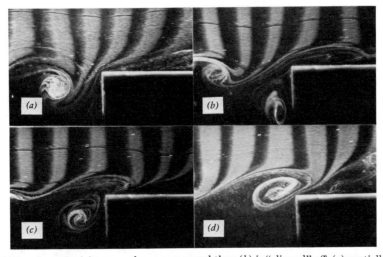

Figure 6.30. A vortex (a) approaches a corner and then (b) is "clipped" off, (c) partially clipped off, and (d) escapes (photographs by D. Rockwell and C. Knisely, Lehigh University, Bethlehem, Pennsylvania).

acoustic field can be justified within the framework of perturbation theory. Details on vortex sound are found in References 73–76.

Oscillations in a jet can also be intensified by means other than solid bodies. Schneider [77] placed a loudspeaker below a burner that produces a diffusion flame. This flame was stimulated by the loudspeaker to sound emission: The vortices in the jet mix the flammable gas and the air and ignite the mixture. The expanding burnt mixture then produces or intensifies sound.

6.5. ROTATING BODIES

According to the statements made in Section 2.1, a rotating body is a vortex, and the fluid dragged along by the rotating body also moves in a rotatory way. The simplest rotation is obtained by an infinitely long circular cylinder that rotates with constant angular velocity in a fluid at rest. The velocity distribution of the fluid around the body is that of a potential vortex (Fig. 3.8).

The rotation of noncircular cylinders like that of an infinitely long plate is more complex. The motion is steady only in a reference frame that rotates with

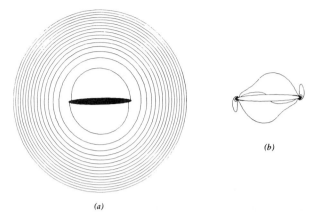

Figure 6.31. Constant rotation of a plate in a fluid at rest at infinity. Re = $\Omega D^2/\nu$ = 400. (a) Steady streamlines in a reference frame fixed to the plate. The direction of rotation is counterclockwise. (b) Equivorticity lines (computer-generated pictures from H. J. Lugt and S. Ohring [78]).

the plate at constant angular velocity, and no vortices separate from the edges of the plate. Figure 6.31 shows steady streamlines and equivorticity lines for the case Re = $\Omega D^2/\nu$ = 400; Ω is the angular velocity and D is the width of the plate. If the plate is abruptly set into constant motion, vortices

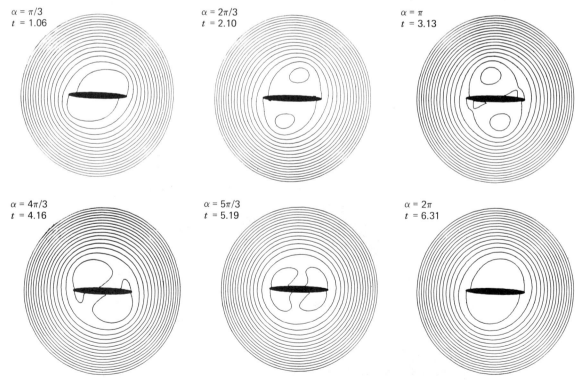

Figure 6.32. Plate abruptly put in constant rotation. Initial state t = time · Ω = 0, α = 0° for Re = 400. Streamlines are plotted in a reference frame fixed to the plate (computer-generated pictures from Reference 78).

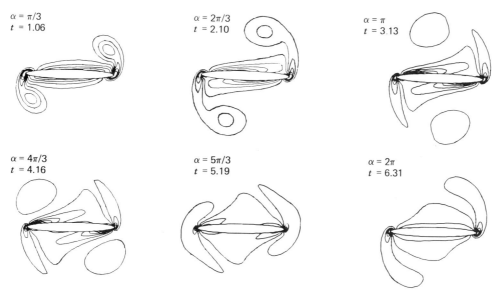

Figure 6.33. Equivorticity lines for the flow in Fig. 6.32. The steady state of Fig. 6.31 is reached when $t = \infty$.

are shed from both edges and, over several revolutions, cause an oscillating torque (Figs. 6.32 and 6.33). After a sufficiently long period of time the steady state of Fig. 6.31 is reached.

For cylinders of finite length a secondary flow is superposed on the rotation (in Fig. 6.34 sketched in the meridional plane). The example of a rotating sphere can be used to explain the characteristic flow properties of rotating bodies with finite length. A very slowly rotating sphere drags the surrounding fluid with it in concentric circles ($Re = \Omega R^2/\nu < 10$). At higher Reynolds numbers a secondary flow is superposed on the rotating field (Fig. 6.35). Fluid is sucked to the poles and pushed away at the equator. With increasing Reynolds number the rotational field and the secondary flow assume the character of a boundary layer: the fluid moves only in the immediate vicinity of the rotating sphere. At about Re = 40,000 the flow starts to become unstable. This instability begins in the region of highest local velocity, that is, at the equator, and migrates in a band to the poles with increasing Reynolds number. The band covers about 10° latitude and consists of about 10 vortices (at the equator) to 26 (at each pole), all rotating in the same direction. The band of vortices separates laminar and turbulent flow (Fig. 6.36). Turbulence dominates from Re = 400,000 on.

Instability near a rotating body can be studied even better with a rotating flat disk. Here, too, centrifugal force pushes the fluid away from the center and toward the edge. A secondary flow develops as

Figure 6.34. Schematic representation of the secondary flow around a rotating circular cylinder of finite length (meridional flow).

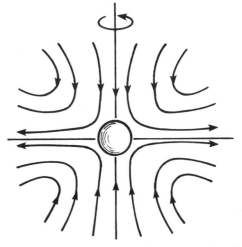

Figure 6.35. Secondary flow around a rotating sphere for Re > 10 (from O. Sawatzki [79]).

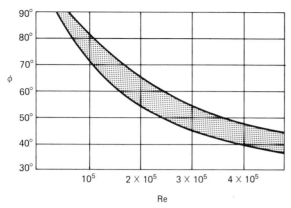

Figure 6.36. Occurrence of instability on the surface of a rotating sphere as a function of the Reynolds number (from O. Sawatzki [79]).

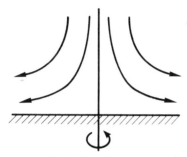

Figure 6.37. Secondary flow near a rotating disk.

sketched in Fig. 6.37. Above Re = $\Omega R^2/\nu$ = 190,000 the flow becomes unstable, and a vortex ring forms, which separates the laminar from the turbulent flow (Fig. 6.38) [80].

If the rotating body is placed in a parallel flow, a force normal to the flow direction and normal to the axis appears, which can be explained in principle by the theory of an airfoil in a potential flow. In Section 4.7 circulation around an inclined plate was introduced as a hypothesis to ensure smooth streaming at the trailing edge. This circulation causes a lift, and the presence of a vortex near a body has the same effect (Fig. 6.14). If a body such as a circular

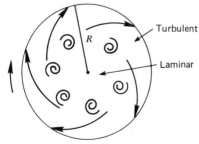

Figure 6.38. Instability vortices on a rotating disk.

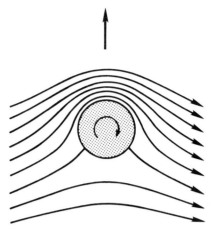

Figure 6.39. A rotating cylinder generates lift in a parallel flow. This phenomenon is called the Magnus effect.

cylinder rotates in a parallel flow, the rotation of the body causes a fluid circulation with the same result: A lift is generated, which tries to divert the body in the direction of this force (Fig. 6.39). This phenomenon is called the "Magnus effect" after the physicist G. Magnus [81].

Three types of flows past rotating circular cylinders are distinguished, depending on whether the points of separation and attachment are at the body surface, coincide there, or are inside the fluid. To describe these types a new parameter is introduced: the ratio of flow velocity V to tangential velocity ΩR. This parameter is called the "Rossby number," abbreviated Ro:

$$\text{Ro} = \frac{V}{\Omega R}.$$

Often the roll parameter $p = 1/\text{Ro}$ is used and is preferred in this chapter.

In potential flow the prescription of the roll parameter is sufficient to determine the flow. Points of separation and attachment occur at the body surface when $p < 2$ (Fig. 6.39). They coincide when $p = 2$, and they are within the fluid when $p > 2$ (Fig. 5.89). These values are approximately accurate for real fluids also.

The Magnus effect has been known for a long time. Newton observed in 1671 that the path of a tennis ball could be influenced by rotation. In 1742 Robins made systematic experiments with a pendulum to determine the magnitude of the aerodynamic side force acting on a sphere [82, 83]. Tennis and golf players know that the flight path of a ball can be changed if the ball spins and that the direction of the rebound will be affected (Fig. 6.40) [84]. In technology the "Flettner rotor," which is based on the Magnus effect, was applied in the 1920s to the "Rotor ship" and the "Rotor airplane" [85].

In reality the motion around a rotating body is more complicated than the potential flow shown in Fig. 6.39. Figure 6.41 shows the real flow past a circular cylinder for Re = 200 and $p = 1$ at a certain time. Both the roll parameter and the Reynolds number must be prescribed here. If periodic vortex shedding occurs behind a rotating cylinder, then the dimensionless frequency (Strouhal number) is a function of both the Reynolds number and the roll parameter.[3] Measurements of the lift force also show deviations from potential-flow theory. For instance, the Magnus force can have a negative direction in a certain range of Re and p. This has been observed for circular cylinders and spheres [86, 87]. Figure 6.42 displays measured data for a rotating cylinder.

So far it has been assumed that an external force, such as a motor, rotates the body, but the flow itself can also rotate the body. Such a "free" rotation is achieved in a shear flow. In Fig. 6.43 a circular cylinder is shown in a parallel shear flow. The larger the shear, the faster is the rotation of the body [88]. Here also a side force occurs that causes a freely moving body to migrate perpendicular to the flow. This phenomenon is important in the study of suspensions in boundary layers and shear flows [89].

The flow past rotating bodies that are neither circular cylindrical nor spherical is extremely complex, because the rotating fluid is not steady but periodic.

Figure 6.40. Spinning of a table tennis ball shortens the flight path.

Figure 6.41. Streamlines about a rotating cylinder at a certain time for Re = 200 and $p = 1$. The constant rotation can be maintained by an electromotor (adapted from D. C. Thoman and A. A. Szewczyk [22]).

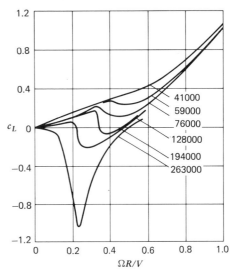

Figure 6.42. Lift coefficients of a rotating circular cylinder in a parallel flow as a function of $\Omega R/V = p$ for various Reynolds numbers (from F. N. M. Brown [87]).

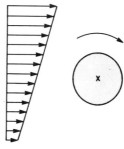

Figure 6.43. A freely rotating cylinder turns in a shear flow.

Again the flow past a rotating plate will serve as an example, and again as for the flow past a rotating circular cylinder, three types of flow can be distinguished. They are recognized for a plate rotating in a real fluid by the direction of the separating vortices. The magnitude of the roll parameter p determines whether the vortex will separate at the retreating edge in flow direction or against it (Figs. 6.44b and 6.44e). Between the extreme values of p a situation exists in which no vortex is attached in the position shown in Fig. 6.44d. The shed vortices can, however, disturb the nascent vortices so that the flow becomes quite complicated several cycles after the start of the rotation (Fig. 6.44c). One may compare the sketches with the computer-generated pictures for pure translation (Fig. 6.12) and for pure rotation (Fig. 6.34). Figure 6.45 shows streamlines and equivorticity lines for $p = 0.25$ and Re = 200, which illustrates the situation in Fig. 6.44d. In analyzing the flow patterns one observes again the advantage of considering the vorticity field rather than the streamlines. The vorticity field remains invariant when changing from an inertial to a rotating system, except for a constant. For very slow rotations of the plate relative to the flow ($p \ll 1$) the frequency of vortex shedding is different from the frequency of rotation of the plate. For $p = 0.25$, Re $= 200$ (Fig. 6.45) vortex shedding and plate rotation are synchronous (lock-in region). Vortices are generated and shed during half a revolution, that is, in the period π. At faster plate rotations ($p \gg 1$) vortices can be trapped by the plate, and the period of vortex generation and shedding is larger than π.

In general a driving torque (averaged over one revolution) must keep the plate in constant rotation. This is certainly true for large and small roll parameters, that is, for $p \gg 1$ and $p \ll 1$. In between a region can exist in which the motor does not drive the plate but has a braking effect (Fig. 6.46). This means that, if the plate could rotate freely in the flow, the plate would be induced by the kinetic energy of the flow to rotate faster until an angular velocity is reached beyond which again a driving torque is necessary for constant rotation. This situation, in which the average torque is zero and which is designated in Fig. 6.46 with A, is called "stable autorotation." It requires a certain moment of inertia and occurs for $200 \leq \text{Re} \leq 400$ at ca. $p = 0.5$ [Reference 35, Chapter 4].

Autorotation was mentioned in Section 4.8 when the Lanchester propeller and the autorotating wing were described. Those bodies rotate about an axis parallel to the flow. In this section the axis of the plate is considered normal to the flow. The essential difference between these types of autorotation is

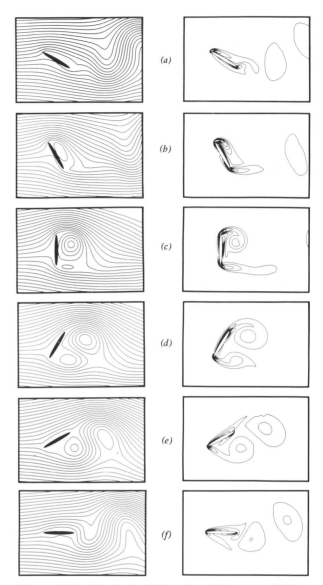

Figure 6.45. Constant rotation of a plate in a parallel flow. To the left are streamlines, to the right equivorticity lines for Re = 200 and $p = 0.25$ (computer-generated pictures from Reference 78).

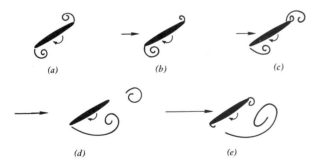

Figure 6.44. Vortex shedding behind a rotating plate in a parallel flow. (a) Initial phase for pure rotation ($p = \infty$); (b) initial phase for $1 < p < \infty$; (c) vortex shedding for $0.5 \leq p < 1$, Re = 200 after several revolutions; (d) no vortex attached at the retreating edge, $p \approx 0.25$, Re = 200; (e) vortex shedding for $p \leq 1.67$, Re = 200 (from Reference 90).

that motion parallel to the flow is essentially steady in the stable state, whereas motion perpendicular to the flow is basically periodic. A body can also autorotate obliquely to the flow, in which case the flow is very complex.

When the rotation of the plate is not powered, a rectangular or round plate, or a cruciform arrangement of plates, may rotate freely about a fixed axis in a parallel flow (Fig. 6.47). If an impulse is given to the plate, it will rotate several times and then come to rest with the broad side normal to the flow. However, the initial rotation can pass over to auto-

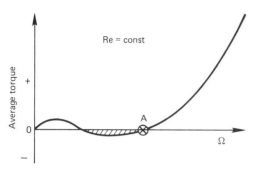

Figure 6.46. Average torque of a rotating plate as a function of the angular velocity. In the positive region of the curve a driving torque is necessary to rotate the plate. In the negative region (shaded) the motor brakes the plate. If the plate is allowed to rotate freely in this range, the angular velocity will increase up to the point A (provided the moment of inertia is sufficiently large). This is the state of stable autorotation.

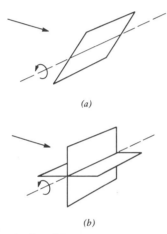

Figure 6.47. A plate (a) can autorotate about a fixed axis perpendicular to the flow. Also a cruciform system of plates (b) can autorotate.

rotation under certain conditions. Then, the torque is partly driving and partly braking during one revolution. To overcome the retarding phase the plate needs a sufficiently large moment of inertia. Autorotation thus depends on the moment of inertia as well as on the Reynolds number. The angular velocity, expressed in dimensionless form by p, is no longer independent of the Reynolds number (as in the case of a plate driven by a motor) but is a function of the Reynolds number and the (dimensionless) moment of inertia [67, 90].

If the moment of inertia is large enough, the plate will autorotate with almost constant angular velocity. The example in Fig. 6.45 for Re = 200 and p = 0.25, which was computed under the assumption of constant angular velocity, can therefore be used to investigate autorotation. This case lies in the shaded region of the curve in Fig. 6.46. However, the explanation for the source of the additional torque is complicated. It may be sufficient here to say that autorotation is the result of a subtle interplay of the following conditions:

1. The (dimensionless) moment of inertia of the plate must be sufficiently large.
2. Vortex shedding and plate rotation must be synchronous. This condition is necessary but not sufficient. For instance, vortex shedding behind a freely wobbling plate is also synchronous (Fig. 6.27).
3. The amount of vorticity produced about the retreating edge in the braking phase (Figs. 6.45d, 6.45e, 6.45f) must be equal on both sides of the plate. Because of this, the pressure differences at the plate's surface are smaller than in the supporting phase (Figs. 6.45a, 6.45b, 6.45c). Hysteresis of the boundary layer, that is, the accelerated or decelerated development of the boundary layer, also contributes to autorotation since in the supporting phase it causes a delay of vortex shedding at the retreating edge.
4. Sufficient vorticity must be produced in the supporting phase at the retreating edge. Blunt bodies autorotate more slowly or not at all. Sharp edges are most advantageous for autorotation.

Conditions 3 and 4 indicate that the strength and position of the vortex at the retreating edge of the plate are decisive for autorotation.

Autorotation is not restricted to small Reynolds numbers. It has been observed in the range $100 <$ Re $< 5 \times 10^5$ [67], and the explanation for small Reynolds numbers is also valid for large ones. Figure 6.48 shows the similarity of the flow patterns for three ranges of Re. The photographs in Fig. 6.49 show, although not as clearly as the movie from which they were taken, the strength of the vortex

Figure 6.48. Flow behind an autorotating plate for various Reynolds numbers.

118   Instability and Turbulence

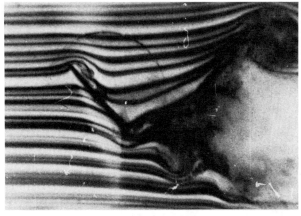

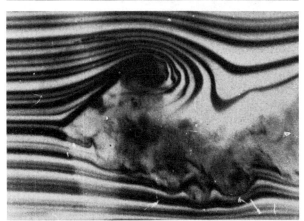

Figure 6.49. Autorotating plate about a fixed axis. Streak lines for Re ≈ 10^5. The plate rotates clockwise and the flow is from left to right (photographs from F. N. M. Brown, University of Notre Dame, Indiana).

behind the retreating edge. With larger Reynolds numbers the roll parameter p increases and is about 0.8 to 1.0 up to Re = 10^5. Below Re = 1000, p decreases to about 0.5, and autorotation ceases when Re → 0. p decreases with smaller aspect ratio of the plate and with blunt edges.

Cruciform plates also autorotate (Fig. 6.50). The vortex behind the retreating edge is clearly visible on flow patterns.

So far, the axis of the rotating body has been kept fixed. If the plate is allowed to fall freely, the axis too will move in a complicated way. To see a falling plate autorotate, anyone can easily make the following experiment: A rectangular strip of paper about 1 cm × 10 cm is held horizontally and then dropped. The strip rotates rapidly about its longer axis and falls obliquely downward. The deviation from vertical fall is caused by the Magnus lift force. The motion of the axis itself (Fig. 6.51) induces an additional torque that supports autorotation. The explanation for this behavior was given by Maxwell in 1853: When the plate passes from the horizontal position 1 to position 2 normal to the flight path, the velocity of fall is greater than the average because the drag of a plate parallel to the flow is smaller than that normal to the flow. In this phase the torque acts in the direction of rotation, and hence the increased velocity of fall supports the torque. However, when the plate rotates from the perpendicular position 2 to the parallel one between 3 and 4, the velocity of fall is smaller than the average and the torque, which in this phase acts as a brake, di-

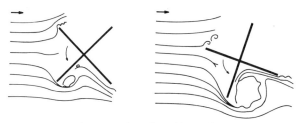

Figure 6.50. Autorotation of cruciform plate arrangements. Sketches of streak lines for Re ≈ 10^5.

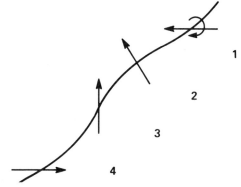

Figure 6.51. Movement of a freely falling strip of paper. When released, the strip must be positioned horizontally.

minishes. Thus, the net torque over a whole cycle supports autorotation [91].[4]

Aerodynamicists have given great attention to autorotating plates because the fins of missiles can cause autorotation and impede the control of the missile. Also reentry bodies of space ships can autorotate in the subsonic range. Astronauts Glenn and Carpenter reported unstable oscillations of their Mercury capsule during the last phase of reentry. Only through opening of the parachute could the capsule be stabilized [92].[5]

6.6. SOME REMARKS ON TURBULENCE

*Big whirls have little whirls,
Which feed on their velocity.
Little whirls have smaller whirls,
And so on to viscosity.*

L. F. Richardson

Every fluid flow will probably become unstable at sufficiently large Reynolds numbers. With further increase of the Reynolds number a state of flow develops that is called "turbulence." The definition of turbulence is not simple and will be postponed for the time being. The following paragraphs describe briefly what is observed in turbulence.

In 1883 Reynolds [93] published the following result of an experiment[6]: If water flows out of a container through a long pipe and the flow is marked at the entrance of the pipe with dye, one observes a clean dye filament at low speed. In laminar flows dye particles diffuse so slowly that they do not have time to spread out across the flow. If the flow velocity, that is, the Reynolds number, is increased, a sudden change occurs above a critical value at a certain distance from the entrance of the pipe. The dye filament now mixes quickly with the fluid over the whole cross section of the pipe. One observes disordered movements of single dye filaments (Fig. 6.52). If the pressure and velocity are measured behind

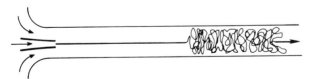

Figure 6.52. Reynolds' experiment: Transition from laminar to turbulent flow near the critical Reynolds number 2300. At the entrance of the pipe the flow is laminar. Small disturbances increase downstream and cause the transition to turbulence.

this transition point, one notes fast spatial and temporal oscillations. This irregular motion is characteristic of turbulence.

A similar process is observed in the water jet of a faucet. At low velocity of the jet the motion is laminar. With increasing velocity this motion changes suddenly to turbulence.

Two kinds of transition from laminar to turbulent flow may be observed [94]. In the "abrupt transition," which is depicted in Fig. 6.52, the flow changes directly to the turbulent state after the first occurrence of instability. In the "transition through spectral development" (Landau 1944 [95]) instability first generates a new type of laminar flow that consists of ordered vortices. With further increase in the Reynolds number new instabilities give rise to an increasing number of vortices, until finally turbulence is reached. An example of the transition through spectral development is the flow around a cylinder. The periodic vortex street in the range $45 \leq \text{Re} < 200$ is the first intermediate step. With increasing Reynolds number vortex shedding becomes more irregular. This means that higher frequencies of the disturbance appear. A better example, in which several intermediate steps are clearly visible is the flow between two concentric cylinders when the inner cylinder rotates faster than the outer one. The so-called "Taylor vortices," which occur after the first instability, will be described in Section 7.4.

According to Reynolds, turbulence can be imagined as the sum of a basic flow determined by average values and a superposed irregular disturbance flow [Reference 35, Chapter 1]. The latter can be represented by a Fourier analysis and interpreted as vortex motion of varying size and intensity. An "eddy" of frequency f is then a disturbance, which contains energy in the vicinity of f. According to Reference 96 a certain "band width" of energy around f must be assumed in order to distinguish the eddy from a wave. There exists thus a whole spectrum of small and large eddies, which are superposed on the basic flow. The largest eddies have the dimension of the entire turbulent flow region. If one assigns a Reynolds number to each eddy and assumes that the angular velocity has the same order of magnitude for all eddies, then the largest eddies have the highest Reynolds number. This means that for large eddies the viscous forces are small in relation to the inertial forces.

This statement is important in understanding the energy transport in a turbulent flow. The largest eddies receive their kinetic energy either from the basic flow through shear forces or directly from sources outside the flow, for instance, through heating of the fluid. The large eddies then transfer kinetic energy to the smaller eddies in a cascading way until the smallest eddies transform their kinetic energy through dissipation into heat. This process is described in the quotation from Richardson at the beginning of Section 6.6.

The transfer of kinetic energy from large to small eddies without dissipation may be imagined as vortex stretching. According to Helmholtz's second theorem, this is possible in a three-dimensional flow along a pathline, but not in a two-dimensional motion. Hence, turbulence can be understood only in a three-dimensional flow.

A plot of the energy against the frequency of the disturbance gives the "energy spectrum" of a turbulent flow (Fig. 6.53). The largest eddies, whose structures depend on the way the flow is generated, are very durable but do not contain the largest amounts of kinetic energy. The next smaller eddies do. If one follows the eddy spectrum to higher frequencies, the structure of the eddies becomes more and more independent of the special form of the large eddies. This statement is the first hypothesis of Kolmogorov [97], who in 1941 studied the eddy spectrum of higher frequencies. Eddies that are almost independent of the form of the large eddies lie in the "universal range." Kolmogorov's second hypothesis says that for high Reynolds numbers the largest eddies in the universal range transport their energy almost exclusively by means of inertial forces and do not dissipate. This part is called the "inertial subrange." Adjacent to this range at still higher frequencies is the range in which the smallest eddies lose their energy through dissipation.

In certain turbulent shear flows, however, energy can also be transported in the opposite direction. This means that energy can go from small eddies to large ones and into the average basic flow. This "energy reversal" probably plays an important role in meteorology, oceanography, and astronomy [98].

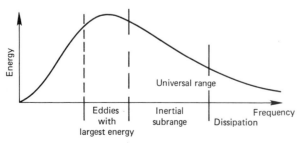

Figure 6.53. Energy spectrum of a turbulent flow.

The essential characteristics of turbulence can be summarized in the following statements (modified from Hinze [Reference 42, Chapter 1] and considered as a kind of a definition):

1. Turbulence is an irregular fluid motion in which physical quantities like velocity and pressure oscillate in space and time.
2. Turbulence is an exchange process, which is orders of magnitude faster than diffusion of vorticity in the laminar region. Without turbulence there would be no life on earth.
3. Energy is transported in general from large eddies to small ones.
4. Turbulence is not an entirely statistical process, since coherent, long-lasting vortical structures exist within it. This item will be discussed later in this section.

Various models for describing turbulence, based on these conceptions, have been developed in the last 100 years. The first attempt goes back to Boussinesq [99] who in 1877 replaced the molecular viscosity coefficient (used in laminar flow) with an "exchange coefficient," a multiple in order of magnitude of the molecular viscosity coefficient. This formal replacement, however, corresponds only to a lowering of the Reynolds number and does not give new insight into the nature of turbulence.

Much more successful was the semiempirical model of Prandtl [Reference 28, Chapter 5]. He introduced, in analogy to the free path of gas molecules, the concept of "mixing length" of turbulent clusters. Although this analogy is not exact since the turbulent clusters influence each other strongly and gas molecules do not, the idea was fruitful. Similar theoretical assumptions by G. I. Taylor and von Kármán led to almost the same results as those of Prandtl. The theory of mixing length is still the basis for most turbulence calculations in engineering today.

Another model, which depends less on empirical data than the hypothesis of mixing length and which is, therefore, more satisfying from the theoretical point of view, is based on statistical arguments. The pioneering work started with G. I. Taylor [100]. He suggested for the first time that the oscillations of the velocity be determined statistically and used to compute the energy spectrum. Taylor's program, which he carried out for a one-dimensional energy spectrum, triggered a flood of publications by such well-known investigators as

Heisenberg, Burgers, von Kármán, Howarth, Obukov, von Weizsäcker, Kovasznay, Batchelor, Lin, and Kraichnan, to name a few. Despite the tremendous effort and gain in insight, the statistical theory of turbulence has not found practical application in engineering nor has it solved the turbulence problem.

More recent investigations have shown that turbulent motion is not as chaotic as had been thought, but that it has coherent structures of deterministic nature. Such coherent structures have been observed for a long time, but their existence was attributed to the geometric configuration of the flow domain (see next section). When coherent vortex structures with a turbulent fine structure inside were observed in a turbulent shear flow (Fig. 6.54), which was considered to be quite homogeneous, the significance of large-scale structures to turbulence and the existence of two scales were recognized [Reference 43, Chapter 1, 101–103].

Turbulent vortices differ from their laminar counterparts in the following way: The molecular velocity in laminar motion (quantitatively expressed in the constant viscosity coefficient) is more or less decoupled from the macroscopic events, but in turbulent flow it is not, and interaction between the two scales of motion is strong.

If one considers molecular viscosity and turbulent "fine-grained" motion as "building blocks" for larger flow structures (see Remark 1), it is actually not surprising that, in general, laminar vortices have turbulent counterparts, and that their behavior is similar, at least in their overall features.

The identification and analysis of coherent structures in turbulent flow are much more difficult than vortices in laminar flow. It may be recalled from Chapter 2 the difficulty of defining and identifying a vortex. In turbulence signals from fluctuating flow quantities have to be analyzed. Through statistical methods important information may get lost. The problem becomes a matter of "pattern recognition" [104, 105].

There are other properties of turbulent flow that are explained by vortices. At the boundary of turbulent flows, fluid free of turbulence is drawn into the turbulent flow. The reverse process is also possible, that turbulent fluid parcels drift temporarily into the turbulence-free outer flow. These phenomena can be explained by means of vortices that are quite large with respect to the turbulent fine structure.

With the aid of computers it will be possible in the future to simulate turbulent flows in detail. The beginning of such an effort is described in References 106–108.

6.7. EXAMPLES OF TURBULENT VORTICES

The occurrence of vortices in turbulent shear layers demonstrated in Fig. 6.54 indicates that in turbulent fluid motion vortices can occur or can be produced in the same way as in laminar flow. There are laminar and turbulent vortex rings, swirling flows, and edge vortices, among others. However, the properties of turbulent vortices can be quite different from those of laminar vortices, and flow phenomena exist that occur only in turbulent motion. In the following paragraphs a few of the different and newly recognized vortex characteristics will be presented.

A turbulent boundary layer separates from a sharp wedge, rolls up, and subsequently detaches as

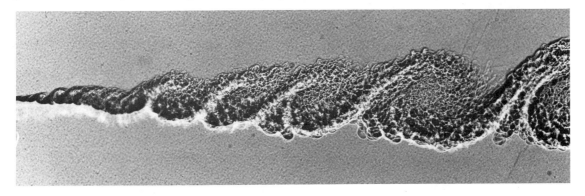

Figure 6.54. Shadowgraph of a mixing layer between flows of helium (upper side) and nitrogen from left to right. The coherent, rollerlike vortical structures and the small-scale turbulence are clearly visible. (Courtesy of A. Roshko, California Institute of Technology, Pasadena, California.)

a free vortex in the same way as its laminar counterpart. In fact, for high Reynolds numbers the model of a discontinuity line (Fig. 5.33) can be applied to both laminar and turbulent flows. However, the turbulent vortex behaves differently in the core and adjacent annular region, where the winding discontinuity lines thicken and smear out through the action of viscosity [Reference 29, Chapter 5]. These properties are described for the completely detached (or free) line vortex and vortex ring decaying in time.

The decaying turbulent line vortex has in common with the laminar one (the potential vortex) that it rotates near the axis like a solid body and far away from the axis like a potential vortex (Fig. 3.9). At both ends of this scale turbulent oscillations die away faster than viscous diffusion slows down the flow. In between, a region of turbulent motion exists that exhibits the phenomenon of "overshoot," that is, between the maximum tangential velocity and the outside flow a region appears over a certain time span, in which the velocity increases with time instead of decreasing [References 4 and 5, Chapter 3; 109]. The generation of such a turbulent line vortex may be envisioned as similar to the generation of a potential vortex (or to the Rankine vortex when a nonzero core is considered, Fig. 3.10): A rapidly rotating rod produces a turbulent boundary layer with an adjacent $1/r$ velocity decrease in a fluid at rest. When the rod is removed, the rotating turbulent and laminar motions decay to a standstill.

At high Reynolds numbers vortex rings generated by the ejection of fluid from an opening become immediately unstable and then turbulent (see Section 5.7). Instability manifests itself in the generation of waves over the circumference of the ring. The thinner the core of the vortex, the larger is the number of waves. This number varies between 5 for thick vortex cores and 13 for thin ones [Reference 26, Chapter 3]. Beautiful photographs were made by Magarvey and MacLatchy [110] and by Didden [111] (Fig. 6.55). Data on the critical Reynolds number for which the flow becomes unstable are sparse and not very exact. According to Maxworthy [Reference 82, Chapter 5] the critical Reynolds number is about 600 when the Reynolds number is based on the velocity and the diameter of the ring.

In the turbulent stage, the mechanism of fluid entrainment into the vortex ring differs decisively from that of entrainment into a laminar ring: While a laminar ring grows by entraining fluid over the whole surface through molecular diffusion (Fig. 5.84), the growth of a turbulent ring is governed by entrainment of turbulent fluid in the core region only. Most of the fluid entrained into the bubble from outside the ring is ejected into the wake. The core, however, gains fluid from the bubble but periodically loses some fluid with large vorticity to the bubble. This fluid is torn from the core by the turbulent motion of the bubble and appears in the wake in the form of distinct small vortex rings (Fig. 6.56) [112, 113]. The slower growth of the turbulent vortex ring, compared to the laminar case, is reflected in the slightly slower decrease in its translational velocity: The velocity of the turbulent vortex ring is proportional to $t^{-4/5}$, that of the laminar one proportional to t^{-1}. This result is surprising, since usually the rate of decay is faster for turbulent motion than for laminar flow.

Turbulent swirling flows, either in a pipe of circular cross section or from tip vortices, decay faster than laminar swirls and vortices [Reference 65, Chapter 5; 114]. In straight pipes of noncircular cross section the laminar flow without initial swirl

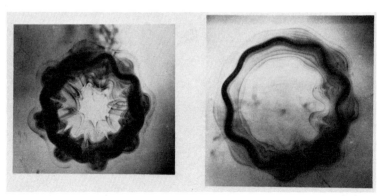

Figure 6.55. Unstable vortex rings (photographs by N. Didden, Max-Planck-Institut für Strömungsforschung, Göttingen).

Figure 6.56. Wake of a turbulent vortex ring with small secondary vortex rings shed from the region close to the core (photograph by P. E. M. Schneider, Max-Planck-Institut für Strömungsforschung. Göttingen).

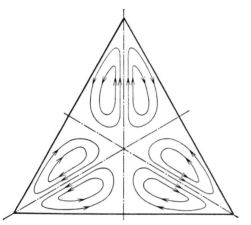

Figure 6.57. Secondary turbulent flow in a pipe of triangular cross section (from J. Nikuradse [115]).

remains that way. In turbulent pipe flow, however, a secondary flow develops, as shown in Fig. 6.57, which with the primary axial flow represents a swirling motion. This phenomenon has also been observed in laminar non-Newtonian fluid flows and indicates that the origin of the secondary flow is to be sought in the nonlinear stress–strain relation.

6.8. FLOW DRAG AND ITS CONTROL

With the last two sections on vortices in turbulent flows, the description of vortices in homogeneous fluids is almost complete. This description has treated flows past spheres, disks, circular cylinders, and plates. The behavior of these flows is fairly well known for the entire range of Reynolds number from zero to extremely high values. In addition, typical flow patterns for each range of Reynolds number have been described in Chapters 4 through 6. They will be summarized now in a discussion of drag curves.

The simplest flow is the constant motion of an incompressible fluid through a straight circular pipe. In the laminar range below Re = 2300 the velocity profile is parabolic, since the nonslip condition causes the velocity to be zero at the wall (Fig. 6.58). This is the oldest known laminar flow. Hagen (1839) and Poiseuille (1840) discovered it independently.

Beyond the critical Reynolds number 2300 the flow changes abruptly to the turbulent state whenever small disturbances are present. The exchange of momentum in the cross section is now larger, and the velocity profile averaged over the temporal and spatial oscillations is flatter.

The pressure loss in a length of pipe is usually made dimensionless by the stagnation pressure. This "friction factor" is a function of the Reynolds number only for a pipe with smooth inner walls[7] [Reference 35, Chapter 1]. Figure 6.59 shows clearly the abrupt transition from laminar to turbulent flow and the sudden increase in the friction factor due to turbulence. The range of the Reynolds number stretches from 0 to 10^7 and is so large that, for plot-

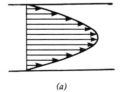 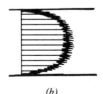

(a) (b)

Figure 6.58. (a) Laminar and (b) turbulent velocity profiles in a straight, circular pipe at the steady state.

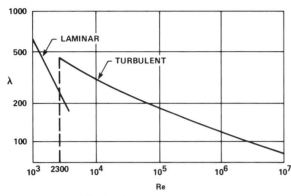

Figure 6.59. The friction factor λ of a pipe flow as a function of the Reynolds number $\mathrm{Re} = DV/\nu$, with D being the pipe diameter and V being the mean velocity.

ting the curve, a logarithmic scale must be used. Because of this distortion, the region of small Reynolds number is presented more accurately.

The flow around a body is more complicated than pipe flow. Here the drag on the body is caused solely by the vorticity in an incompressible fluid. To describe this drag the dimensionless drag coefficient c_D is introduced.[7] Figure 6.60 shows the relation between c_D and Re for flows past spheres and circular cylinders. For the sphere the following flow regions are distinguished:

$0 \leq \mathrm{Re} < 20$	No flow separation
$20 \leq \mathrm{Re} < 400$	Vortex ring attached to the body (Fig. 5.7)
$400 \leq \mathrm{Re} < 4 \times 10^5$	Periodically shed vortices of double-helix form (Fig. 6.19). Irregular frequencies appear with increasing Reynolds number
$4 \times 10^5 < \mathrm{Re}$	Turbulence

The drag curve for a circular cylinder looks similar. The flow regions here are:

$0 < \mathrm{Re} < 5$	No flow separation
$5 \leq \mathrm{Re} < 45$	Vortex pair attached to the cylinder
$45 \leq \mathrm{Re} < 200$	Purely laminar, stable vortex street (Fig. 6.12)
$200 < \mathrm{Re} < 4.5 \times 10^5$	Formation of vortex layers. Vortex street with superposed irregular frequencies
$4.5 \times 10^5 < \mathrm{Re}$	Turbulent vortex street

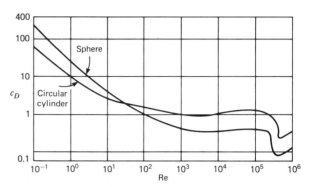

Figure 6.60. Drag coefficient c_D for sphere and circular cylinder as a function of the Reynolds number $\mathrm{Re} = DV/\nu$.

The sudden drop of the drag curve in Fig. 6.60 at the occurrence of turbulence can be explained by the fact that the wake region drastically shrinks. This strange flow behavior also contains the solution of the so-called "Eiffel–Prandtl paradox." At the end of the last century the French engineer Eiffel experimented with spheres dropped from the tower in Paris named after him. He found that the drag on a sphere decreases with increasing terminal velocity (in the region around $\mathrm{Re} = 4 \times 10^5$). Later Prandtl resolved this seeming paradox on the basis of the drop of the c_D curve at the transition to turbulence. He confirmed this conclusion experimentally by inducing turbulence artificially below the critical Reynolds number by means of a trip wire around the sphere.

The drag coefficient, thus, depends on the form of the body as well as on the Reynolds number. With these parameters the drag of a body can be adjusted for special functions in nature and technology. The sinking of a body can be delayed by increasing the drag (parachute principle), and an increase in the velocity for a constant propulsion force can be obtained by reducing the drag (streamlined bodies of fast swimmers and flyers).

Refined methods of drag control can be used to keep the flow laminar or to accelerate the transition to turbulence.[8] Streamlined bodies have a smaller drag in laminar flow. For blunt bodies the situation can be reversed, as the Eiffel–Prandtl paradox reveals.

Remarks on Chapter 6

1. The tendency of nature to form ordered structures seems to contradict the second law of thermodynamics which states that in a closed system entropy increases, or, in other words, that mechanical energy dissipates. Fluid flows past obsta-

cles or fluids heated from below are irreversible processes in an "open" thermodynamic system, and these processes are not in equilibrium. Prigogine [Reference 44, Chapter 1] has shown that thermodynamic nonequilibrium can be a source of order. The tendency toward larger ordered structures may be exemplified by the roll up of an infinitely long discontinuity line, approximated by point vortices (Fig. 3.29). The row of point vortices rolls up to a new row of concentrated vortices of larger size (clustered point vortices) and wider spacing as determined by the initial disturbance. The new row is unstable, too, and may roll up to a still larger vortex arrangement. This is, of course, an idealistic situation. In reality, a finite shear layer will break up after the first generation of vortices (Fig. 6.2). Other examples of apparent increased ordering are the Bénard cells (Section 8.3) and the Taylor vortices (Section 7.6).

2. Large changes in a vorticity field occur in a nascent vortex at the border of the vortex core. It is of interest that the model of point vortices or a band of point vortices fails in the description of sound radiation. The roll up of a chain of point vortices in Fig. 3.29 to vortex clusters causes nonphysical sound at the center of the developing vortex cluster through the greatly accelerated rotation of point vortices about that center (Fig. 6.61). In reality friction quickly causes the core to rotate like a solid body.

3. Nominally, a simple relationship exists between roll parameter (or Rossby number) and Strouhal number so that the flow past a rotating body could also be designated by the pair of parameters Re and St. The frequency of a periodic event is denoted by f, and its unit is $1/s = 1$ Hz (Hertz). The angular velocity is $\Omega = 2\pi f$. The roll parameter $p = \Omega R/V$ and the Strouhal number St $= fR/V$ are thus related by $p = 2\pi$ St or Ro $= 1/(2\pi$ St$)$. In this book the Strouhal number is applied to oscillating flows and bodies; the roll parameter or the Rossby number is applied to rotating phenomena.

4. The statement that Maxwell explained the falling, autorotating plate must be qualified. Maxwell recognized that the center of rotation does not coincide with the center of pressure and that, therefore, a torque exists. The explanation of Fig. 6.51 was also provided by Maxwell. However, he believed that the produced torque causes a net torque in the direction of rotation, which is in equilibrium with an opposite torque due to rotation. In fact, it is only the rotation that under certain conditions produces a supporting torque. Details are described in Reference 85.

5. One finds in nature autorotating tree seeds (for instance, *Ailanthus altissima* and *Liriodendron tulipfera*) that can spread over a wide range. Hail stones in the form of spheroids can also autorotate [116].

6. As mentioned in Section 1.3, Leonardo distinguished between laminar and turbulent flows. In modern times Hagen in 1854 was probably the first to point out this difference in his study of pipe flow [117].

7. The use of the curves in Figs. 6.59 and 6.60 may be illustrated with the following example. The friction factor λ is defined by

$$p_1 - p_2 = \lambda \frac{L}{D} \frac{\rho}{2} V^2,$$

with $p_1 - p_2$ = pressure loss between 1 and 2
L = length of pipe from 1 to 2
D = diameter of the pipe
V = average flow velocity
ρ = density of the fluid

The following quantities are prescribed with three significant numbers:

$V = 1.31$ m/s
$L = 100$ m
$D = 0.100$ m
$\rho = 102$ kp·s^2/m^4 for water at 10°C
$\nu = 1.31 \times 10^{-6}$ m^2/s for water at 10°C

How large is the pressure loss in such a flow of water in a horizontal, smooth pipe?
Solution: First the Reynolds number Re = DV/ν is obtained, which is Re = 1.00×10^5. With the aid of this number the friction factor λ is computed from Fig. 6.59 (or from a more accurate table in Reference 35, Chapter 1: $\lambda = 0.0180$). From the above formula the pressure loss is then $p_1 - p_2 = 1580$ kp/m^2 or 1.58 m water height.

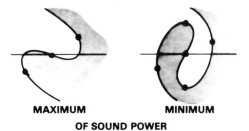

Figure 6.61. Constellation of point vortices in the immediate vicinity of the center of rotation of a rolled-up row of point vortices. In (*a*) a maximum of sound power is created, in (*b*) a minimum. However, the model of point vortices is unrealistic. The shaded area is the upper side of the vortex chain in Fig. 3.29.

Similarly, the curves in Fig. 6.60 may be used to compute the drag of spheres and circular cylinders. The drag D is related to c_D by $D = c_D \rho V^2/2$.

8. Suction through pores and slots and the injection of non-Newtonian fluids into the boundary layer are examples of the drag reduction on bodies by more sophisticated methods. The basic idea of suction is simple, although the engineering realization is difficult. Fluid particles near the wall, which have been slowed down through adherence to the wall, are sucked away into pores or slots so that particles with greater momentum come closer to the wall. Through this stabilization of the boundary layer the transition to turbulence is avoided or delayed. The injection of non-Newtonian fluids into the boundary layer can easily be carried out practically, but a satisfactory explanation of its effect has not yet been developed. So far, all fluids mentioned in this book have been considered to be "Newtonian," that is, the deformation of a fluid particle is proportional to the shear stress. This "constitutive" relation is valid for water, air, and oil, but not for the dough mentioned in Remark 6 of Chapter 3. Honey and polymers also behave in a non-Newtonian way. Fluids with long chain molecules are well suited for drag reduction. In 1948 Toms [118] found that a weak solution of polymethylmethacrylate drastically decreases the pressure loss in a turbulent pipe flow. In the 1960s the "Toms phenomenon" was carefully investigated. Polyox [polyethylene(oxide)] was most frequently used because it was cheap and very effective. A drag reduction of 33% is reached if 18 parts of polyox with a molecular weight of 10^6 is mixed with 10^6 parts of water [119, 120]. For short-time engineering tasks (as the drag reduction of torpedoes) a painting of polyox is sufficient. For drag reduction over longer periods of time, the non-Newtonian fluid must be injected into the boundary layer, for instance, through pores or slots. Fish apply a similar technique when they secrete slime on the skin. However, not every slime is drag reducing. In general, slime serves as protection against infection, as lubricant between scales, or as food for the young. However, experiments have shown that in many cases slime is used for drag reduction. According to Reference 121 the Pacific barracuda *(Sphyraena argentia)* produces the most effective slime. This predatory fish has, in addition, a very streamlined body and reaches high velocities.

PART TWO

7. Fluid Flow in a Rotating System

7.1. ABSOLUTE ROTATION AND MACH'S PRINCIPLE

The heliocentric system is simpler, not truer than the geocentric system.

According to Leibniz

In the first part of this book it was assumed that the earth's rotation has no influence on vortical motion. For local flows such as those in laboratory experiments, this assumption is often justified, but not always. Foucault's pendulum[1] and the disturbance-free vortex in the bathtub are examples of rotation explained by the influence of the earth's rotation and even serve as a proof for the earth's rotation.

Large-scale vortices in the atmosphere and ocean are significantly influenced by the earth's rotation. The variable densities of air and water, caused by the influence of temperature, pressure, and material concentration, also have an effect on vortices. The second part of this book deals with vortices in stratified fluids and with vortices in a rotating system with or without stratification.

As an introduction, some remarks are necessary on the space in which the movement of matter takes place, for the description and interpretation of motion, especially of rotation, depend on the physical properties of space. Two conflicting theories of space will be contrasted in the following text through their historical development.

It makes sense to talk about the movement of a body only in relation to another body, that is, a "reference frame" is needed. In Aristotle's conception of the world and in that of the medieval scholastics, the earth was the reference frame, and in relation to it the stars moved on natural (i.e., circular) orbits (Section 1.2). Interpretation of the epicyclical movement of the planets, however, caused difficulties, and the epicycles were thought to be composed of circular orbits (Fig. 7.1). In 1507 Copernicus found an explanation: If the sun became the reference frame (heliocentric system), the epicycles became simple, closed orbits. Copernicus still thought the orbits to be circular in the traditional sense, a conception which was still held by Galileo (1564–1642). It was Kepler who in 1609 discovered that the planetary orbits are elliptical. This finding required abandoning the idea of natural circular orbits and searching for a new explanation [1].

For the scholars of the Middle Ages the space between the heavenly bodies was not empty in the sense of a vacuum but filled with a light substance, the aether. Descartes (1596–1650) went a step farther and taught that matter and space are identical [Reference 27, Chapter 1]. Then a body can be surrounded only by another body (Section 2.2). For example, the stars revolve within gaseous vortices.[2] This model contained the idea that stars remain in their orbits because of a certain inertia. Descartes understood inertia in the sense that a body resists a change of motion. About the same time Kepler developed for his newly found elliptical orbits a concept of inertia that differs from that of Descartes in that a body resists movement from rest [1].

Newton (1642–1727) built on these new ideas of Descartes and Kepler and developed a systematic theory with which he created the general foundation of classical mechanics. According to his first law of motion, the state of a body at rest cannot be distinguished from that of a body moving with constant

130 Fluid Flow in a Rotating System

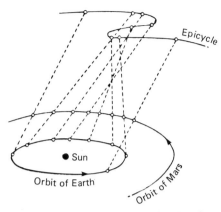

Figure 7.1. For an observer on earth, the planets move along complicated epicyclic orbits. In a reference frame fixed to space, these orbits simplify to ellipses.

speed. Through this relativity the distinction between the inertia concepts of Descartes and Kepler became meaningless. According to Newton's second law of motion, a force is necessary to change the velocity of a body, that is, to accelerate it. The state at rest and the constant straight movement are, thus, force-free states. The fact that a force is required for a change in velocity is also expressed in this way: the body has inertia. Newton's second law then reads: Force is equal to inertial mass times acceleration.

This law contains a problem of principle: It does not state the reference frame in which a body is accelerated. Newton had to postulate for this the concept of "absolute space," which itself need not be related to a reference system. This space has the physical property that a body can be accelerated in it without the presence of other matter. Inertia is thus an inherent property of matter and is not dependent on the presence of other matter. To prove the existence of absolute space and absolute rotation, that is, rotation relative to absolute space, Newton carried out his famous bucket experiment. A bucket filled with water was hung on a rope and turned so that the rope became twisted. When released the bucket rotated around its own axis. This rotation was transmitted to the water and produced a parabolic surface (Section 3.3). When the bucket was suddenly stopped, the water continued to rotate until it finally came to rest and the surface became flat again. Newton concluded from this experiment that the water rotates absolutely if the surface is curved, whether the bucket moves relative to the water or not. In the same way, Newton would have interpreted Foucault's pendulum experiment and the bathtub vortex as proofs that the earth rotates absolutely.

In addition to the laws of motion Newton formulated his famous law of gravitation. It says that two bodies attract each other inversely proportionally to the square of the distance between them and directly proportionally to the product of their masses. The masses are basically different from the inertial mass in the second law of motion. The masses in the law of gravitation owe their existence to the presence of other bodies and are called "gravitational masses." Newton's theory of gravitation confronted Descartes' vortex theory, which could not explain Kepler's planetary laws; but since Descartes' ideas were in fashion in the French salons, Newton's new thoughts were accepted only slowly at first. In 1736–1737 Maupertuis, a French physicist and president of the Prussian Academy, organized an expedition to Lapland to measure latitude and thus to determine whether the earth is flatter at the poles or at the equator. According to Descartes' vortex theory the earth would be more curved at the poles; according to Newton's theory of gravitation, it would be flatter. Maupertuis confirmed Newton's view, and with this success and with the correct interpretation of Kepler's planetary laws, Newton's authority was established.

The second theory on the nature of space can also be traced to the time of Newton. During his lifetime, the idea of absolute space did not remain unchallenged. About 20 years after the publication of Newton's *Principia* in 1710, the Irish philosopher and bishop, Berkeley (1685–1753), argued that the movement of a body makes sense only if it is measured relative to another body. As an empiricist from Locke's school he rejected the existence of absolute space as an unjustified abstraction and proposed to consider inertia not as a property of matter but attributable to the influence of the other matter in the universe. In this context, space without the presence of matter becomes meaningless. Berkeley supported his view by showing that Newton's bucket experiment can be interpreted in the following way: The curvature of the water surface during rotation is not independent of the other matter in the universe but is caused precisely by its presence. Without the fixed stars the water surface would always be flat [2]. Euler rejected Berkeley's idea, whereas Leibniz (1646–1716) arrived at the same conclusion as Berkeley. Leibniz's correspondence of 1715–1716 with Newton's friend Clarke reveals the theological difficulties of both parties. Newton's theory of gravitation needs the omnipresence and continuous interference of God (so Newton thought) in order to maintain the rotation of the stars. Leibniz could not accept God's intervention. The struggle ended with Leibniz's death in 1716 [Reference 27, Chapter 1].

As already mentioned, Newton's theory prevailed in the ensuing years. Except for philosophical considerations (Kant, Hegel), no scientific discussion on space took place in the following 150 years.

In 1872 Ernst Mach again took up Berkeley's idea that inertia comes into existence through the presence of the remaining matter of the universe. Space without matter is meaningless, and motion can be only relative. The earth rotates in relation to the fixed stars, but it does not matter whether the earth rotates relative to the stars or the stars revolve about the earth. This is the content of "Mach's principle" [Reference 57, Chapter 1; 3], which may also be expressed in the following way: The inertial force, which accelerates the earth with respect to the total mass of the universe, may also be interpreted as a force that acts on the earth at rest through the presence of the accelerated masses of the stars. Centrifugal force, thus, does not appear in absolute rotation but in rotation relative to the remainder of the universe. Einstein [4] used Mach's idea in his theory of relativity. He postulated that inertial and gravitational masses are equivalent, but he could not show that inertia depends completely on the remaining matter. Although space is influenced by the presence of matter (space becomes curved) within the framework of general relativity, no solutions have so far been found that are zero in the absence of matter. New approaches to overcome these difficulties are found in the work of Brans and Dicke [5] and of Hoyle and Narlikar [6].

What significance has the discussion of space and matter for the description of vortices, other than the perception that Mach's principle is more satisfying from the natural philosopher's point of view than Newton's absolute space? In describing events on earth there is no appreciable difference between the two theories. Newton's perception may thus be used from here on as a basis, since a satisfying quantitative formulation of a theory that rests on Mach's principle is still lacking. However, if one considers the evolutionary process of the universe, in particular, the constancy of the angular momentum of galaxies and metagalaxies, then the consequences of Mach's principle may be decisive. Today it is believed that the gravitational constant decreases in an expanding universe [7].

For the discussions that follow this statement is sufficient: The earth rotates absolutely according to Newton, relative to the fixed stars according to Mach, once around itself in 23 hours, 56 minutes, and 3 seconds (or in 86,163 seconds).[3] The angular velocity of the earth can be considered as approximately constant. This angular velocity, denoted by Ω, is then

$$\Omega = \frac{2\pi}{86{,}163 \text{ s}} = 7.292 \times 10^{-5} \text{ s}^{-1}.$$

Thus, the earth is a solid-body vortex with a vorticity of 2Ω, provided one neglects its elastic and plastic properties as well as the convection inside the earth.

7.2. CENTRIFUGAL FORCE AND CORIOLIS FORCE

Newton's laws of motion change in a reference frame that is accelerated in relation to absolute space. The description of events on the rotating earth, when the earth itself is the reference frame, is an example. Newton's laws must then be extended by additional terms. Absolute velocity is the vector sum of relative velocity and the earth's rotational velocity. The change of velocity with time, that is, acceleration, differs in a rotating system from that in absolute space by two additional portions: the centrifugal and the Coriolis accelerations. Multiplied by mass, these contributions appear as additional forces in the momentum equation. They are also called fictitious forces because they are kinematical quantities. The centrifugal force always acts perpendicular to the axis of rotation (Fig. 3.14) and increases linearly with the distance from the axis and quadratically with the angular velocity. The Coriolis force (named after the physicist Coriolis, 1835) acts perpendicular to the axis of rotation and perpendicular to the relative velocity. The appearance of this force may be explained in the following way: If a particle moves from the equator to the north, it crosses latitudes with decreasing circular radius. To preserve its angular momentum, the particle must rotate faster than the earth and thus will be deflected to the right (Fig. 7.2). A par-

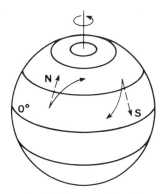

Figure 7.2. The Coriolis force acts in such a way that objects moving horizontally to the surface of the earth are deflected in the northern hemisphere to the right and in the southern hemisphere to the left.

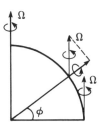

Figure 7.3. The angular velocity of the earth has, on the surface of the earth, a vertical component that is zero at the equator and largest at the poles.

ticle which travels from the north to the equator will be retarded relative to the earth and, hence, will also be deflected to the right. On the southern half of the earth the deviation is to the left. An observer on the earth concludes from the deviation of the particle that a force exists, the Coriolis force [8].

On a rotating sphere the component of the angular velocity perpendicular to the sphere's surface changes sinusoidally with latitude. Twice its value is the vorticity component perpendicular to the surface of the sphere and is called the Coriolis parameter f (Fig. 7.3):

$$f = 2\Omega \sin \phi, \quad \phi = \text{latitude}.$$

The rule that the Coriolis force causes a deviation to the right in the northern hemisphere explains Shapiro's bathtub experiment (Section 3.4): A fluid that would flow radially into a sink without the influence of the earth's rotation takes on a counterclockwise rotation in the northern hemisphere due to the Coriolis force (Fig. 7.4a). This rotary direction is called "cyclonic." A source flow will be deflected clockwise in the northern hemisphere. It is "anticyclonic" (Fig. 7.4b). Low-pressure regions are cyclonic; high-pressure regions are anticyclonic.

Thus the earth's rotation is capable of generating vortices in the atmosphere and ocean through a redistribution of vorticity. These vortices differ essentially from those in a nonrotating system. Whereas in the latter the pressure always has a minimum at the center (Fig. 3.16), the pressure in the vortex of a rotating system can have a minimum or a maximum, depending on the direction of rotation. In cyclonic rotation the pressure has a minimum; in anticyclonic rotation, it has a maximum. The concepts of "high" and "low" are familiar from the weather map [9].

The relation between streamlines and isobars is, under certain conditions, a simple one, which is important in the interpretation of weather maps. If only pressure and Coriolis forces are acting in atmospheric motions, the isobars coincide with the streamlines for a steady-state flow. This state, which represents equilibrium between pressure and Coriolis forces, is called "geostrophic" [9]. Although it seems to contradict the law that flow particles move toward lower pressure, this paradox can be resolved as follows: A mass of air is accelerated in the direction of the pressure gradient (Fig. 7.5a), but the wind direction changes under the influence of the Coriolis force (Fig. 7.5b). The wind will then be accelerated until equilibrium is reached between pressure and Coriolis forces (Fig. 7.5c). In this case the wind direction is parallel to the isobars. Since the pressure can easily be measured, isobars are drawn on the weather map. If the wind is geo-

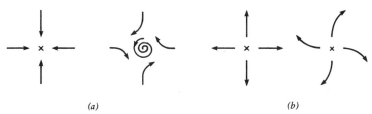

Figure 7.4. Deviation of (a) a sink flow and (b) a source flow through the earth's rotation in the northern hemisphere.

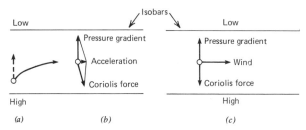

Figure 7.5. Development of a geostrophic wind. In a geostrophic state, pressure and Coriolis forces are in equilibrium, and the streamlines coincide with the isobars.

strophic, these lines are also streamlines. Hence, isobars on the weather map give an indication on the movement of air masses. The assumption of a geostrophic state does not hold, however, near the earth's surface, where friction forces cause a boundary layer, nor for strongly curved streamlines.

For strongly curved streamlines (as in vortices) a state similar to the geostrophic one can be reached if the centrifugal force is considered (Fig. 7.6). Here, too, streamlines and isobars coincide when the pressure, Coriolis, and centrifugal forces are in equilibrium. Such a flow is called "gradient flow."

A simple fluid rotation is possible if only inertial and Coriolis forces, but not the pressure force, act on a fluid particle. The particles then follow circular paths whose rotary direction is clockwise in the northern hemisphere and counterclockwise in the southern hemisphere. The period is $\pi/\Omega \sin \phi$, which is one-half the period of the Foucault pendulum.[1] This special kind of rotation is called an "inertial vortex." Figure 7.7 shows the path of such motion observed in the Baltic Sea.

Near a solid or liquid surface the assumption of a geostrophic or a gradient flow is not valid because the adherence of the fluid at the surface causes a boundary layer (Fig. 4.16). Without the influence of

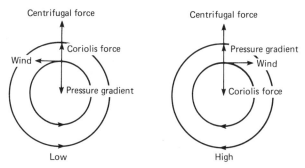

Figure 7.6. Regions of high and low pressure. The isobars are streamlines for gradient flow. The wind direction drawn is for the northern hemisphere.

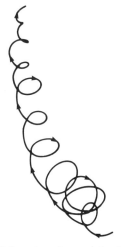

Figure 7.7. Inertial vortex observed in the Baltic Sea by Gustafson and Kullenberg [10]. The circles are drawn apart through superposition of a translational motion. Friction decreases the radii of the circles.

rotation, streamlines do not change their direction in a parallel flow; the boundary-layer motion is in the same direction as the outer flow. However, under the influence of the Coriolis force, the boundary-layer profile will be twisted. If the axis of the rotating system is perpendicular to the wall, the boundary layer is twisted to the right[4] in the northern hemisphere (Fig. 7.8). This boundary layer is called "Ekman boundary layer" or "Ekman layer" for short after the oceanographer Ekman (1905) [11]. If the axis of rotation is parallel to the wall, a

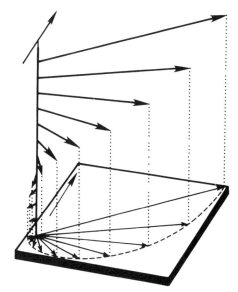

Figure 7.8. Ekman boundary layer.

"Stewartson layer" develops, named for the fluid dynamicist Stewartson (1957) [12].

For many flow phenomena in nature or in the laboratory the total motion can be thought of as composed of geostrophic flow and Ekman and Stewartson layers. The conditions for this motion will be described in Section 7.5.

7.3. DYNAMIC SIMILARITY IN A ROTATING SYSTEM

In all motions on earth, centrifugal and Coriolis forces caused by the earth's rotation are always present. These forces, however, are so weak in local events that they can be neglected in most cases. According to the descriptions in Section 7.1, this situation can also be expressed in such a way that the local reference frame agrees with the absolute one (in the Newtonian sense) or with the nonrotating one of the distant masses of the stars (in Mach's sense). When can the influence of the earth's rotation on fluid motion be neglected, or, in other words, what is a local event? Since the centrifugal force and the pressure gradient are usually considered together (acting in the same direction), the discussion will first be restricted to the Coriolis force.

The circumstances under which the Coriolis force causes a noticeable effect depend on the magnitude of the other forces. In the ocean and atmosphere inertial and frictional forces are present as in a nonrotating system. The ratio of forces is described by such parameters as the Reynolds number, introduced in Section 4.2 as the ratio of inertial to frictional forces. In the presence of the Coriolis force two new parameters can be defined: The ratio of inertial to Coriolis force, and the ratio of frictional to Coriolis force. The first parameter is called the "Rossby number," the second the "Ekman number" [Reference 40, Chapter 1]:

$$\text{Rossby number} = \frac{\text{inertial force}}{\text{Coriolis force}},$$
$$\text{Ekman number} = \frac{\text{frictional force}}{\text{Coriolis force}}.$$

In most atmospheric and oceanic motions the inertial force exceeds the frictional force, and thus the Rossby number is of greater significance than the Ekman number. The Rossby number is therefore used to find a criterion for the influence of the earth's rotation on fluid motion. For this reason it is necessary, just as in the case of the Reynolds number, to define the Rossby number (and for later use the Ekman number also) mathematically through characteristic quantities. For the Rossby number Ro and the Ekman number Ek the relations are

$$\text{Ro} = \frac{V}{Lf} \quad \text{or} \quad \text{Ro} = \frac{\omega}{f}, \quad \text{Ek} = \frac{\nu}{L^2 f}.$$

Here, V designates a characteristic velocity, L is a characteristic length, f is the Coriolis parameter, ν is the kinematic viscosity, and ω is a characteristic vorticity related to the rotating system. The definition of the Rossby number ω/f is useful for the description of vortices. The Rossby number is here merely the ratio of relative (in the rotating system) vorticity to the vorticity component of the earth's rotation perpendicular to the earth's surface. It may be mentioned that in Section 6.5 the Rossby number was introduced in general form as the ratio $V/\Omega R$, where Ω is the angular velocity of the rotating body.

The influence of the earth's rotation vanishes if the Coriolis force becomes so small relative to the inertial and frictional forces that it can be neglected; in mathematical terms, this means that the Rossby and Ekman numbers become infinitely large. These parameters, however, become infinite only if either $f = 0$ or $L = 0$, or if $V = \infty$ or $\omega = \infty$ or $\nu = \infty$. This condition can be realized exactly only at the equator, where $f = 0$. On the remainder of the globe the rotation-free state can only be approximated but never reached. Still, for many practical purposes the influence of the Coriolis force can be neglected because local disturbances outweigh the Coriolis force. One may recall the example of the bathtub vortex in Section 3.4. Disturbances due to asymmetrical initial conditions, such as the influence of temperature, side walls, or already existing fluid motions, determine the direction of rotation of the vortex when water is discharged. Only if all disturbances are carefully avoided and the Coriolis force has time to act is the Coriolis force strong enough to determine the direction of the bathtub vortex. These considerations mean that approximate data based on experience can be given for the Rossby and Ekman numbers, whether the Coriolis force is insignificant or not.

For vortices, the maximum velocity and the radius of the vortex may be chosen as the characteristic velocity and length, respectively. The following approximate Rossby numbers are obtained for some typical vortices on earth (with $f = 7.29 \times 10^{-5} \approx 10^{-4}$ s^{-1} for the latitude $\phi = 30°$):

	L	V	Ro
Bathtub vortices	1 cm	0.1 m/s	10^5
Dust devils	3 m	10 m/s	3×10^4
Tornadoes	50 m	150 m/s	3×10^4
Hurricanes	500 km	50 m/s	1
Low-pressure systems	1000 km	10 m/s	10^{-1}
Ocean circulations	3000 km	1.5 m/s	5×10^{-3}

As will be shown later, the rotary direction of tropical whirlwinds (hurricanes, typhoons) depends on the Coriolis force, while tornadoes are local events.[5] The critical Rossby number, which gives the limit for the noticeable influence of the earth's rotation, is thus about Ro = 10^4.

A clue to the time needed for the Coriolis force to have an appreciable effect on fluid motion is obtained for the special case of the inertial vortex. Its period is one-half as large as that of the Foucault pendulum. A flow thus would be deflected by the Coriolis force by 1° in 12 h/360° sin ϕ, which corresponds to 2 min/sin ϕ.

7.4. HYPERBOLICITY

A reason for the fascination which radiates from rotating media may be sought in the fact that the human mind—developed and formed in a locally non-rotating environment—cannot interpret motions intuitively in a rotating system.

Vorticity can spread in the form of waves within a rotating incompressible fluid. This sounds incredible because the wave concept in fluid dynamics is usually connected with the existence of a free surface (ship waves) or with the compressibility of the medium (shock waves in supersonic flows). In general, waves occur in a physical system if the disturbed system is brought back into equilibrium by a restoring force. For instance, surface waves are generated when gravity tries to bring fluid particles, which are not in equilibrium, back to their original positions. The neighboring particles are affected by the oscillation, and the disturbance spreads in a wavelike way. In a rotating system the Coriolis force functions as gravity does for water waves. This may be imagined in the following way: If the motion in a rotating system is related to absolute space, "absolute vorticity" is governed by Helmholtz's laws in the absence of frictional forces (Section 4.6). For instance, absolute vorticity in a two-dimensional flow is constant on a particle path and is bound to the particles. Absolute vorticity is now the sum of twice the angular velocity of the rotating system and the vorticity relative to the rotating system. This relative vorticity can spread out in wave form. Rossby [13] demonstrated this with the following argument. He assumed that the absolute vorticity component ω_A of a mass of air perpendicular to the earth's surface remains constant:

$$\omega_A = f + \omega_R = \text{const}$$

with ω_R being the relative vorticity. Frictional forces are excluded and the flow has no appreciable shearing. The air moves, as is usual in midlatitudes, from west to east. If this air has a velocity component in the north–south direction, the following situation occurs (Fig. 7.9): In position 1 a mass of air moves from southwest to northeast. Its absolute vorticity ω_A will be equal to f. As it moves northward, however, f increases ($f = 2\Omega \sin \phi$) and ω_R must assume values of opposite sign to keep ω_A constant (position 2). This ω_R value enforces a curvature of the streamline toward the south[6] until the latitude ϕ is reached, at which $\omega_A = f$ (position 3). As the air mass continues to travel south, f decreases, and ω_R of the same rotary direction must be generated (position 4). This ω_R value requires a curvature of the streamline to the north. At position 5 the cycle starts again. The mass of air hence moves in waves from west to east around the globe. These waves are called "Rossby waves."

In mathematics the basic equations for describing waves are called "hyperbolic." Hyperbolicity is

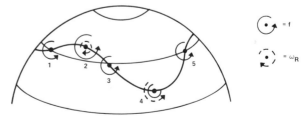

Figure 7.9. Rossby waves can be explained by the constancy of the absolute vorticity.

the key to understanding meteorological and oceanic motions. Not only rotating fluids in a rotating reference frame[7] can behave hyperbolically, stratified flows do also, and they are described in Chapter 8.

Waves, however, cannot spread indefinitely, but are enclosed by lines (or areas). These boundary lines are called "characteristics." Examples are the characteristics of ship and shock waves (Fig. 7.10).

The hyperbolic behavior of rotating fluids was pointed out by Lord Kelvin [14] and by Görtler [15]. Oser [16] was the first to show the existence of characteristics experimentally: He oscillated a horizontal circular disk perpendicular along the axis of a rotating mass of water. At the edge of the oscillating disk velocity jumps occur (in a frictionless flow), and these jumps propagate along the characteristics. The characteristics, shown in Fig. 7.11, divide the whole flow field into four regions: No disturbances can penetrate into regions I and IV. Whereas region I is at rest relative to the rotation, the fluid in region IV takes part like a solid body in the perpendicular movement of the disk. Regions II and III are penetrated by the disturbances, and secondary flows develop as shown in Fig. 7.12.[8]

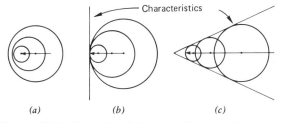

Figure 7.10. Example of the spreading of a disturbance. The source of a weak pressure wave moves (a) at a speed less than the speed of sound, at which the pressure itself is spreading; (b) at a speed equal to the speed of sound, and (c) faster than the speed of sound. In the latter case the disburbance can spread only within the characteristics. In the special case (b) ("parabolic"), the source of the disturbance moves with the disturbance itself.

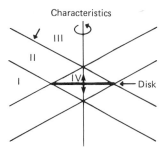

Figure 7.11. Characteristics caused by a disk oscillating in a rotating fluid (from H. Oser).

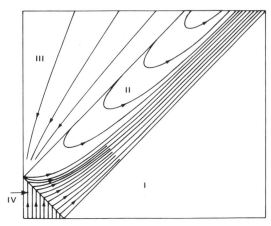

Figure 7.12. Streamlines between the characteristics of Fig. 7.11 during the upward moving phase of the oscillation (from H. Oser).

If the frequency of the disturbance approaches zero, the characteristics are aligned parallel to the axis of rotation, and the situation is identical with the geostrophic flow described in Section 7.2. This fluid motion, in which pressure and Coriolis forces are in equilibrium, has a strange property: because the characteristics are parallel to the axis, the motion cannot change in that direction. This means that the velocity can change only in the direction perpendicular to the axis. A streamline along the bottom forces streamlines above to be exactly parallel to it, and hence the flow must be parallel to lines of constant depth. Expressed another way, a column of fluid must keep a constant height during its movement. If the contour of the bottom changes due to a local elevation, the fluid must move around this bump as if it were a column of the height of the fluid. Proudman [17] formulated this phenomenon of the "two-dimensional constraint" in 1916 mathematically. In 1921 G. I. Taylor [18], inspired by this result, made the following experiment: He pulled a body along the bottom of a rotating container filled with water (Fig. 7.13). He observed that the fluid, in agreement with Proudman's prediction, did not flow over the body but sidewise around it as if an imaginary perpendicular cylinder had been placed in the container. This phenomenon is called the "Taylor–Proudman column," and the mathematical formulation of the two-dimensional constraint is known as the "Taylor–Proudman theorem."

This theorem holds only under the assumption that the pressure and Coriolis forces are in equilibrium in a steady flow. Inertial and frictional forces

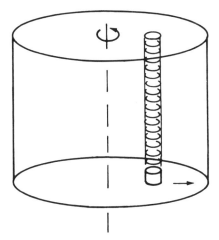

Figure 7.13. Taylor–Proudman column.

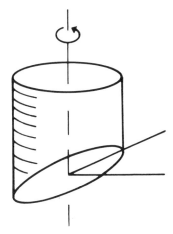

Figure 7.15. Container for which a geostrophic contour does not exist.

are neglected, that is, the Rossby and Ekman numbers must be vanishingly small. If the Rossby number is significant, the two-dimensional constraint disappears (Fig. 7.14).

Hide [20] interpreted the Great Red Spot on the planet Jupiter as a Taylor–Proudman column. Today, however, it is believed that the Great Red Spot is a soliton (Section 12.3).

The Taylor–Proudman theorem gives an answer to the question of why vortices with small friction are columnar. Changes in the velocity parallel to the axis are suppressed, which means that the particle paths are approximately circular or helical.

Geostrophic flow is possible only in spaces or containers in which a column of fluid (parallel to the axis of rotation) can keep a constant height. The movement of such a column must take place along a closed bottom line of the container [Reference 40, Chapter 1]. The closed bottom line is called the "geostrophic contour." An important example of a

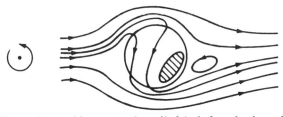

Figure 7.14. Movement of a cylindrical obstacle along the bottom of a container perpendicular to the axis of rotation of the fluid. The observer moves with the obstacle. With increasing Rossby number the Taylor–Proudman column diminishes (shaded area). Ro = $V/R\Omega$ = 0.014, where V is the velocity of the obstacle, Ω is the angular velocity of the container, and R is the radius of the container (from R. Hide and A. Ibbetson [19]).

container that does not possess a geostrophic contour is a cylinder with an oblique bottom (Fig. 7.15). Since the fluid cannot move in a geostrophic way, it must undergo wave motions. Such waves are similar to the Rossby waves described at the beginning of this section. It can be shown theoretically that the role played by the angle of inclination in a rotating container corresponds approximately to the change of the Coriolis parameter with latitude ϕ [Reference 40, Chapter 1]. This analogy is even exact if the Coriolis parameter f is approximately $f = \text{const} + \beta y$, where y is the coordinate in the northern direction. The phase velocity U of the Rossby wave can then be written, with V as the mean west–east flow velocity and λ as the wavelength:

$$U = V - \frac{\beta \lambda^2}{4\pi^2}.$$

The phase velocity may also become zero or negative, depending on the magnitude of V.

The course of motion is changed by frictional forces, and this will be discussed further in Section 9.1. There the great significance of the Rossby waves for the general circulation of the atmosphere will be appreciated.

Yet another special case of Taylor–Proudman columns occurs if the axis of symmetry of an axisymmetric body coincides with the axis of a rotating fluid [21–23]. In Fig. 7.16 the flow of a rotating fluid about a fixed sphere is sketched. For frictionless flow the Taylor–Proudman column forms an infinitely long circular cylinder (dashed lines). Viscosity changes this column to separation regions in front of and behind the sphere. Several different

Figure 7.16. Taylor–Proudman effect in the flow of a rotating fluid around a fixed sphere for small Ro and large Re. Dashed line: Taylor–Proudman column in a frictionless flow (adapted from T. Maxworthy [24]).

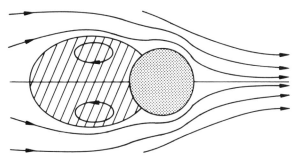

Figure 7.17. Taylor–Proudman effect in the flow of a fluid rotating around a fixed sphere for small Ro and small Re (adapted from S. Singh [26]).

flow types exist depending on the magnitude of the Reynolds and Rossby numbers. Figure 7.16 shows a typical flow field for small Ro and large Re [24, 25] and Fig. 7.17 is the flow for small Ro and small Re. The wake region behind the body has vanished in this case.

7.5. CIRCULATION IN ROTATING VESSELS

In a vessel rotating with constant angular velocity the fluid moves like a solid body (Fig. 3.6). To reach this state the vessel must first be set in motion. After an abrupt start, the solid-body rotation develops as follows: Immediately after the sudden start of rotation, the walls of the vessel revolve relative to the fluid at rest. Between the walls and the fluid a thin shear layer forms. Within a few revolutions rotation influences this shear layer, that is, boundary layers of the Ekman and Stewartson types develop. This means that in the Ekman layers at the bottom and the top of the vessel fluid is deflected in the radial direction toward the side wall, in the Stewartson layer at the side wall in the direction parallel to it (Fig. 7.18). A secondary flow develops, which consists in the meridional plane of two circulation cells. A particle path then is composed of the primary rotation about the axis and of the meridional circulation. In addition, inertial waves can occur at the start, which slowly vanish

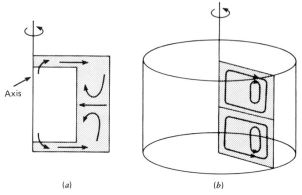

Figure 7.18. The "spin-up process": A cylindrical vessel is suddenly put in constant rotation. (a) Ekman and Stewartson layers develop in which the angular momentum is brought into the nonrotating fluid. (b) In the meridional plane a secondary circulation is visible, which dissipates in the course of time. Finally, the fluid rotates like a solid body.

with time. In the boundary layers the angular momentum of the vessel is transferred to the fluid. With advancing time more and more fluid inside the vessel assumes the angular velocity of the vessel, and the secondary flow vanishes slowly through dissipation. Finally, solid-body rotation remains. This transient process—from rest to solid-body rotation, or from a certain rotation to one with higher angular velocity—is called "spin-up" [Reference 40, Chapter 1; 27]. Of course, one may also stop a vessel in which a fluid is rotating like a solid body. Then, again, a secondary flow develops in the transient stages until the fluid has come completely to rest (Fig. 7.19) [28]. This decay of rotation differs essentially from that of pure vorticity diffusion (Fig.

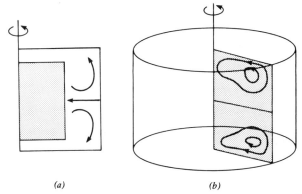

Figure 7.19. The "spin-down process": If a rotating vessel is suddenly stopped in which the fluid spins like a solid body, the state at rest is also reached over a secondary circulation. In distinction to the spin-up, the spin-down process can easily become unstable and can produce Taylor vortices (Section 7.6).

3.12). In "spin-down" the secondary flow takes care of the redistribution of angular momentum so that the rotation is stopped more quickly than it would be by friction alone. For instance, the motion in a cup of tea stops after stirring in about 1 minute. In contrast, the diffusion process alone would require almost half an hour to stop the fluid [Reference 40, Chapter 1].

For the understanding of atmospheric and oceanic circulations the general case will now be studied in which a cylindrical vessel rotates with a constant angular velocity Ω while the lid revolves with an angular velocity of Ω_c (Fig. 7.20). The ratio $(\Omega - \Omega_c)/\Omega$, that is, the relative difference in angular velocity of the lid and of the vessel, is introduced as the Rossby number for this arrangement:

$$\mathrm{Ro} = (\Omega - \Omega_c)/\Omega.$$

For the solid-body rotation $\Omega = \Omega_c$ is Ro = 0. If the lid is fixed ($\Omega_c = 0$), then Ro = 1. If the vessel is at rest ($\Omega = 0$) and only the lid rotates, Ro = ∞. For ocean basins, which rotate according to the earth's rotation with vorticity f, and in which the wind shear stress at the surface is roughly represented by the lid, Ro $\ll$ 1. Flow patterns for the three cases Ro $\ll$ 1, Ro = 1, and Ro = ∞ will be presented. The meridional plane of the vessel is, for simplicity, assumed to have a quadratic shape.

The development of the flow may be imagined as similar to the spin-up process: First, the vessel and the lid rotate together with the constant angular velocity Ω. The fluid rotates like a solid body, or, in other words, the fluid is at rest in relation to the spinning vessel. If suddenly the angular velocity of the lid changes to Ω_c, a shear layer is produced at the boundary between lid and fluid and develops to an Ekman layer in which the fluid is transported radially to the side wall (in some cases to the axis of rotation). With the aid of a Stewartson layer at the side wall and an Ekman layer at the bottom, a meridional circulation is generated which, in general, consists of a single cell.[9] In contrast to the spin-up or spin-down process, however, and after a sufficiently long time, a steady meridional circulation and an additional azimuthal velocity are present, which obtain their energy from the vessel and the lid.

For very small Rossby and Ekman numbers the inertial forces are small, and the frictional forces are restricted to the boundary layers along the walls. In the range Ro $\ll$ Ek $\ll$ 1, which is important in geophysics, the vessel can be divided into three areas: the Ekman layers at the lid and bottom of the vessel, the Stewartson layer at the side wall, and the geostrophic interior (Fig. 7.21). Typical data for the ocean are: Ro $\approx 10^{-5}$, Ek $\approx 10^{-4}$ with Ek = $\nu/R^2\Omega$. Ocean basins, of course, have a stretched rectangular shape, and the flow regions are distorted correspondingly.

It may be mentioned that the boundary layer at the earth's surface for large-scale atmospheric motions is divided into an Ekman layer and a turbulent layer very close to the surface. In this latter layer, the Coriolis force can be neglected (Fig. 7.22).

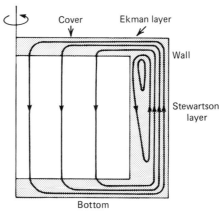

Figure 7.21. For Ro $\ll$ Ek $\ll$ 1 the flow in the vessel can be divided into three regions.

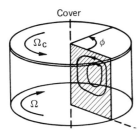

Figure 7.20. Steady meridional flow in a rotating vessel with the angular velocity Ω. The lid rotates with Ω_c.

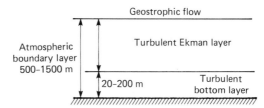

Figure 7.22. Order of magnitude of the atmospheric boundary layers.

140 Fluid Flow in a Rotating System

The temporal development of the meridional circulation is displayed in the computer-generated pictures of Fig. 7.23. The motion begins in the gap between lid and side wall at time $t = 0$, after the abrupt change of the lid's rotation. Inertial waves become visible in the form of oscillating streamlines and bubbles (trapped waves) around the axis of rotation. These bubbles appear and vanish twice in the transient period. Separation regions of this kind have been described in Section 5.5 for swirling flows and have been labeled vortex breakdown. In Fig. 7.23 the bubbles occur only temporarily and have vanished in the steady state.

If the lid is kept fixed and only the vessel revolves, then Ro = 1. Inertial forces can no longer be neglected, and the flow in the interior is not geostrophic. In the steady-state case the magnitude of the Ekman layer determines whether the streamline patterns are monotonic or wavy (Fig. 7.24). The Ekman layer which separates the two types is about 0.005.

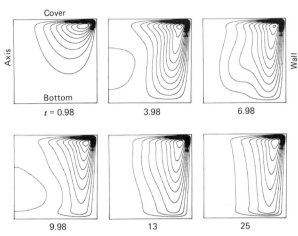

Figure 7.23. Temporal development of the laminar meridional circulation (streamlines) for Ro = 10^{-5}, Ek = 10^{-3}. The dimensionless time is time · Ω (computer-generated pictures by H. J. Lugt and H. J. Haussling [29]).

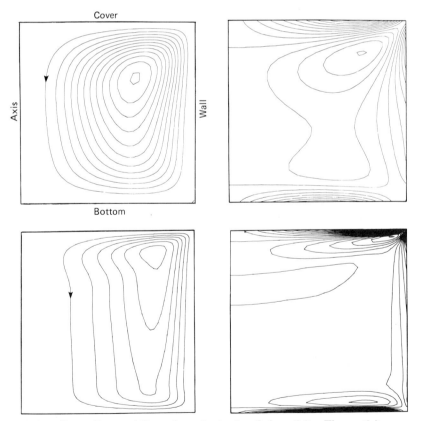

Figure 7.24. Streamlines and lines of constant azimuthal vorticity. The vorticity curves have their maximum value at the walls. Ro = 1, Ek = 0.01 above, Ek = 0.001 below (computer-generated pictures by H. J. Lugt and H. J. Haussling [29]).

The flow behavior for Ro = 1 explains the heaping up of tea leaves at the center of the bottom of a tea cup. If Fig. 7.24 is inverted, the fixed lid represents the bottom of the cup and the stirring tea spoon is approximated by the rotating vessel. At the bottom of the cup the tea leaves move with the fluid particles toward the center. Since the tea leaves are a little heavier than the water, they are not carried upward but heap up at the bottom. This can best be observed after stirring, when the fluid motion slows down (Fig. 1.5).

With the vessel at rest and the lid rotating, the Rossby number is infinite. Figure 7.25 shows the meridional flow for Ek = 0.0025 or Re = 400, where the Ekman number is now defined by $\nu/R^2\Omega_c$ and the Reynolds number by Re = 1/Ek. The flow direction in Fig. 7.25 is opposite to that in Fig. 7.24 for Ro = 1. The flow is monotonic in this case but changes to wavy patterns for Re = 1000. Bubbles do not appear, but Vogel [30] found in experiments that, with increasing relative height H/R of the vessel, steady-state bubbles can occur. The inertial waves are "frozen" or "trapped." Figure 7.26 shows the limiting curve in the (Re, H/R) plane, which separates the regions in which bubbles do and do not occur. Computer calculations confirm the existence of the bubbles and the accuracy of the limiting curve [31]. Figure 7.27 shows the steady-state situation for Re = 1350, H/R = 1.58. With increasing Re the bubble moves downward, that is, upstream. This tendency agrees with previous results for swirling flows (Section 5.6).

The flow direction in Figs. 7.25 and 7.27 is opposite to that in Fig. 7.24 for Ro = 1. This different

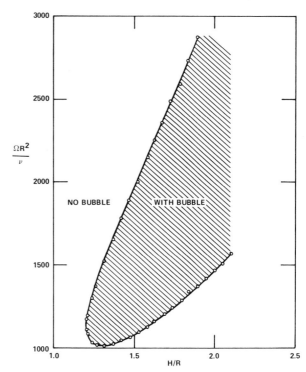

Figure 7.26. Limiting curve for vortex breakdown according to experiments by H. U. Vogel [30].

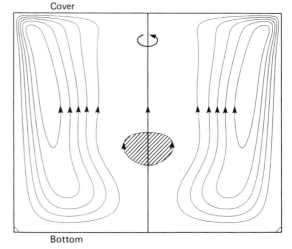

Figure 7.27. Vortex breakdown produced by the rotating lid of a fixed vessel: Re = 1350, H/R = 1.58 (computer-generated picture by H. J. Lugt and H. J. Haussling [31]).

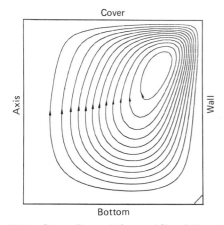

Figure 7.25. Streamlines of the meridional flow for Ro = ∞, Re = 400 (computer-generated picture by H. J. Lugt and H. J. Haussling [29]).

behavior is extremely important in understanding vortex flows. In general, the following rule is valid for solid surfaces (and liquid surfaces in gas motions): If the fluid rotates over a fixed bottom, no matter what the velocity distribution of the spinning fluid, the meridional flow is always directed toward the center of rotation. If, however, the bottom rotates and the fluid far away from the bottom is at rest, then the meridional flow is directed away from the center (Fig. 7.28).

The flow shown in Fig. 7.28a, which is generated by a rotating bottom, was first studied by von Kármán [32]. The explanation of the radial flow outward is quite simple: The fluid that adheres to the bottom is carried away from the center by centrifugal force. More complicated is the explanation of the meridional flow caused by the rotating fluid. Here, the equilibrium between pressure and centrifugal forces within the rotating fluid is disturbed at the bottom. The angular velocity is diminished by the nonslip condition, and the centrifugal force is reduced near the bottom. The new dominant pressure force generates a radial flow toward the axis. The type of meridional flow, either monotonic or oscillating (in space), depends on the velocity distribution of the rotating fluid. Bödewadt [33] was the first to investigate the flow field produced by a solid-body rotation. The corresponding potential-vortex problem is difficult because of the singularity at the axis of rotation. Goldshtik [34], Kidd and Farris [35], Schwiderski [36], and Burggraf et al. [37] have studied this case.

Depending on the boundary conditions, for instance through additional inflow of fluid far away from the axis, separation can occur at the axis (Fig. 7.29). An example for this is the wake region behind a sphere in a rotating fluid (Fig. 7.16).

Of particular significance in nature and technology are rotating fluids in containers with a hole in the bottom for discharge. Such flows have been described in Section 3.4 but without mention of the secondary flow.

This kind of fluid motion is quite easy to understand if the bottom rotates and the fluid is at rest far away from the center of rotation (Fig. 7.28b). With a center opening at the bottom, the meridional flow changes as shown in the sketch in Fig. 7.30.

More complicated is the flow above an opening if the fluid rotates. According to Long [38] and Shih and Pao [39] there are two types of fluid flows, depending on the magnitude of the Rossby number Ro $= Q/2\pi\Omega R^3$, where Q is the efflux, R is the radius of

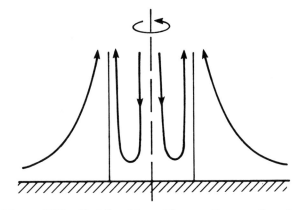

Figure 7.29. Meridional flow with separation near the axis.

Figure 7.28. Direction of the meridional flow (a) in a rotating fluid with fixed bottom and (b) for a rotating bottom and a fluid at rest far away from the bottom.

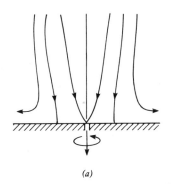
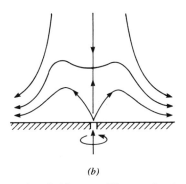

Figure 7.30. Meridional flow over a rotating bottom in a fluid at rest. The opening forms a sink in (a), a source in (b) (from C. G. Richards and W. P. Graebel [40], The American Society of Mechanical Engineers).

the container, and Ω is the angular velocity of the container. Above Ro = 0.26 the motion is similar to a potential flow, that is, to a potential vortex with a superposed sink. Near this critical Rossby number the outflow is concentrated in the vicinity of the axis and below the critical value is restricted to a core region around the axis. For the last case (Ro < 0.26) Shih and Pao distinguished three phases: In the beginning the motion is similar to potential flow. In the second phase the core flow develops with an almost stagnant ring area. In the final phase the core becomes a fast rotating jet. The second phase is sketched in Fig. 7.31. The core region I is surrounded by an almost motionless annular region IV. These are separated by a thin shear layer II. At the bottom and at the side wall are the Ekman layer III and the Stewartson layer V. The axial velocity W and the tangential velocity V are drawn in Fig. 7.31. V is compared with the circumferential velocity Ωr of the container. In the shear layer, V has a local maximum.

In the final phase, shortly before the last fluid is discharged, all the fluid motion takes place in the Ekman layer at the bottom. During this period the flow direction can change [41].[10]

The behavior of a rotating fluid in a container depends also on the position of the openings for inlet and outlet. For instance, the inlet angle can determine whether a separation region occurs near the axis of rotation [42].

A concentrated columnar vortex can be produced

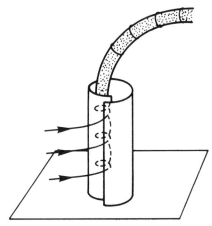

Figure 7.32. Apparatus for the generation of a columnar vortex after F. O. Ringleb.

with a cylinder that is sliced open on one side and whose cross section is a spiral (Fig. 7.32). When air is blown against the cylinder, it enters the interior and is put into fast rotation. Simultaneously, the axial velocity at the outlet transforms this motion into a swirling columnar vortex [43].

In technology various forms of containers with diverse arrangements of inlets and outlets have been designed. Ring-type entrance and exit chambers for turbomachines and centrifuges, conical vessels for spraying, and cylindrical containers for cyclone combustion chambers have been studied [44–46]. Physically, these confined vortex flows can exhibit typical vortical phenomena of engineering importance: helical oscillations; vortex breakdown; roll up of twisted discontinuity sheets; large differences in temperature, pressure, and concentration; cavitation; and noise. A few examples of engineering applications are given in the paragraphs that follow.

Containers with tangential inlets and axial outlets are used as centrifuges. The cyclone in Fig. 7.33 serves, for instance, as a dust separator and collector [47, 48]. In the nuclear industry, gas centrifuges play an important role in the enrichment of uranium [49–51]. In cybernetics vortex valves and amplifiers are used to control flux without moveable parts. In Fig. 7.34 a vortex valve is sketched. The flux to be controlled enters radially into a circular-cylindrical box and leaves it axially. At the same time a control fluid (of the same medium) is tangentially injected into the box. It mixes with the main fluid and generates a vortical flow. The larger the control fluid flow, the smaller is the main flux until a state is reached in which the main flux is zero. The enlarged pressure drop in the vortex increases the

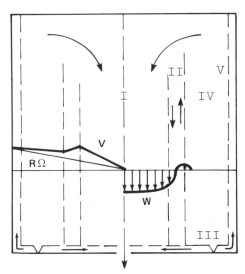

Figure 7.31. Schematic presentation of the flow in a rotating container with an outlet for Ro = 0.26 (adapted from H. Shih and H. Pao [39]).

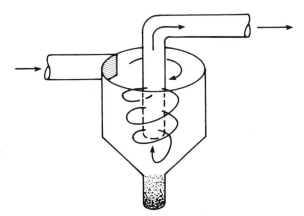

Figure 7.33. Dust cyclone.

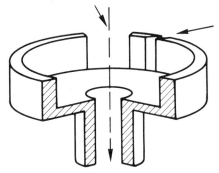

Figure 7.34. Sketch of a vortex valve. The lid is removed.

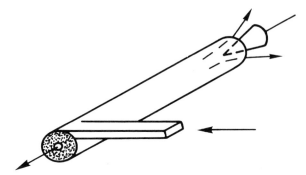

Figure 7.35. Vortex tube as a cooling device (Ranque–Hilsch tube).

flow resistance in the valve and thus throttles the flux [52].

A vortex chamber increases the time a particle stays in a confined space since the helical path from entrance to outlet is much longer than the straight distance between them. This fact is cleverly exploited in devices for chemical reactions, in particular for combustion [Reference 49, Chapter 1; 53, 54]. Axial return flow is sometimes enforced to increase further the delay through the generation of helical vortex filaments in the shear layer of the axial flow and its counterpart [55].

In other devices and designs the properties of rotating fluids in containers are technically exploited. Examples are vortex pumps [56], vortex whistles [57], magnetohydrodynamic (MHD) generators [58], heat exchangers [59], vortex reactors [60], and devices for stabilizing gas discharges [61]. Moreover, the "vortex tube" or "Ranque–Hilsch tube" is mentioned because of its surprising physical property. With compressed rotating air, it can be used as a cooling device [62]. The tube has at one end a tangential inlet nozzle and a circular orifice as outlet (Fig. 7.35). At the other end of the tube is a second opening in the form of a valve. When compressed air flows into the tube, a vortex is produced whose core cools down while the air close to the wall heats up. The cold air escapes from the circular orifice and the warm air from the valve, which regulates the ratio of cold to warm air. According to Reference 63 compressed air at several atmospheres pressure at room temperature can be cooled to $-50°C$.

There are numerous interpretations in the literature for the temperature difference produced in the vortex tube [62]. All agree that the air in the core is cooled through adiabatic expansion and the air in the outer part of the vortex is heated through centrifugal pressure. Different opinions exist, however, in the explanation of how the energy is transported from the core of the vortex to the wall of the tube and in the axial direction. This behavior is determined by the change in the velocity and temperature profiles along the tube's axis [64].

In nature cyclone-type flows can be found in fast-moving mountain brooks, rivers, and tidal currents at sea. Vortices in corners and rough spots hollow out the ground when the secondary flow pulls sand and stones upward (Fig. 7.36). They grind off walls and bottoms in the course of time and form almost circular holes in the stone of the bed. In the rocky banks of the Potomac river north of Washington, D.C. there are such "potholes" up to 3 m in depth and diameter (Fig. 7.37a).

In the same manner potholes are formed by stones in the melting water of glaciers. Larger whirlpools occur behind waterfalls, for instance, beneath Niagara Falls and at the Iron Gate east of Belgrade, Yugoslavia, where the Danube squeezes through the Carpathian Mountains. These whirlpools, called "Lepenski Vir," vanished in 1970 when a dam was built. In the sand of shallow coastal waters "kolks" are generated by the tides (Fig. 7.37b). They too

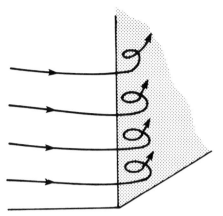

Figure 7.36. Corner vortex.

have their cause in the secondary flow of vortices, which are created at irregular spots in the ground. Such vortices can also be produced by the sudden break-up of an irregular or oscillating turbulent boundary layer. Violent upward motion is connected with the break-up, and the local "boiling" at the surface of rivers and tidal currents is a sign of this phenomenon [65]. Larger tidal vortices in the ocean will be described in Section 10.6.

7.6. INSTABILITY OF ROTATING FLUIDS

As in flows of nonrotating systems, instability can occur in rotating media, and the variety of forms and the beauty of vortices are here far greater. Lord Rayleigh [66] showed in a simple way the conditions under which a rotating fluid becomes unstable. If the equilibrium of a particle on a circular path is disturbed, it follows from the law of angular momentum that the motion becomes unstable when

$$(r_o V_o)^2 < (r_i V_i)^2.$$

This means that the fluid becomes unstable if the angular momentum of the particle in the inner circle with radius r_i and velocity V_i is larger than that of the particles in the outer circle with r_o and V_o. The particles on the inner circle push to the outside harder than those on the outer circle. The fluid between two rotating coaxial circular cylinders is an example (Fig. 7.38). This problem was first studied by G. I. Taylor [67]. He found that the flow between rotating cylinders becomes unstable if the inner cylinder rotates faster than the outer one. For creeping flow a pure shear motion is observed. With increasing angular velocity of the inner cylinder the flow becomes unstable. Alternating vortex rings are formed which are called Taylor vortices. In Fig. 7.39 curves are drawn that separate the stable from the unstable flow region. This new state, rotation with Taylor vortices, does not last. If the rotation of the inner cylinder continues to increase, the Taylor vortices also become unstable, and oscillations in the azimuthal direction appear. The new stability curve is also drawn in Fig. 7.39. The instability of the Taylor vortices is sketched in Fig. 7.40.

Taylor vortices occur not only between concentric circular cylinders; they can form during the spin-down of a container (Fig. 7.19) near the wall and lid where the largest differences in velocity occur. At every concavely curved wall, Taylor vortices can form parallel to the flow direction. In this case they are called "Görtler vortices" after H. Görtler, who first studied them (Fig. 7.41) [68]. Note that they protrude beyond the boundary layer.

The twisted Ekman layer of a rotating fluid on a solid surface (Fig. 7.8) can become unstable exactly as does the plane boundary layer. This problem is similar to that of an unstable boundary layer on a rotating disk (Fig. 6.38). Within the Ekman layer, Faller, Kaylor, Tatro, and Mollö-Christensen [Reference 40, Chapter 1] have found two types of cylindrical instability vortices (vortex rolls). Vortices of the first type, which occur first when the flow becomes unstable and are called vortices of "class A," move with a wavelike motion in the radial direction toward the center of rotation. The axes of the vortices form an angle between 0° and −8° with the tangential velocity V of the fluid (Fig. 7.42). Vortices of class B also migrate radially to the center but with an angle of 14°. The velocity of propagation of class B vortex rolls and their distances from each other are smaller than for the vortices of class A.

In vessels whose geometry deviates from the straight circular cylinder and flat disk, complicated flow instabilities of the Taylor and Ekman types can appear. An example is the flow between two concentric spherical shells rotating with different velocities. The vicinity of the equator is similar to the region between two rotating cylinders. One thus expects Taylor instability if the inner sphere rotates faster than the outer shell. The polar region is similar to the area between two parallel rotating disks. Here, the occurrence of Ekman instability is expected. Since the velocity of the inner sphere $V = R_i \Omega_i \cos \phi$ has its largest value at the equator, instabilities always occur first in this region. In the stable range below the limiting curve (which is similar to the lower curve in Fig. 7.38) fluid particles move spirally toward the outer shell from the equator to the poles; near the inner sphere they move in the opposite direction. The meridional flow for a quadrant is shown in Fig. 7.43.

According to Sawatzki and Zierep [70] five different flow configurations are distinguished beyond the limiting curve. They can occur, surprisingly, with equal boundary conditions and equal Reynolds number, and depend according to Wimmer [71] only on the initial acceleration, that is, on the acceleration from rest to constant rotation Ω_i. These flow states are briefly described as follows:

State I: At a sudden acceleration the fluid has apparently no time to form vortices. A state without vortices is established which is similar to the subcritical state in Fig. 7.4. Beyond $Re_i = 5500$ vortices appear around the poles and their axes form an angle of ca. 12° with the tangential velocity. These vortices probably belong to class B of the Ekman instability (Fig. 7.41). At about $Re_i = 150,000$ the flow becomes turbulent.

State II: The polar vortices in State I are labeled State II according to [70]. They also occur in all other states as transient forms (superposition) for $Re_i \geq 5500$.

State III: At very slow acceleration, two Taylor vortices are generated north and south of the equator. In the remaining region fluid flows as in the subcritical state (Fig. 7.44a). It may be mentioned that the main flow and not the centrifugal force determines the rotational direction of the Taylor vortex. State III is very stable up to turbulence.

State IV: If the initial acceleration is increased (compared to State III), the number of Taylor vortices doubles (Fig. 7.44b). This state passes over to State III at higher Reynolds numbers.

State V: This flow configuration appears in an acceleration range, which lies between that of States I and IV. It is unsteady and consists of periodic vortex rolls that migrate from the equator to the poles.

For a small width of the gap between the spherical shells, that is, for $(R_o - R_i)/R_i \leq 0.1$, the flow patterns simplify. Instability generates first Taylor vortices of State III. With increasing Reynolds

Figure 7.37. (a) Circular "potholes" and (b) kolks formed by periodic vortices at the Irish coast (copyright of Fig. 7.37b → H. Glatz, Neustadt, Germany).

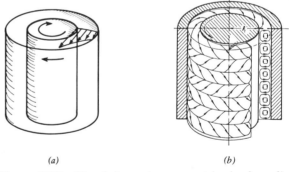

Figure 7.38. Flow between two concentric circular cylinders. The inner cylinder rotates faster than the outer one. (a) For creeping flow a pure shear motion is observed. (b) Through instability Taylor vortices occur.

number the Taylor vortices multiply but do not reach the polar region. With further increase of the Reynolds number wavelike disturbances in the Taylor vortices occur, as sketched in Fig. 7.40. Finally, the turbulent state is approached.

The stability of these states (at least, states I, III, IV, and V) at one and the same Reynolds number

Instability of Rotating Fluids 147

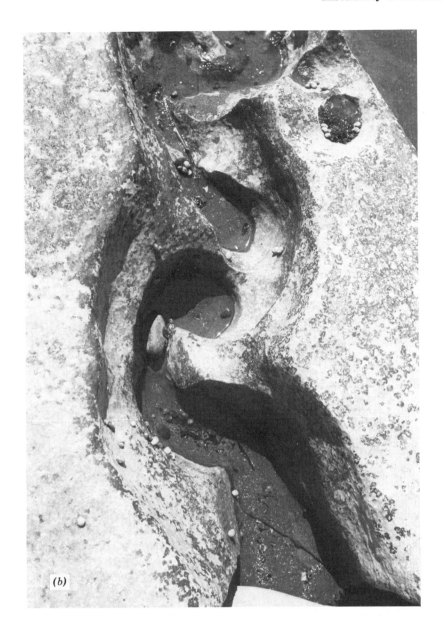

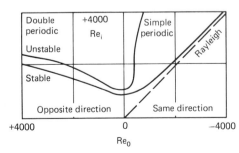

Figure 7.39. Stability curves for two rotating concentric cylinders.

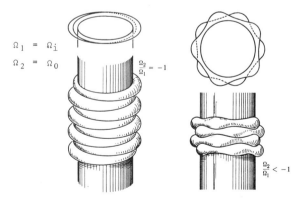

Figure 7.40. Taylor's original sketches for the instability of Taylor vortices (from Reference 67).

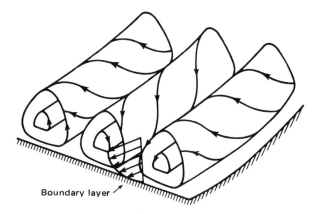

Figure 7.41. Görtler vortices.

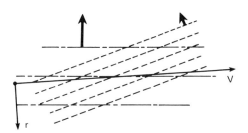

Figure 7.42. Vortex rolls of classes A and B that are generated by instability of the Ekman layer. In the schematic representation, the V arrow is directed toward the angular velocity, the r arrow is directed away from the center of rotation (from H. P. Greenspan [Reference 40, Chapter 1]. *The Theory of Rotating Fluids*, Cambridge University Press, 1968).

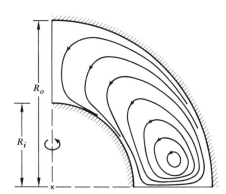

Figure 7.43. Meridional flow between two concentric spherical shells, the outer one being at rest and the inner one rotating. Re = Ek^{-1} = $R_i^2 \Omega / \nu$ = 250 with $R_o = 2R_i$ (adapted from C. E. Pearson, *Journal of Fluid Mechanics*, Cambridge University Press [69]).

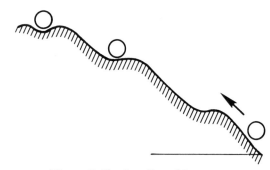

Figure 7.44. Meridional flow with Taylor vortices: (*a*) in State III, (*b*) in State IV (from O. Sawatzki and J. Zierep [70]).

Figure 7.45. Locally stable states.

may be explained by the fact that the states are locally stable from the thermodynamic point of view (see Remark 1 in Chapter 6) in a state diagram. This may be imagined in the following naive way: A sphere is rolled up a steep hill in which there are troughs. The "state level" (Fig. 7.45) the sphere will reach will depend on the initial conditions (see also Roesner [72]).

In containers other than those of cylindrical and spherical shapes the fluid motion can become quite involved. Dolzhanskii et al. [73] and Roesner [74] have investigated the spin-up and spin-down processes of fluids in ellipsoids (fluid gyros). Here again, the final state depends on the way the vessel is accelerated.

Turbulence decays faster in a rotating fluid than in a nonrotating one. This may probably be explained by the transport of energy in inertial waves to the edge of the rotating fluid [75]. For weak turbulence the two-dimensional constraint acts in such a way that isotropic turbulence becomes anisotropic and in such form that the turbulent eddies align themselves parallel to the axis of rotation. During this process strong discrete vortices are generated that are statistically distributed. This phenomenon probably plays a large role in the development of atmospheric vortices (Section 11.3) [76].

Remarks on Chapter 7

1. Foucault's pendulum (1851) is a simple pendulum that can swing freely in all directions. If it is swung in one plane at the start, this plane changes during the day, and that in 24 hours by

$2\pi \sin \phi$, where ϕ is the latitude. At the poles, thus, the plane rotates once a day. In "absolute space" the direction of the plane remains unchanged while the earth rotates below it.

2. Here, gas is not meant in today's physical sense. Descartes distinguished three kinds of matter that have no relation to the state phases of matter.

3. The absolute value of the earth's rotation has a fundamental influence on human life. Not only is life pulsating in a rhythm of 24 hours, the life expectancy depends also on this value (according to J. Aschoff, Max Planck Institut at Erling—Andechs, Germany).

4. Sand dunes in the desert, which are not aligned perpendicular to the wind direction, can be explained by the Ekman layer [77].

5. The rotational direction of tornadoes depends in most cases on the vorticity of larger systems that are under the influence of the Coriolis force. Therefore, preferred rotational directions are also found in tornadoes (Section 10.3).

6. If one neglects the shearing of the flow, the vorticity is equal to the curvature of the streamline times flow velocity. Thus when the sign of the curvature changes, the vorticity also changes its sign.

7. A physical event like the occurrence of waves and discontinuity lines in rotating media is, of course, independent of the choice of the reference frame in which it is described. Wavelike disturbances on earth, however, would be difficult to study in an inertial system. It is thus advantageous to chose a reference frame rotating with the earth and then to treat the disturbance separately from the earth's rotation. A perturbation technique for the equations of motion leads to a hyperbolic equation that can describe atmospheric and oceanic waves directly.

8. In Fig. 7.12 the flow field is drawn without frictional effects. Friction causes tip vortices during the up-and-down motion and shear layers along the characteristics.

9. Depending on the magnitude of the Rossby number and the ratio of bottom length to height, the formation of more than one cell is possible. For instance, the meridional flow for Ro = 4, Ek = 0.01 has two cells [29].

10. Theories on the cause of this reversal are controversial. See also Reference 78.

8. Stratification in Ocean and Atmosphere

8.1. THE EARTH'S AIR AND WATER ENVELOPE

The layer of air and water that surrounds the earth is relatively so thin that on a desk globe it would correspond to a coat of paint. All motions of air and water take place in this thin atmospheric and oceanic layer, and because it is so thin, vortices of global extension are disklike (Fig. 2.4a), as photographs of the earth from satellites confirm (Fig. 8.1).

It follows that convection flows and turbulent eddies are correspondingly distorted, that is, their horizontal extensions and velocities are larger than those in the vertical direction. However, the thinness of the layers does not mean that vertical flows in global circulations are of only secondary importance (Section 9.1). In local flows they are decisive and can be of considerable strength, as thunderstorms and tornadoes show.

The vertical structure of the oceans and atmosphere is of great significance in understanding water and air motions. This structure is determined essentially by the gravitational field of the earth, which prevents water and air from escaping from earth and presses them against the earth's surface. Gravitation thus dominates over the centrifugal force of the earth.

At sea level a column of air presses on a square centimeter with an average weight of 1.033 kp, corresponding to the pressure of a 760-mm high column of mercury. Torricelli demonstrated the balance between the column of mercury and the atmospheric air pressure in 1643 in Florence with his "barometer." In meteorology the millibar (mb) is also used as a unit of pressure. At sea level 1013 mb or 1.033 atm correspond to a 760-mm column of mercury.

With increasing height the pressure diminishes by a factor of 10 every 10 miles. In water the pressure increases by 1 atm for every 10 m. At the deepest place in the ocean, the Mariana trench, which is 10900 m deep, the pressure is enormous—1090 atm.

The density of the atmosphere and oceans depends not only on the pressure but also on the temperature and the composition of matter. These two state variables do not change monotonically with height (or depth) as does the pressure. Figure 8.2 shows a temperature profile of the atmosphere, which of course represents only certain average values (standard atmosphere). On this profile, four different layers are distinguishable.

The sun's rays in the visible and infrared range are absorbed by the earth's surface and there the air temperature reaches a relative maximum. With height the temperature decreases by 0.6°C per 100 m up to the tropopause, which forms the boundary between troposphere and stratosphere. The troposphere is about 16 km thick at the equator and 8 km at the poles. Within the stratosphere the temperature is constant at the lower levels and then increases up to an altitude of about 50 km, where it reaches the temperature of the earth's surface. This increase can be explained by the absorption of the sun's ultraviolet radiation.

In the mesosphere the temperature decreases again and then increases in the thermosphere through absorption of short-wave radiation (e.g., X rays). Here, however, the air is so thin that the high temperature is of minor significance. Thus, it can be seen that the atmosphere, which becomes thinner

The Earth's Air and Water Envelope 151

Figure 8.1. Earth as seen from an Apollo spaceship (NASA photograph).

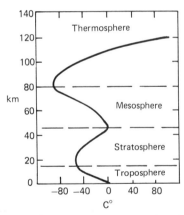

Figure 8.2. U.S. standard atmosphere (1962). More than 75% of the atmospheric air in the midlatitude and 90% in the tropics are in the troposphere.

with height, filters out certain wave ranges of solar radiation, resulting in an uneven temperature distribution.

Locally and with time, temperatures can deviate more or less from those of the standard atmosphere. In the troposphere the temperature decreases on the average by 0.6°C per 100 m, but if it drops more rapidly, the upper air layer becomes unstable. With increasing temperature ("inversion") the air layer becomes so stable that it forms a barrier to upward motion.

Pressure and temperature fields are related to each other. If the isobars are parallel to the isotherms, or, equivalently, if the density is a function only of the pressure, the state of the air is called "barotropic." The nonbarotropic case is called "baroclinic."

The gaseous composition of air is of minor importance in the study of vortices. In the few cases in which it is significant, for instance in the generation of vortices in dusty air, special attention will be given.

As is well known, water has its highest density at +4°C. This "anomaly of water" is of great significance in nature since it hinders the solid freezing of lakes and ponds. At the freezing point water is lighter than at +4°C so that water of +4°C sinks to the bottom and ice is formed first at the surface. If the bottom is frozen and the water surface is warmer than +4°C, then water with the largest density will be between the bottom and the surface, and the water will be unstable between the bottom and the layer of +4°C (Fig. 8.3).

152 Stratification in Ocean and Atmosphere

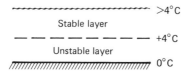

Figure 8.3. Water above a bottom of 0°C is unstable because of the anomaly of the water.

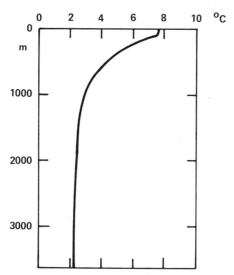

Figure 8.4. Typical temperature profile of the ocean.

A typical temperature profile of the ocean is given in Fig. 8.4. Layers of greatest changes in temperature are called "thermoclines." They owe their existence to convection flows that are generated through wind or temperature differences at the water surface [1].

Material composition plays a larger role in the development of currents in seawater than it does in the atmosphere. This will be pointed out in Section 9.3. The salinity of the ocean averages $35^0/_{00}$, that is, 35 g per 1000 g water.

8.2. UPWARD AND DOWNWARD MOTIONS

Changes in density in a fluid can occur in various ways: through change in temperature, pressure, and material composition. Warm air is lighter than cold air; seawater with high salinity is heavier than that with lower salinity. Density changes can lead to an unstable state, and the fluid will then not remain at rest but will start to move to reestablish stability. The resulting exchange of heat or matter is called "convection" to distinguish the process from "diffusion" (Section 4.1). Heat convection, first discovered by Count Rumford in 1797 [2], is the most important fluid motion in nature. The global circulation of the atmosphere is maintained by the differential heating of the earth by the sun. Another example is the slow convection in the mantle of the earth.

It may first be assumed that the fluid changes temperature only locally. When one makes an open fire, a ring-shaped circulation is generated by the buoyancy of the heated air (Fig. 2.8). Depending on the magnitude of the temperature difference between heat source and surroundings (the appropriate parameter will be given later), the buoyancy will be correspondingly large. During a heavy fire storm in Tokyo in 1923 velocities of 70 m/s were estimated. In more recent memory are the area fires after air raids in World War II. One of the most terrible fire storms occurred in Hamburg on the 27th and 28th of July 1943. Dry summer weather with unstable air stratification supported the fire [3].

Less violent but of potential danger are smog layers over big cities. The sun heats urban areas more than the surrounding country. Hence, the air rises over the city and circulates in the form of a vortex ring (Fig. 8.5). Dust and exhaust gas accumulate in this ring until the polluted air is carried away by strong winds or cleaned by rain. Particularly dangerous are smog clouds which are stabilized by a low inversion and thus have a high concentration of dirt and exhaust gases [4].

Of a more peaceful nature are the upward and downward motions of the air over the countryside. Rivers, wooded areas, and open land are extended heat sources or sinks and generate rolls instead of circulation rings.

The daily change of land and sea breezes along coasts is another phenomenon caused by convection. During the night, the land cools more than the sea, and during the daytime the land is heated more than the sea. The developing temperature differences generate breezes toward the sea or the land. The monsoons are large-scale convections with semiannual periods (Section 9.1).

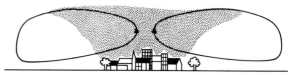

Figure 8.5. Air circulation in an urban layer of haze.

An upward flow is easily recognized if the humidity of the air is sufficiently high that water vapor can condense into droplets that form clouds. Impressive are the huge cumulus clouds before a thunderstorm. On sultry days in summer the heating of the ground by the sun can be sufficient to produce unstable air stratification. The ensuing convection can lead to a local thunderstorm, which is not the same as a "frontal thunderstorm" (Section 10.2). Local thunderstorms, thus, do not cause a change in weather, that is, they bring no change of air mass.

When the Coriolis force has an opportunity to act on the flow (for $Ro < 10^4$), an upward motion is twisted in such a way that a horizontal circulation with its axis perpendicular to the earth's surface is generated. Near the ground an Ekman layer forms. The rotation is cyclonic there because the fluid moves toward the center. Above a certain height the direction of rotation changes when the fluid moves away from the axis. The rotation is then anticyclonic. A particle path is sketched in Fig. 8.6. In the fire storm in Hamburg cyclonic rotation occurred after about 2 hours. The change in rotational direction with height is observed in hurricanes (Section 11.4).

There is an essential difference between the horizontal rotation in heat convection and the circulation in a rotating vessel. In the latter case the wall generates the meridional flow and the rotation of the vessel is the primary flow; in convection the horizontal rotation must first be created by the Coriolis force and is here the secondary flow.

Mathematically, a measure for the buoyancy relative to the inertial force is the Grashof number Gr:

$$\mathrm{Gr} = gb(T_1 - T_0)\frac{L^3}{\nu^2},$$

where g is the gravity, b is the expansion coefficient, and $T_1 - T_0$ is the temperature difference between two bodies 0 and 1. Another important parameter for the heat transfer is the Prandtl number, which is a measure of heat convection relative to heat conduction:

$$\mathrm{Pr} = \frac{\nu}{a}.$$

The coefficient a is the thermal diffusivity. The Prandtl number depends only on the material properties. For air it is approximately $\mathrm{Pr} = 0.7$, for water $\mathrm{Pr} = 7$. Instead of the Grashof number the Rayleigh number Ra can be used:

$$\mathrm{Ra} = \mathrm{Gr}\cdot\mathrm{Pr}.$$

Heat convection in a cylindrical box may be used as an example. The walls of the box have a temperature T_0, a plate in the middle of the bottom a temperature $T_1 > T_0$. The height of the box is L. For the special case $\mathrm{Gr} = 4 \times 10^6$ and for air, the steady laminar streamlines were determined with the aid of a computer and are displayed in Fig. 8.7.

The study of buoyancy flows over a horizontal plane without side walls is important for the description of convection in the atmosphere and in the ocean [7]. When a local heat source exists, various types of convection may be distinguished (Fig. 8.8).

Figure 8.6. The Coriolis force generates a horizontal circulation in an upward flow. The direction of rotation is valid for the northern hemisphere (from Reference 5).

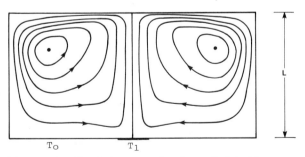

Figure 8.7. Steady buoyancy flow in a cylindrical box, generated by a flat plate in the middle of the bottom. Meridional cross section of streamlines for $\mathrm{Gr} = 4\times 10^6$ and $\mathrm{Pr} = 0.7$ (computer-generated picture by K. E. Torrance and J. A. Rockett [6]).

154 Stratification in Ocean and Atmosphere

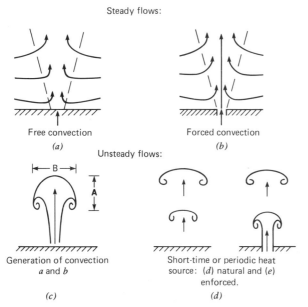

Figure 8.8. The various types of steady and unsteady convection flows over a local heat source.

Examples of steady natural convection flows (by buoyancy only) are given in Figs. 8.5 and 8.6. In "forced" convection heat is not transported by buoyancy alone but also by an independently produced jet, for example, by a fan.

The steady convection flows in Figs. 8.8a and 8.8b are initiated by a starting vortex (Fig. 8.8c) Short-time or periodic heat sources generate vortex rings (Fig. 8.8d and 8.8e) (Fig. 8.9). Natural vortex rings are also created in unstable fluids over uniformly heated horizontal planes with local disturbances of the fluid. These vortices are observed during the formation of ice, in gas absorption in liquids, and in the atmosphere on hot days when the ground is strongly heated. It is believed that the main part of heat transfer from a uniformly heated horizontal area takes place through warm air vortices, at least in the Rayleigh number range 10^5–10^7 [8].

Birds cleverly exploit these warm air vortices for soaring [9]. The birds circle inside the vortex ring in the ascending jet and chose their position in such a way that they are carried upward with the vortex

Figure 8.9. Photograph of a smoke ring formed at the opening of a chimney (courtesy of G. F. Stork, Chevy Chase, Maryland).

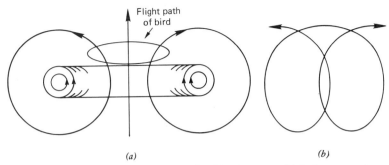

Figure 8.10. (a) The flight path of a soaring bird in the upward jet of a warm-air vortex. The vortex itself moves upward through buoyancy and carries the bird with it. The diameter of the flight path of vultures is of the order of 30–40 m. It decreases with the strength of the side wind. (b) In a side wind the flight path of soaring birds is a trochoid as seen from the earth.

ring (Fig. 8.10). Numerous data have been collected from observations of soaring vultures in the eastern United States and in India. In a side wind the flight path of these birds is not circular, but trochoidal when seen from the ground (Fig. 8.11).

Steady and short-time convection flows differ in the following properties: The fluid in jetlike flows mixes with the surrounding fluid resulting in an increase in the width and mass of the jet. The flow velocity, however, decreases (Figs. 8.8a and 8.8b). The boundary of the jet is sharper in the turbulent state than in the laminar case. In vortex rings, however, mixing with the surroundings is small and is essentially confined to the developing stage (Section 5.7). In Fig. 8.12 isotherms are drawn at various times which indicate the sucking-in of surrounding fluid. The upward velocity of the "front" is plotted against time in Fig. 8.13. According to Shlien [10] four different phases may be distinguished:

1. The heat transfer in the fluid starts by diffusion.
2. Convection takes over the transport. The acceleration is almost constant.
3. A period of approximately constant velocity follows.
4. The vortex decays. The various types of decay will be illustrated further in Fig. 8.15.

It may be mentioned that the form of the vortex ring differs from that of the starting vortex in Fig. 8.8c. The starting vortex in steady, laminar convection is not as flat as the warm-air vortex. The ratio $A:B$ in Fig. 8.8 is about 0.9 for the starting vortex, about 0.4 for the warm-air vortex [10].

Owing to the influence of the Coriolis force, a horizontal vortex can develop in a steady buoyancy flow (Fig. 8.6). Other influences can also transform the convection jet in Figs. 8.8a and 8.8b into a strong columnar vortex. For instance, previously existing background vorticity can concentrate to a vortex (as in the case of the bathtub vortex, Section 3.4), or obstacles can create a rotational component.

An interesting way to produce rotation of a fluid in a container is to move a heat source, for instance a flame, underneath it on a circular path. The temperature wave in the fluid, generated by the heat source, induces the fluid to rotate. This "moving-flame experiment" has been used to explain the fast rotation of the upper atmosphere of the planet Venus [12, 13].

Flows generated by the density difference of two media can behave in a way similar to heat convection. Easy to observe are immiscible unstable fluid layers separated by a distinct boundary line. This situation is called "Rayleigh–Taylor instability." Figure 8.14 shows the development of a downward flow of heavier liquid into a lighter one. The smaller the density difference, the more distinct the vortex ring forms.

Four cases may be distinguished for vortex motion between fluids of different densities:

1. The "heavy vortex" moves downward due to its own weight.
2. The "light vortex" moves upward due to buoyancy.

In forced motion through ejection of fluid from a nozzle two additional cases occur:

3. The heavy vortex moves upward against gravity so that it decelerates.

Figure 8.11. Leonardo da Vinci. Flight of birds in an updraft. *Codex Atlanticus*, folio 308 r-b. Ambrosian Library, Milan.

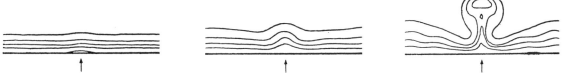

Figure 8.12. Isotherms at various times during the formation of a warm-air vortex (from J. W. Elder, *Journal of Fluid Mechanics*, Cambridge University Press [11]).

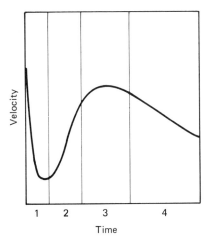

Figure 8.13. The four phases of a warm-air vortex. The upward velocity is plotted over the time (adapted from D. J. Shlien [10]).

4. The light vortex moves downward against buoyancy. It too decelerates.

The properties of the vortex rings are different in these four cases. For instance, the heavy vortex in case 1 is more unstable than the light one in case 2, since in case 1 the heavier fluid rotates, a situation which is more unstable. The same is true for the vortex ring in case 3 compared to that in case 4 [15].

The decay of a vortex ring depends not only on the density difference but also on the Reynolds number. Chen and Chang [15] have studied case 1 vortex rings produced by ejection of gas from an orifice. The density of the gas was 1.5 times greater than that of the surrounding air. Three ranges of Reynolds numbers were distinguished; the Reynolds number is defined by the exit velocity of the gas, the diameter of the orifice, and the kinematic viscosity of the air.

1. **Re < 1500.** After its development behind the orifice the vortex ring moves about 10 diameters downward without a noticeable change in its diameter. Wave-type disturbances in Fig. 8.15a develop to smaller vortex rings, which again decay in a cascading way.
2. **1500 < Re < 2000.** In this range disk-type vortices perpendicular to the original vortex appear after the first phase. They then decay in a cascading way (Fig. 8.15b).
3. **Re > 2000.** The decay of the vortex ring results in turbulence (Fig. 8.15c).

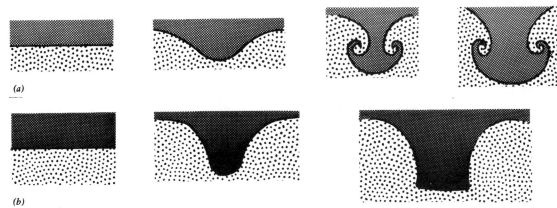

Figure 8.14. Temporal sequence of an unstable downward motion of heavy liquid into a lighter one. (*a*) The density ratio of heavy to light liquid is small and (*b*) the ratio is large (adapted from B. J. Daly [14]).

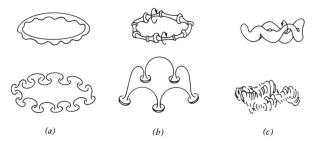

Figure 8.15. The decay of vortex rings for three different Reynolds number ranges (from C. J. Chen and L. M. Chang [15], The American Society of Mechanical Engineers).

Figure 8.16. Cascading decay of a sinking vortex ring.

For asymmetric or disturbed vortices, the cascading decay is irregular (Fig. 8.16) [16].

The sinking of fluid through the formation of vortex rings explains an interesting phenomenon observed years ago in Lake Zurich. The annual deposits at the bottom of the lake, which can be distinguished from each other like annual tree rings, showed that the deposit of the most minute particles still took place within a year, although according to the law of the terminal velocity of single particles they should have taken much longer. Such rapid deposit can be explained by vertical density flow. At the lake's surface dust and microorganisms collect in the water to give a higher density than that of pure water. This layer, of course, is unstable and following a small disturbance moves down like a jet. During this process the jet dissolves into vortex rings. The particles contained in the jet and the rings reach the bottom together much more quickly than would single particles [17].

Jetlike density flows are known not only from the fast deposit in lakes; they also play a large role in volcanic eruptions and in the development of tornadoes. These examples will be discussed in Section 10.3.

8.3. CELLULAR CONVECTIONS

When a thin horizontal fluid layer is heated uniformly from below, a stratified fluid develops whose density (in general) increases with decreasing temperature. This state is usually unstable. For small temperature differences between the lower and upper boundaries, viscosity and heat conduction are sufficient to maintain the state at rest. However, if the temperature difference is increased, a state is reached in which a small disturbance is sufficient to trigger a convection. Steady, stable roll-like vortices form as sketched in Fig. 8.17 and shown on the interferograms of Fig. 8.18. With further increase of the temperature (that is, at higher Rayleigh number), the rolls become unstable, too, and three-dimensional convection flows develop [18]. Their properties, however, have not yet been fully investigated. Rolls have been observed perpendicular to the original ones; Whitehead [19] called these rolls "bimodal." In addition, time-dependent states exist, in particular at still higher Rayleigh numbers, before the flow becomes turbulent. Best-known among the steady, three-dimensional flow configurations are hexagonal vortex cells that look like honey combs (Fig. 2.10) and are called "Bénard cells." Palm [20] considers them generated from the superposition of three rolls displaced by 120°. Two disturbance rolls are superposed on a steady roll, and the three join to form a hexagonal configuration. Bénard cells are preferred in nature when the fluid at rest has asymmetric properties below and above the horizontal middle line [19]. Such asymmetry occurs with temperature-dependent viscosity coefficients. In gases the viscosity increases with temperature; in liquids, it decreases. This difference explains the fact that in gases matter sinks at the center of the Bénard cells and rises at the walls, whereas in liquids the flow direction is reversed. In

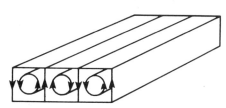

Figure 8.17. Straight rolls which form in an unstable fluid.

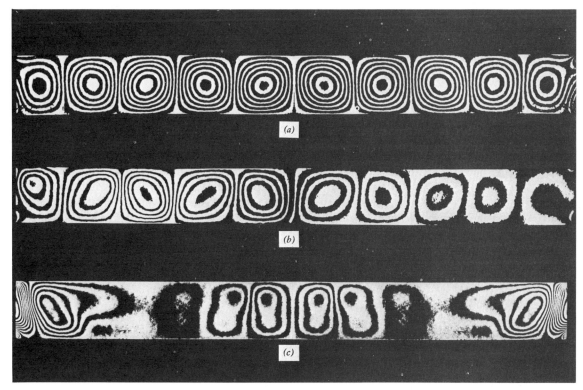

Figure 8.18. Interferograms of convection rolls: (a) with uniform heating of the lower plate, (b) with horizontal temperature gradient, and (c) with rotation (from H. Oertel, jun., University of Karlsruhe).

convection with surface tension Bénard cells can form even below the critical Rayleigh number. This behavior is explained by the Marangoni effect, which is described at the end of this section.

Bénard cells are easily made visible in a cup of coffee. If boiling water is poured on instant coffee and immediately afterward cream is added, the cell structure is clearly observable at the surface of the liquid (Fig. 8.19). The cells have a diameter of about 1 cm, and the downward motion occurs at the walls of the cells. The unstable stratification is generated by evaporation at the surface.

Weber [21] probably first mentioned the cell structure of the fluid in 1855 but he interpreted it incorrectly. Careful experiments were done and more profound knowledge was obtained by Henri Bénard [22] whose name is connected with the hexagonal cells. The first theoretical studies were made by Lord Rayleigh [23] in 1916. Further historical data may be found in Reference 24.

The transition from the fluid at rest to convection is expressed quantitatively by the Rayleigh number. The following critical values are obtained for different kinds of boundaries [25]:

1. Solid walls at the lower and upper boundaries: Ra = 1700.
2. Both boundaries are free surfaces: Ra = 658.
3. The lower boundary is solid, the upper one is free: Ra = 1100.

For case 1, the regions of the Rayleigh numbers for the various flow configurations are presented in Fig. 8.20 as a function of the Prandtl number.

The structure of the convection flow can be changed in various ways. The influence of surface tension has already been mentioned. Convection in a linear shear flow (Couette flow) causes longitudinal

Figure 8.19. Bénard cells at the surface of hot coffee.

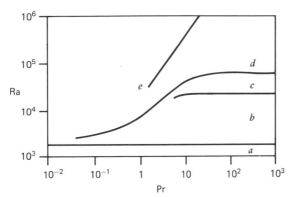

Figure 8.20. The various forms of convection between two solid plates, the lower warmer than the upper. The following flow regions are distinguished: (a) motionless; (b) two dimensional; (c) three dimensional, steady; (d) three dimensional, unsteady; and (e) turbulent (adapted from R. Krishnamurti [26]).

rolls (Fig. 8.21). In a parabolic shear flow (Fig. 6.58a) heat is transported in longitudinal rolls for Pr = 1 up to Re = 5400, above those values in transverse rolls. In the first case, the instability is of thermal origin; in the second case, the shear layer becomes unstable [24].

The magnitude and rotational direction of the rolls are also influenced when a horizontal temperature gradient exists in addition to the vertical temperature difference. Then, a basic circulation in the whole flow domain occurs, and the direction is clockwise if the higher temperature difference occurs at the left boundary, as in Fig. 8.22 (see also Fig. 8.18). Superposed on this basic circulation are rolls that are weakened where the rotation is opposite to the basic flow direction. The number of possible convection states increases when such horizontal temperature gradients are considered [20, 27].

In nature rolls and Bénard cells occur frequently, but they are in most cases invisible. In the upper layers of the ocean, in lakes, and in the atmosphere unstable fluid layers can form, and there rolls and Bénard cells occur. In the atmosphere they may become visible through the formation of clouds. Mackerel clouds and cloud bands are typical evidence of

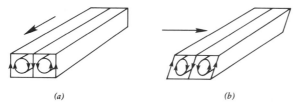

Figure 8.21. (a) Longitudinal rolls and (b) transverse rolls in a shear flow.

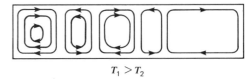

Figure 8.22. Steady rolls between two plates with horizontal temperature gradient (from J. Zierep [27]).

this kind of instability [28]. In land and sea breezes rolls of about 2 km thickness are also observed [29].

In the earth's mantle, convection cells are created by the temperature difference between the hot interior and the cooled surface of the earth. The influence of these cells on the earth's crust, particularly on the continental drift, will be discussed in Section 9.5.

Fig. 8.23 is a photograph of the hexagonal cell pattern formed during the dry-up of a salt lake.

Granulation at the surface of the sun is merely a system of Bénard cells (Fig. 12.3).

Similar to the cascading decay of sinking vortex rings in a lake caused by the accumulation of dust and microorganisms and described at the end of Section 8.2, microorganisms can concentrate to an unstable layer, and cellular convection can be initiated. This fluid motion is called "bioconvection" [Reference 14, Chapter 4].

A special kind of convection cell develops in unstable water layers below a sheet of ice. These cells are not hexagonal but circular. They melt cavities in the ice sheet in the form of spherical caps (Fig. 8.24). The cells are not steady but grow in time by eating into the ice.

In this context a type of cellular flow is mentioned that originates in a local change of surface tension (Marangoni effect). In the presence of temperature or concentration differences, a gradient of the surface tension appears that represents a driving force for the liquid and that is balanced by viscous stress in the steady state. The generated liquid circulation is called "Marangoni convection" or "Marangoni vortex" and plays a significant role in the engineering process of crystal growth, in which undesirable liquid motion and heat transfer of the molten material must be controlled.

In investigating the Marangoni convection experimentally, the influence of buoyancy must be minimized. Figure 8.25a shows an arrangement by Chun [31] to verify the existence of such a phenomenon. The apparatus consists of two coaxial cylindrical copper rods of 3 mm diameter which can be heated. The gap between the rods is filled with a liquid (in

Figure 8.23. Arid salt lake in East Africa (Afar). (Copyright G. Gerster, Zumikon-Zürich, Switzerland.)

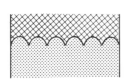

As seen from above

Figure 8.24. Convection cells in an unstable water layer under an ice sheet. The cells are not hexagonal but spherical (from Y. Yen [30]).

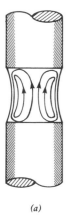 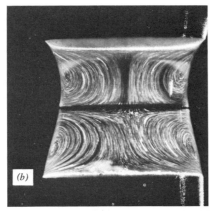

(a) (b)

Figure 8.25. (a) Marangoni vortex ring between two coaxial rods depicted in the meridional plane. (b) Photograph of two Marangoni vortex rings produced by a ring heater between the two rods (from C. H. Chun, DFVLR, Göttingen).

Chun's experiment with silicone oil). In Fig. 8.25a the upper rod is warmer than the lower one, and the direction of the Marangoni vortex is depicted in the meridional plane. Changing the sign of the temperature gradient reverses the flow direction. Beyond a critical temperature difference the steady flow circulation becomes oscillatory.

Two Marangoni vortex rings can be produced by a ring heater located at the midplane between the rods (Fig. 8.25b). This arrangement is of advantage in the technology of crystal growth and will briefly be discussed in Section 8.6.

8.4. VORTICES AND VORTICITY IN STRATIFIED FLOWS

Density changes in a fluid, created through differences in temperature, pressure, or material composition, may cause motions that do not occur in homogeneous media. Such fluids and flows with local density variations are called "stratified"; the atmosphere and ocean are examples of stratified media. The most important property of stratified fluids is their ability to transport energy in the form of waves.

The simplest and perhaps oldest perception of wave motion is that of a disturbed water surface. At the border between water and air, the density changes abruptly. If water particles are brought out of their vertical equilibrium position through a disturbance, they oscillate due to the action of gravity and transfer energy to neighboring particles. This process of energy transport is called a "wave."

At this point an essential difference between a wave and a vortex may be mentioned. A wave transports (almost) no matter but does transport energy; a vortex carries matter as well as energy with it. This concept of a wave must be distinguished from that of the streamline wave, which follows from the superposition of a vortex on a parallel flow (Fig. 2.14).

First a homogeneous fluid is considered with a free surface on which a disturbance generates waves. If the distance L from the surface to the bottom is small compared to the wavelength, the propagation velocity of the wave is $\sqrt{gL}$, where g is the gravitational constant. The ratio of the flow velocity V to this propagation velocity is called the "Froude number." It can be shown that

$$(\text{Froude number})^2 = \frac{\text{inertial force}}{\text{gravity force}}.$$

The flow over an obstacle can be characterized as either of two types of motion, depending on whether the Froude number is smaller or larger than unity. If it is smaller, one speaks of a subcritical motion; if larger, of a supercritical motion.

In the atmosphere and ocean the density differences are not abrupt as they are at a free water surface, but are gradual. Such a continuously stratified fluid was tacitly assumed in Sections 8.1 through 8.3, where natural upward motions were mainly considered; now the forced flow of stratified media will be described. These media will be considered as stable for the time being. If from the equilibrium state the position of a fluid element shifts in the direction of gravity, and if the particle is not constrained, it will oscillate. Hence, the fact that during the flow past an obstacle a perpendicular velocity component appears explains qualitatively already the occurrence of waves in a stratified medium [32]. The frequency N of the oscillation, called the "Brunt–Väisälä frequency," is $N^2 = -(d\rho/dz)\,g/\rho$. ρ is the density of the medium, and $d\rho/dz$ is the density change with height z. The critical velocity $\sqrt{gL}$ of the surface waves corresponds here to the expression NL, where L is a characteristic vertical length. The "internal" Froude number is V/NL [33].

Waves downstream of a mountain or ahead of a cold front are called "lee waves." In the 1920s and 1930s glider pilots participated in the study of these waves and discovered that upwinds occur not only in front of but also behind a mountain. The first investigations in Germany are connected with the names of Georgii and Koschmieder. Also mentioned are the soaring flights of Küttner, who in 1937 reached an altitude of 8000 m in the Riesengebirge, and of Klöckner, who 2 years later soared to 11,400 m in the eastern Alps [34]. Birds as well as gliders use the upwinds of lee waves.

The internal Froude number allows the characterization of five different kinds of flows over an obstacle, as sketched in Fig. 8.26 For a homogeneous fluid is $N = 0$ and hence Fr = ∞. Here, of course, no waves occur (Fig. 8.26a). At high Froude numbers, lee waves appear whose behavior depends on the velocity profile upstream. It is remarkable that the wake vortex vanishes, and that, instead, a strong downward flow occurs immediately behind the obstacle (Fig. 8.26b). These downward currents can cause great damage in the mountains [35]. With further decrease of the Froude number vortices appear in the lee waves, which are called "rotors." Under the rotors at ground level the flow direction is reversed (Fig. 8.26c). At small Froude numbers a jetlike flow develops, which is similar to the water jump of surface waves (Fig. 8.26d). For very small Froude numbers "blocking" occurs (Fig. 8.26e). This behavior is similar to the Taylor–Proudman column in rotating fluids (Section 7.4).

An idea of the magnitude of lee waves is provided in Figs. 8.27 and 8.28. The maximum upward velocities in lee waves are of the order of 20 m/s. Lee waves can also occur in the ocean behind sea mountains.

Stratified flows have another important property in addition to their ability to produce waves: They

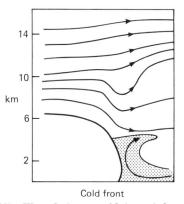

Figure 8.27. Wave before a cold front (adapted from W. Georgii [36]).

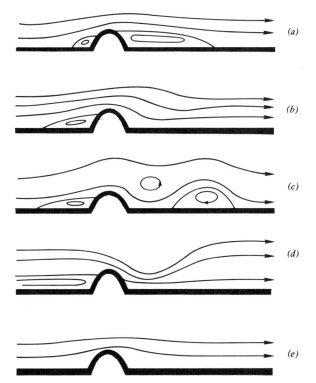

Figure 8.26. The various types of flows around a mountain range. Besides the Froude number, the wind profile upstream influences the motion.

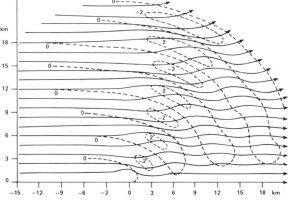

Figure 8.28. Streamlines (solid) and equivorticity lines (dashed) for lee waves in a baroclinic, inviscid fluid (adapted from P. Queney [32]).

can generate vorticity. So far, discussion of the creation of vorticity has been restricted to surfaces (Section 4.1); moreover, relative vorticity can be generated in homogeneous rotating fluids at the expense of vorticity of the rotating system. Then the whole process is only a redistribution of vorticity; the absolute vorticity remains constant (Section 7.4). In stratified flows, however, vorticity can also be generated inside the fluid, provided that the fluid is baroclinic [37]. This process is called "baroclinic generation of vorticity."[1] In Fig. 8.28 streamlines and equivorticity lines are displayed for lee waves in an inviscid medium.

In a stratified viscous fluid vorticity is generated both at a solid wall and in lee waves (and also through Helmholtz instability). Local extrema of the vorticity are no longer unique indications of the presence of vortices but may also mean the existence of lee waves. The interpretation of a vorticity field thus becomes quite difficult. An example of the combined surface-generated and baroclinic vorticity is given in Fig. 8.29.

Additional differences between homogeneous and stratified flows are briefly enumerated in the following paragraph.

Jets in a stable stratified medium are narrower and longer than in a homogeneous one if the density of the jet's fluid is greater than that of the surrounding medium, as for example, a water jet in air. The decay of vortices is delayed in a stratified fluid. The behavior of turbulent jet flows in stratified media is important for the study of air pollution. At small Froude numbers, that is, for strong inversion in the atmosphere, vertical turbulent mixing of chimney smoke is prevented, and in weak winds the smoke is carried away horizontally. With unstable stratification, on the other hand, strong vertical turbulent heat convection takes place (Fig. 8.30).

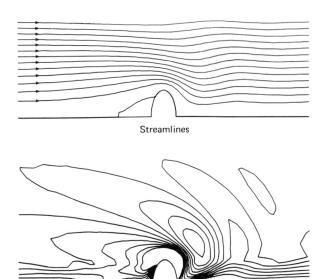

Figure 8.29. Lee waves in a viscous fluid (computer-generated picture by H. J. Haussling [38]).

8.5. ANALOGY BETWEEN ROTATING AND STRATIFIED FLUIDS

The attentive reader may have noted that a number of properties typical of stratified fluids are remarkably similar to phenomena in rotating systems. The analogy can under certain conditions be pushed even to the same mathematical formalism. The reason for this close relationship is that in both cases restricting forces of similar kinds act on the fluid. In a rotating medium it is the Coriolis force; in a stratified fluid, gravity.[2] These two forces act to restore the disturbed equilibrium of a fluid element. Like a spring, the restoring force generates oscillations (Fig. 8.31). Waves can, therefore, be generated in rotating and stratified fluids. They behave hyperboli-

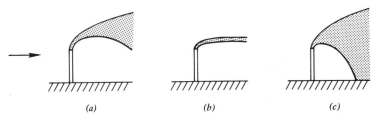

Figure 8.30. Schematic representation of air pollution from smoke stacks under various conditions of atmospheric stratified flows: (a) turbulent spreading in homogeneous air, (b) at strong inversion, (c) at unstable stratification (adapted from H. A. Panofsky [39]. Reprinted by permission, *American Scientist*, Journal of Sigma Xi, the Scientific Research Society).

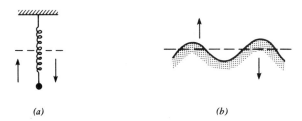

Figure 8.31. Oscillating systems: (*a*) spring and (*b*) surface wave. The Rossby wave is drawn in Fig. 7.9.

cally (Section 7.4). Lord Rayleigh [23] first pointed out this analogy between rotating and stratified fluids in 1916.

The restriction that a hyperbolic system puts on fluid motion is of great significance in meteorology, oceanography, and astrophysics. The following important characteristics of rotating and stratified fluids will, therefore, be discussed.

1. Disturbances in rotating systems are subject to the two-dimensional constraint (Taylor–Proudman theorem). The corresponding phenomenon in stratified fluids is the blocking effect.
2. Instability of rotating fluids in cylinders leads to the development of Taylor vortices. The analogous phenomenon in stratified media is the Bénard convection. This analogy is also valid for spherical Taylor instabilities, which in stratified fluids correspond to convection rolls between two plates with a horizontal temperature gradient [27].
3. The spin-up process (Fig. 7.18) in rotating fluids has an analogy in stratified fluids in the "heat-up" process [40].

During the spin-up process in an abruptly rotated vessel, Ekman layers develop in which (in connection with a Stewartson layer) angular momentum is transferred to the part of the fluid not yet rotating (Fig. 7.18). This process takes place more rapidly than the diffusion of vorticity. The heat-up process can be understood in a similar way. The temperature gradient at the side walls of a vessel, which causes the stratification, is suddenly increased. A "buoyancy layer," analogous to the Ekman and Stewartson layers, is formed, which draws fluid from the interior and carries it upward at the side walls. The developing circulation heats the interior of the fluid more quickly than would pure heat conduction.

8.6. ROTATION IN STRATIFIED FLUIDS

In meteorology, oceanography, and astrophysics global movements of matter are dominated by the effects of both rotation and stratification. Because of the complexity of such systems the initial discussion will be restricted to simple examples. This discussion will conclude the theoretical part of this book. The remaining chapters will interpret natural phenomena with the aid of this theoretical knowledge.

In the first simple example, which can easily be solved theoretically, warm fluid is located over a cold one with higher density, and hence the system is stable. Both fluids are assumed to be frictionless and immiscible, and they rotate with different angular velocities. In a steady low-pressure vortex (which rotates counterclockwise in the northern hemisphere), the curvature of the separation line between the fluids depends on whether the warm or the cold fluid rotates faster. In a high-pressure vortex the separation line curves in the opposite way (Fig. 8.32).

In rotating fluids the two-dimensional constraint delays the start of instability. Since this constraint is completely valid only for frictionless fluids and because it relaxes with viscosity, one arrives at the surprising result that viscosity can encourage instability in a rotating fluid (as it happens in boundary layers of nonrotating fluids).

For this reason the generation of Bénard cells in a rotating fluid requires a larger temperature difference between the plates than in a nonrotating fluid, and the cells are smaller in their horizontal extension than those in a nonrotating fluid [41]. The streamlines are twisted under the influence of the Coriolis force as sketched in Fig. 8.6 for convection over a solitary heat source (Fig. 8.33). Convection rolls are shown in Fig. 8.18.

In containers stratification has essentially no effect on the horizontal Ekman layer but it does affect the vertical Stewartson layer. This layer may be considered as divided into two parts, a "buoyancy" layer and a "hydrostatic" layer [42]. In the spin-up process stratification retards the exchange of angular momentum in the Ekman layer so that diffusion of vorticity dominates [43].

The axis of a rotating fluid need not be parallel to the direction of gravity. This situation occurs in rotating stars. With sufficiently fast rotation the gravitational force (and the centrifugal force) is ap-

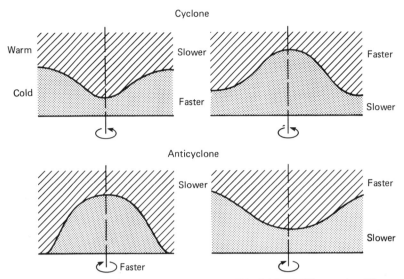

Figure 8.32. Steady vortices in two immiscible fluids of different densities.

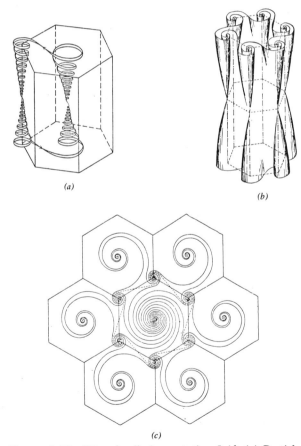

Figure 8.33. Bénard cells in a rotating fluid. (a) Particle paths in a cell, perspective view; (b) sketch of a cell which is distorted through rotation; (c) seven cells seen from above. The solid spiral lines represent particle paths, the dashed line is the border of the middle cell (from S. Chandrasekhar [25], Oxford University Press).

proximately perpendicular to the axis. Since in this case gravity acts parallel to the centrifugal force, relatively simple laboratory experiments are possible. For this purpose a rotating spherical shell is used whose inner surface is kept cold and its outer surface warm. At a sufficiently high Rayleigh number the fluid becomes unstable, and rolls develop, which are oriented parallel to the axis of rotation because of the two-dimensional constraint (Fig. 8.34). In a narrow spherical shell the rolls look like bananas.

At this point the Marangoni convection is brought up again (Fig. 8.25). If a rotation is superposed on the Marangoni vortex ring by rotating the rods, the Marangoni convection can be reduced and its oscillation avoided. Both these effects are desired in the engineering process of crystal growth. Figure 8.35 shows the influence of rotation on the Marangoni vortex. If both rods rotate in the same direction with the same angular velocity, that is, $\Omega_1 = \Omega_2$, the streamlines are displaced toward the upper rod (Fig. 8.35a), and oscillations are suppressed. If the upper rod is fixed, that is, $\Omega_1 = 0, \Omega_2 \neq 0$, the Marangoni vortex weakens and is pushed toward the free surface (Fig. 8.35b). Above the lower rod a thin circulation due to boundary-layer effects appears. Counterrotation ($\Omega_1 = -\Omega_2$) with simultaneous heating of the midplane by a ring heater creates two Marangoni vortex rings. Their strengths can be greatly reduced by sufficiently large rotation (Fig. 8.35c). Further increase in the rotation results in the development of a secondary circulation due to the

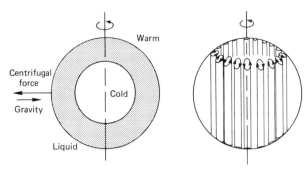

Figure 8.34. Rolls generated by instability in a rotating spherical shell (adapted from F. H. Busse and C. R. Carrigan [44], Copyright 1976 by the American Association for the Advancement of Science).

above-mentioned boundary-layer effect (Fig. 8.35d).

The study of heat convection in a rotating annular container is useful in understanding atmospheric motions. This study, known as the "dishpan experiment," leads to the problem discussed in Chapter 9. In this experiment the outer wall of the ring is warmer than the inner one. The fluid adheres to the bottom but is able to slip along the surface. The temperature difference generates a meridional circulation that is directed toward the center at the surface and outward at the bottom (Fig. 8.36). Because of the slip flow at the surface, the center of the meridional circulation is shifted toward the surface. In a rotating annular container the fluid is pushed outward by centrifugal force. This means that the fluid is supported in its movement at the bottom but hindered at the surface, so the center of the meridional circulation moves closer to the outer wall. At large rotation, two cells may even occur. In addition, the meridional flow changes the azimuthal velocity component. Instead of solid-body rotation (straight line in Fig. 8.37) one observes a velocity distribution that indicates an increase of the veloc-

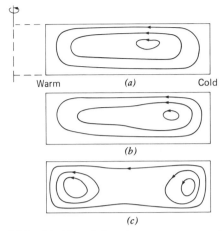

Figure 8.36. Meridional flow in a rotating annular container. Ra = 10,800 with $R_o - R_i$ as characteristic length, Re = $\Omega R_i^2/\nu$, Pr = 1: (a) without rotation, Re = 0; (b) with rotation, Re = 10; (c) with strong rotation, Re = 40 (adapted from M. R. Samuels [45]).

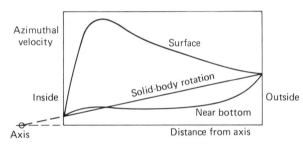

Figure 8.37. Azimuthal velocity distribution in a rotating annular container. The velocity is jetlike at the surface near the inner wall (adapted from M. R. Samuels [45]).

ity toward the center at the surface, a decrease at the bottom. The jetlike increase in the velocity at the surface is similar to the "jet stream" in the atmosphere. This situation will be discussed in more detail in Section 9.1.

Williams [46] has made computer calculations with parameter values and boundary conditions

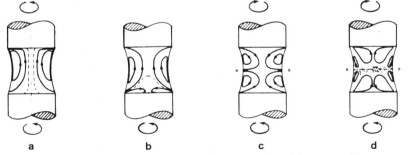

Figure 8.35. The influence of rotation of Marangoni vortices: (a) $\Omega_1 = \Omega_2$, (b) $\Omega_1 = 0, \Omega_2 \neq 0$, (c) and (d) $\Omega_1 = -\Omega_2$ (adapted from C. H. Chun, DFVLR, Göttingen).

that correspond more closely to atmospheric conditions. In particular, he minimized the influence of the boundary layers on the meridional flow by using slip conditions at the side walls. An additional important condition for global atmospheric circulations is the elimination of the total angular momentum. Then azimuthal countercurrents are generated (Section 9.1).

Flows in a rotating pan will be further described in Section 9.4 after the general atmospheric circulation has been discussed. Then, also, more recent computer calculations of atmospheric motions will be mentioned.

Remarks on Chapter 8

1. The generation of baroclinic vorticity is proportional to $\nabla p \times \nabla \rho$, where p is the pressure and ρ is the density. One recognizes that this outer product is zero for a barotropic change of state. Physically, the generation of vorticity may be imagined in the following way: When a fluid element is accelerated perpendicular to the stratification, it behaves like a solid body. No vorticity can be produced. If, however, an element is accelerated parallel to the stratification, then (with equal pressure gradient) the upper part of the fluid element is more accelerated because of the smaller density than the lower one. The fluid element rotates and gains vorticity [37].

2. Also, in the direction of the force a constant magnetic field causes analogous restrictions in the fluid motion [25].

9. Circulations in Atmosphere, Ocean, and Earth

9.1. THE GENERAL CIRCULATION OF THE ATMOSPHERE

Cyclones are an essential part of the general circulation which could not exist without them.
Harold Jeffreys, 1933

The sun is the source of almost all energy on earth. Since the earth as a whole is neither heated nor cooled, the solar energy absorbed must equal the energy released by the earth. However, the radiation of the sun is not the same at all places on the earth's surface. Figure 9.1 shows that at the equinoxes on March 21st and September 23rd the area around the equator is heated most while almost no rays hit the polar regions.

The earth, on the other hand, radiates more energy in the infrared regime at the poles than enters through solar radiation, and radiates less energy at the equator. Hence, the earth's surface is heated at the equator and is cooled at the poles (Fig. 9.2). This temperature difference causes a flow of heat from the equator to the poles, transported by winds and ocean currents.

The distribution of energy on earth is a complicated process, which is shown schematically in Fig. 9.3 [1, 2]. Thirty percent of the total energy that reaches the earth from the sun is immediately reflected into outer space. The remaining 70% is transformed into heat, two-thirds of it on the surface of the earth, one-third in the atmosphere. Surprisingly, most of the heat energy absorbed by the earth is used for the evaporation of water and is stored in the atmosphere as latent heat, whereas the rest of the heat is transferred directly from the earth to the atmosphere. Temperature differences in the heat reservoir of the atmosphere (Fig. 9.2) transform heat into kinetic energy, which is noticeable in the air motions and ocean currents. Air and water flows dissipate and the frictional heat produced is returned to the atmosphere. Finally, the atmosphere releases heat into outer space in the form of infrared radiation, that is, the amount which is absorbed in solar energy from the atmosphere and the surface of the earth.[1]

Local winds are of no significance in the exchange of heat between equator and poles; only such large-scale flows as trade winds and monsoons are important. The totality of these wind systems, which form a complicated three-dimensional circulation, is called the "General Circulation." This is probably the most important concept in meteorology, but its exact definition is difficult; and there are many definitions. Any model of the general circulation will depend on the task for which it is to be used and will, accordingly, incorporate air movements of various orders of magnitude in space and time. For instance, a model for studying climatic changes would use monthly or semiannual or annual averages of observational data, but such a model would be unsuitable for weather prediction, which deals with short-time oscillations in the atmosphere. In the following text these two points of view, that of long-time average and that of short-time changes, will be investigated. As an introduction some historical remarks are useful.

Aristotle mentioned in his book *Meteorologica* that winds are produced through heating of the air

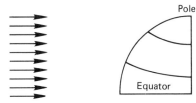

Figure 9.1. The intensity of solar radiation is larger at the equator than at the poles. The figure shows the direction of rays at the equinoxes.

by the sun. In 1686 the famous astronomer Halley (who predicted the return of the comet named after him) also stated in his work on trade winds and monsoons that the sun is the cause of air movements [3].

Hadley was the first to recognize, in 1735, the necessity of air circulation between equator and poles [4]. His perception is represented in Fig. 9.4 (although he himself has not left any sketch). The meridional circulation is the simplest one may imagine: one steady vortex ring in each hemisphere, caused and maintained by the unequal heating of the earth. Hadley also recognized that the movement of air is deflected from the north–south direction by the earth's rotation. Air coming from the poles is carried into regions with a greater west–east velocity, which deflects the air to the west (trade winds). The frictional drag of the trade winds, however, would slow down the earth's rotation if a counterforce did not act on the other parts of the surface. Hadley argued that the poleward flow of air contains a velocity component toward the east because the rotational velocity of the earth decreases toward the poles. This air sinks to the earth at the poles and

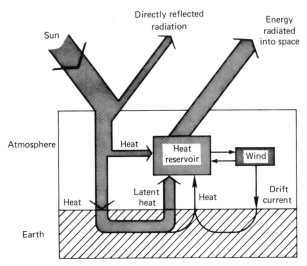

Figure 9.3. The energy flux from the sun to the earth and from there into outer space. (Adapted from A. H. Oort, "The Energy Cycle of the Earth," *Scientific American,* Sept. 1970, 54. Copyright © 1970 by Scientific American, Inc. All rights reserved.)

forms the westerly winds. Their friction on earth balances the frictional force of the trade winds. Hadley's result was outstanding for his time. His contribution may be considered the birth of the theory on general circulation.

Hadley's model, however, has grave shortcomings, which have been overcome only slowly in the course of time. Hadley did not know of the Coriolis effect, and he assumed that the absolute velocity must be preserved. His model thus could not quantitatively describe the east–west deviation. A more serious objection is the fact that the observed cir-

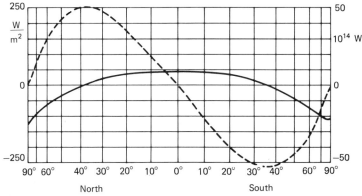

Figure 9.2. Difference between absorbed solar radiation and infrared radiation released by the earth into outer space (solid line, units on the left). The energy transport caused by this difference from the equator to the poles reaches an extremum at about 35° northern and southern latitudes (dashed line, units on the right) (after W. D. Sellers from Reference 28, Chapter 1).

Figure 9.4. Hadley's circulation model of 1735 (from E. N. Lorenz [Reference 28, Chapter 1]).

culation deviates in the midlatitudes. The westerlies do not flow in this region from northwest but from southwest.

This observation was interpreted in 1837 by H. W. Dove when he assumed winds in changing directions for the midlatitudes [5]. This change was also considered to be the reason for lows and highs in the midlatitudes (Fig. 9.5). One recognizes here the first indication of a cyclonic theory, which later was to become critical in explaining the general circulation.

Dove did not know of the Coriolis force either, and he too assumed the conservation of absolute velocity. The discovery of the Coriolis effect on the general circulation is credited to Ferrel [6]. Ferrel's

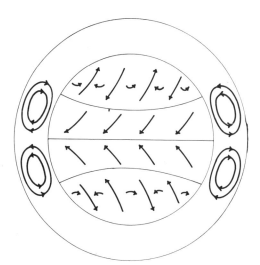

Figure 9.5. Dove's circulation model of 1837 (from E. N. Lorenz [Reference 28, Chapter 1]).

The General Circulation of the Atmosphere 171

model, indeed, comes remarkably close to the actually observed circulation averaged over the year. His model (Fig. 1.6) shows three vortex rings in each hemisphere. This circulation explains the pressure distribution on the earth's surface correctly, in particular the highpressure belt in the horse latitudes. Still, Ferrel's model has shortcomings too. It does not explain the cyclones in midlatitudes, whose existence (as will be shown later) is vital for the balance of energy and angular momentum.

In 1888 Helmholtz pulbished a paper [7] in which he pointed out that, between the equator and poleward winds in the middle of the atmospheric layer, discontinuity areas must occur which cause the exchange of energy and vortices. Unfortunately, his text is not clear as to the magnitude of the vortices, whether they are local vortices or of the order of low-pressure systems. Helmholtz could have become the founder of modern cyclonic theory. Nevertheless, his work inspired Bjerknes to his cyclonic theory, which became the basis for the famous meteorological school in Bergen [8], mentioned again in Section 10.2.

Two more contributions to cyclonic theory led to the circulation model that today—at least qualitatively—is generally accepted. In 1921 Defant interpreted the cyclones in midlatitudes as large turbulent eddies, and with this idea explained the exchange of energy in this region [9] (Fig. 9.6). A little later, in1926, Jeffreys recognized the necessity of cyclones for the exchange of angular momentum [10]. This idea was further pursued in particular by Starr [11].

Today's conception of the general circulation, its colorful history, and the work in developing a good description have received new meaning through the spectacular discoveries of the atmospheres of other planets. They reveal both similarities to the earth's atmosphere (Mars) and striking differences (Jupiter and Saturn). These similarities and differences fur-

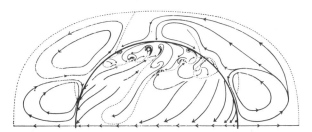

Figure 9.6. Bjerknes' circulation model of 1921 (from E. N. Lorenz [Reference 28, Chapter 1]).

nish additional evidence for the validity of planetary theories (Section 12.3).

The main reason for the obstacles to describing correctly the earth's atmosphere lies in the difficulty of measuring the state variables (in particular the wind velocities) in the presence of local disturbances. From the fluid dynamical point of view, the general circulation is so complicated that this problem—together with the turbulence problem—belongs to the most difficult in classical physics.

From the present state of knowledge (1983) the following model of the general circulation for the annual average can be developed. The earth is here idealized as homogeneous and symmetric, that is, the motions are independent of the longitudes and are mirror images in the northern and southern hemispheres [12].

The steadiest wind system, which forms the basis for all circulation models, is the trade winds. They consist of two vortex rings north and south of the equator. At the equator, in the doldrums, air heated by the sun rises and flows away from the equator to the horse latitudes (30°), where it sinks and returns to the equator, deflected from the north–south direction, however, by the Coriolis force (Fig. 9.7). The meridional circulation is called the "Hadley cell," which is not to be confused with the Hadley circulation in Fig. 9.4.

In the midlatitudes between 30° and 60° the air motion is variable as compared to the steady trade winds. Highs and lows develop and decay in the west-to-east air currents. In these latitudes the temperature gradient between equator and poles is largest (Fig. 9.2), and the exchange of energy and momentum can no longer be handled by a simple circulation with steady zonal velocities like those of the Hadley cell. The flow becomes unstable, and vortex systems develop that take over the transport of energy and momentum.

However, a plot of the average yearly flow in the midlatitudes shows a meridional circulation, which exists only in the statistical sense. This circulation is called a "Ferrel cell" (Fig. 9.8). The weak polar cell seen on the figure has not yet been measured directly.

The average yearly data, which describe the state of the atmosphere, must correctly reflect the sensitive balance on the whole earth. The equilibrium of the local angular momentum can be used as an example. If there were only easterly winds on earth, the effective angular momenta would slow down the rotation of the earth. Conversely, the westerlies would accelerate the rotation. Experience over thousands of years of human history shows, however, that the earth's rotation is nearly constant. Hence, the integration of the local angular momentum over the whole earth must result in zero (Fig. 9.9).

Plotting lines of constant wind velocities in a similar way in the easterly direction ("zonal" velocity) reveals a strange phenomenon. In the midlatitudes close to the tropopause the westerlies reach a maximum (Fig. 9.10). Although the data are averaged, the result indicates a phenomenon known in meteorology as the "jet stream." The occurrence of the maximum is explained by the fact that the atmosphere is baroclinic, that is, the isobars are not parallel to the isotherms (Fig. 9.11), because the earth's surface between equator and poles is not heated evenly. Under the assumptions of geostrophic motion and hydrostatic equilibrium it can be shown that the change of wind velocity with height depends on the horizontal temperature gradient (in Fig. 9.11 along the isobars). The change of the wind velocity is largest near the ground and decreases to zero at a certain height (in Fig. 9.11 at ca. 250 mb). The westerly has its highest velocity there (Fig. 9.10). This locally restricted high-velocity range is called the "jet stream."

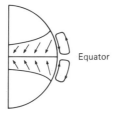

Figure 9.7. The trade winds form two belts north and south of the equator. The meridional motion is called a Hadley cell.

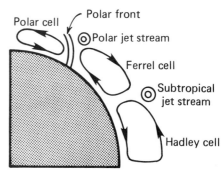

Figure 9.8. Streamlines of the meridional circulation in the annual average. The height scale is much increased.

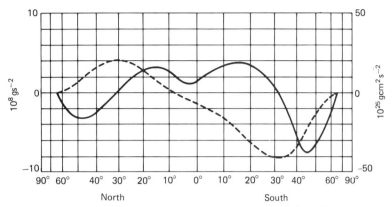

Figure 9.9. Average angular momentum due to surface friction in 10^8 g·s^{-2} (solid line, left scale). Transport of absolute angular momentum to the poles to maintain balance (dashed line, right scale, in units of 10^{25} g·cm^2·s^{-2}). The curves may be considered only as rough estimates (from C. H. B. Priestley [Reference 28, Chapter 1]).

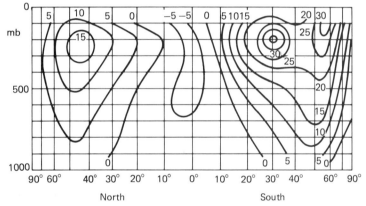

Figure 9.10. Zonal wind velocity, averaged over time and longitudes, for April to September. Positive values in m/s correspond to westerlies, negative values to easterlies (from H. Buch and G. O. P. Obasi [Reference 28, Chapter 1]).

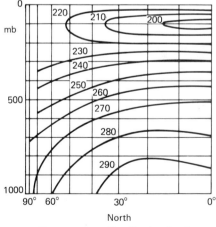

Figure 9.11. Temperature distribution in degrees Kelvin, averaged over time and longitudes, for April through September in the northern hemisphere (from J. P. Peixoto and L. Peng [Reference 28, Chapter 1]).

The average values drawn in Fig. 9.10 are much too coarse to describe an atmospheric situation subject to daily oscillations. Jet streams are in reality more fascinating than the broadly drawn average values indicate. The characteristics of the jet stream are apparent in the official definition of the World Meteorological Organization (WMO) [13]: "A jet stream is a strong narrow current, concentrated along a quasi-horizontal axis in the upper troposphere or in the stratosphere, characterized by strong vertical and lateral wind shears and featuring one or more velocity maxima."

Jet streams, as narrow meandering currents several hundred kilometers wide and a few kilometers high, stretch for thousands of kilometers. They were first discovered during World War II when reconnaissance planes and heavy bombers, forced by flak and interceptors to high altitudes, encountered them. How does the narrow current develop? The basic explanation has already been given, but in reality much stronger temperature differences occur that make the jet stream narrower. In Fig. 9.12 such a situation at a polar front is sketched (see Section 10.2).

Data averaged in space and time do not depict irregularities in the general circulation as those caused by the distribution of continents and oceans. The semiannual oscillations of the wind circulation between continent and ocean (e.g., the Indian monsoon) occur in the same way that the wind direction in a sea breeze changes every half day, because the land is cooler at night and the sea during the day. In winter dry, cold continental air comes from the northeast, in summer humid maritime air from the southwest. Unlike the sea breeze, the monsoon is influenced by the Coriolis force [14].

The seasonal cycle of the monsoon is part of the overall changes of the general circulation in the equatorial region. The ensuing longitudinal circulation patterns (perpendicular to the meridional Hadley cells) are called "Walker circulations" (Fig. 9.13) [15].

There are also nonseasonal changes in the general circulation that are caused by anomalies of the sea-surface temperature (whose origins are not yet fully understood), volcanic eruptions that influence the atmospheric transmission of solar energy, and perhaps changes in the solar-energy output itself. These climatic changes influence the strengths and appearances of the Walker circulations and the monsoons in particular [15]. An example is the occurrence of El Niño, a tongue of warm water off the Peruvian coast that coincides with the disappearance of the upwelling of cold water and the reduction in supply of nutrients for the anchovy fish. The catch drops sharply during the appearance of El Niño.

Short-time or periodic air motion, which determines the weather, cannot be described even with semiannual or monthly models. Weather prediction is actually the temporal extrapolation of an instantaneous state into the future. If one neglects local disturbances like thunderstorms and tornadoes, the temporal global changes of the atmosphere give a picture of the large-scale weather. The most important periodic disturbances in the midlatitudes, the Rossby waves, have already been introduced in Section 7.3. The formula given for the velocity in that section shows that this velocity can be positive, negative, or zero, depending on the average east-west current. This means that the Rossby waves can migrate toward the west (which is rare) or toward the east. They can also be stagnant for a certain time.[2] With these waves high- and low-pressure regions occur; their development will be described in Section 10.2. Here it is sufficient to remark that the description of vortex motion is an essential part of weather prediction.

In Fig. 9.14 streamlines are plotted for an instantaneous state on the 500 mb level (about 6 km high) over the northern hemisphere. The streamlines show clearly the wavy air motion and the distribution of highs and lows. If one looks at the streamlines at higher altitudes, say at about the 100 mb level (ca. 16 km), the closed streamlines of the highs and lows have vanished, and the "circumpolar vortex" with waves of smaller amplitudes predominates (see, e.g., the global weather maps in Reference 17).

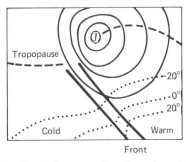

Figure 9.12. Vertical cross section through a cold front with jet stream (adapted from *Introduction to the Atmosphere* by H. Riehl. Copyright © 1965. Used with the permission of McGraw-Hill Book Company).

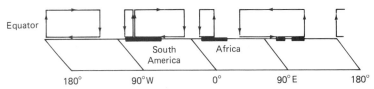

Figure 9.13. Walker circulations over equatorial regions in January (after R. E. Newell [15]. Reprinted by permission, *American Scientist,* Journal of Sigma Xi, The Scientific Research Society).

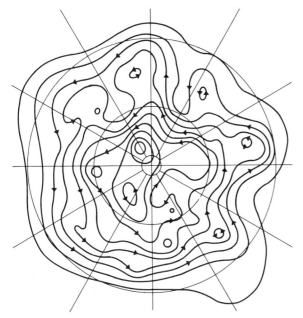

Figure 9.14. Streamlines in the 500 mb layer for the northern hemisphere with the north pole in the center (from P. D. Thompson [16]).

These factors contribute to the great difficulties of weather prediction. Atmospheric events are not periodic. Largescale changes in the general circulation are often triggered by small local disturbances, which can only be approximated in the initial data for the prediction. In addition, the grid size of the observational net (ca. 100 km) and the many parameters to be considered in a physical model put almost insurmountable limits to an exact prediction for more than 2 weeks. This problem will be addressed again in Section 9.4 on numerical weather prediction.

9.2. OCEAN CIRCULATIONS

The general circulation of the atmosphere is not the only mechanism by which the heat flow from the equator to the poles is maintained. The waters of the oceans also participate, and here atmosphere and ocean influence each other. Two essential distinctions must be considered, however, in comparing atmosphere and ocean. The ocean basins have natural boundaries in the continents (except for the Antarctic circulation, which is generated by the westerlies). A global circulation averaged over the longitudes as in the atmosphere is thus not possible. The second difference lies in the generation of the circulation. Whereas the general circulation of the atmosphere results from thermal convection maintained by the temperature difference between equator and poles, ocean circulations come from both drift currents, which are wind-generated motions, and convection flows, produced by differences in density and temperature. Wind-generated water vortices have not yet been mentioned, so some general remarks are introduced here.

Although it offends good table manners to cool a cup of coffee by blowing on it, sometimes it is tempting to overlook etiquette. Beautiful surface vortices can be generated by blowing, particularly if they are made more visible by the addition of milk. Because water and air adhere at the interface, the water is dragged along by the air passing over it. Since the water is restricted by the edge of the cup, it is forced to circulate (Fig. 2.7). In nature, fluid motions generated in this way are called drift currents. Examples are the tidal storm currents at coasts and wind-generated ocean circulations, which are discussed here. Drift currents depend on the strength of the winds that produce them, and they can change in short periods of time. As with the general circulation, however, average values can be obtained that also represent an instantaneous situation when the winds are steady, as in the case of the trade winds [18]. Figure 9.15 shows the average surface circulations of the world's oceans. In the following text the North Atlantic circulation is described in more detail, and its similarity to the North Pacific circulation is pointed out.[3]

The shearing effect of the trade winds and westerlies generates the North Atlantic circulation (Fig. 9.16). The North-equatorial current, driven by the trade winds, forms the southern border of the circulation. It branches before the West Indies and flows partly into the Gulf of Mexico, partly along the Bahamas. The two parts combine again off the coast of Florida and flow as the Gulf Stream along the coasts of Georgia and the Carolinas. This part of the North Atlantic circulation is remarkable for its high flow velocity (up to 3 m/s at the surface) and its limited width (only 100 km). As one approaches the Gulf Stream from the coast, the temperature and salinity increase within a few kilometers by 10°C and 1‰ respectively. The transport of water off the coast of Florida—say at Cape Canaveral—is about 30 million cubic meters per second, which is 22 times as much as all the rivers on earth together. This mighty ocean current leaves the coast of North America at Cape Hatteras and meanders toward

176 Circulations in Atmosphere, Ocean, and Earth

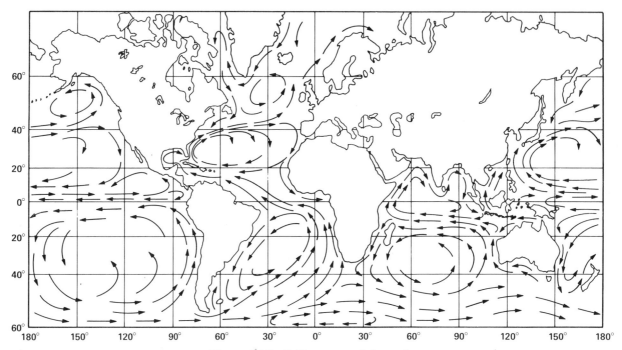

Figure 9.15. Drift currents of the earth.

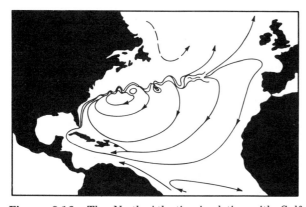

Figure 9.16. The North Atlantic circulation with Gulf Stream.

Europe, where part of it flows along the British and Norwegian Coasts to provide, as the "warm water heater of Europe," the mild climate. The other part flows in southerly direction along the North African coast and starts a new cycle in the circulation as part of the North-equatorial current. Although water is exchanged with neighboring currents at the northern and southern boundaries of the North Atlantic circulation, it is useful for theoretical considerations to regard the circulation as a single big vortex without this exchange. This vortex has two unusual properties: its strong asymmetry, which is related to the fast current at its western boundary, and its meandering path from Cape Hatteras downstream. A theoretical explanation will be discussed a little later.

As in all vortices, the center is a region of relative calm.[4] In the North Atlantic circulation it is the Sargasso Sea. This by no means implies, however, that this part of the circulation is less fascinating than the Gulf Stream, for the Sargasso Sea has its peculiar biological characteristics. There is an influx and downward motion of water at the rotation center since the bottom layer rotates faster than the upper region, the latter retarded by the surface winds (Fig. 7.28b). The water thus sinks into deeper layers from which it flows over the ocean bottom partly to the African west coast, partly to the east coast of South America. Since water at the surface has a larger oxygen content than in the depth, the sinking water carries oxygen with it, and flora and fauna of the Sargasso Sea are thus particularly rich. The large amounts of seaweed gave this part of the ocean its name (seaweed in Portuguese is "sargaço"). The Sargasso Sea is also the spawning ground of the American and European eels. When the larvae reach the border of the circulation, that is, the North American and European coasts, they have already developed to elvers, and they then migrate up the rivers. They finish their cycle when they return as adult eels to the spawning grounds in the Sargasso Sea [19]. Unfortunately, the influx of surface water has made the Sargasso Sea the dumping place for tar, plastic material, and other debris.

Surprisingly similar is the situation in the North Pacific circulation (Fig. 9.15). Here, too, at the western boundary is a current corresponding to the Gulf Stream, the Kuroshio off the Japanese coast. At Cape Nojima, south of Tokyo, the Kuroshio leaves the Asiatic continent and meanders toward North America. Exactly as Europe is warmed by the Gulf Stream, Alaska is warmed by the Kuroshio. Approximately in the center of the North Pacific circulation are the islands of Hawaii. It is remarkable that the Spaniards maintained a lively trade for over 200 years between Mexico and the Philippines, that they sailed north and south of Hawaii, but never discovered the islands. Hawaii was finally discovered by Cook in 1778.

In the southern hemisphere there is also a "westward intensification" of the circulation, although it is less pronounced than that in the north. Off the South American coast is the Brazil current, in the Indian Ocean the Agulhas. A possible explanation for the weakening is the density current in the deep ocean, which runs from the north over the equator toward the south along the whole American coast. In the north it supports the westward intensification, in the south it weakens it [18].

The westward intensification of the ocean circulations is explained theoretically by the Coriolis force. One may think for the time being of a rectangular ocean basin with solid side walls representing the continents, and with free upper and lower boundaries. The depth of the basin is assumed constant. The wind blows from the horizontal middle line downward to the left (trade wind) and upward to the right (westerlies) (Fig. 9.17a). In a nonrotating basin in which friction dominates, a drift current occurs that is completely symmetric in the east–west and north–south directions. If the Coriolis force is considered in this model as a constant force, the circulation pattern in Fig. 9.17a does not change. In reality, however, the magnitude of the Coriolis force changes from zero at the equator to a maximum at the poles. If this change is considered in the model, the westward intensification is apparent (Fig. 9.17b). This conception was presented by Stommel [18] and Munk [20], and as a simplification the Reynolds number is assumed by Munk to be nil, but the model can be improved in various ways. According to Bryan [21] inertial force causes

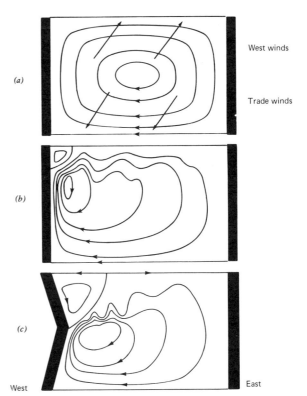

Figure 9.17. Schematic representation of the drift currents in the North Atlantic and North Pacific: (a) without and (b) with influence of the change of the Coriolis force; (c) "Cape Hatteras effect."

an intensification in the north–south direction, that is, toward the north (Fig. 9.17b). Moreover, wavy currents occur at the northern boundary. The model can also be improved by considering the special geography of the western continental border (Fig. 9.17c). The corner schematically represents Cape Hatteras. There the Gulf Stream separates from the coast and migrates toward Europe. In reality, however, the process of separation and the subsequent oscillations of the Gulf Stream are more complicated (Section 9.3) [22].

Exactly as in the general circulation of the atmosphere, there are semiannual and short-time changes in the ocean which correspond to the weather in the atmosphere. The short-time changes are described in Section 9.3. The Somali current is an example of a semiannual oscillation whose generation can be attributed to the monsoon. A theory on such interaction by means of oceanic Rossby waves was proposed by Lighthill [23].

It may be mentioned that wind-driven flows can occur not only in the ocean but in every body of water. It has been observed that in the northern hemisphere almost all lakes and semiopen oceans circulate counterclockwise if averaged over a long period of time. Examples are the American Great Lakes, Lake Constance, Lake Geneva, The Caspian and Black Seas, the Adriatic Sea, and the Dead Sea. Emery and Csanady [24] ascribe this rotation to the combined action of Ekman drift currents and temperature stratification.

9.3. THE GULF STREAM

The Nantucket whale-men being extremely well acquainted with the Gulph Stream, its course, strength and extent, by their constant practice of whaling on the edges of it, from their island quite down to the Bahamas,...

Benjamin Franklin, 1769

Presumably Columbus had some knowledge of westward ocean currents; otherwise the choice of route to "India" would have been merely a lucky chance. He was also tempted to sail west, rather than east, basing his decision on the errant calculation of Ptolemy, who underestimated the circumference of the earth. At any case, Henry the Seafarer had at the beginning of the 15th century founded a kind of oceanographic institute in Lisbon in which sea charts were drawn. A few years after Columbus' voyages in 1513 the Spaniard Ponce de Leon observed an unusually strong current along the coast of Florida. A few years later the Gulf Stream was so well known to the Spaniards that they used the North-equatorial current from Spain to reach the West Indies, and on their voyage back they sailed the Gulf Stream [25]. In the following 200 years, during which the colonization of the North American east coast took place, the Gulf Stream was crossed and traveled innumerable times. The data observed then were astonishingly accurate in contrast to the confused and sometimes grotesque theories. Sea maps of the time contained a mixture of accurate data and fantasy: Happelius' map from 1685 shows as a curiosity two surface currents crossing each other and a large whirlpool off the Lofoten in Norway, the legendary Maelstrom (see Section 10.6). Kepler fancied the reason for the equatorial current was that the ocean cannot follow the rotation of the earth fast enough and hence lags behind. This idea is easy to ridicule from today's point of view, but except for the data collected by experienced sea captains, scientific research on the Gulf Stream started only about 100 years ago with systematic measurements by the U.S. Coast Guard. In the years 1925–1927 the German Meteor expedition participated to a great extent by collecting data in the Atlantic. Since 1930 research has passed almost completely into the hands of the newly established Oceanographic Institute in Woods Hole, Massachusetts.

As can be seen in Fig. 9.16, the Gulf Stream is part of the North Atlantic circulation and transports large quantities of heat to northern latitudes. Vortices play a particular role in this process so that a special section on the Gulf Stream appears justified. Before the current reaches the east coast of Florida and becomes the Gulf Stream proper, it has already had the opportunity at various places to generate large-scale vortices. In the Yucatán Channel the current is squeezed between Cuba and Mexico. There, off the coast of Yucatán, is the island of Cozumel (Fig. 9.18), behind which the current forms an oceanic vortex street [26]. Further downstream, between Yucatán and Florida, the current bends into the Gulf of Mexico. The size of this bend varies with the season. When the current extends farthest into the Gulf of Mexico, large-scale vortices can develop along the west coast of Florida. From the southern tip of Florida up to Cape Hatteras the Gulf Stream behaves as a steady jet. Flat steady vortices through friction at the coast are generated which rotate counterclockwise. These vortices have formed the shape of the coasts of Florida, Georgia, and the Carolinas (Fig. 9.19). But the situation changes abruptly off Cape Hatteras. Here the Gulf Stream leaves the coast and the continental shelf and meanders eastward.

To the disadvantage of Europe-bound vessels, the meandering movement of the Gulf Stream is in

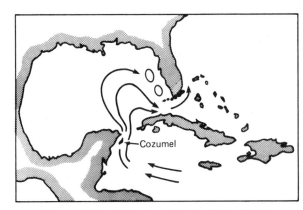

Figure 9.18. The current in the Gulf of Mexico.

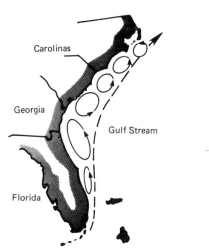

Figure 9.19. Steady, weak vortices at the east coast of the United States between Florida and Cape Hatteras.

no way steady, but irregular in space and time (Fig. 9.20). The Gulf Stream can change its course within a day by 25–30 km and over a longer period up to 400 km. The wavelength of the meandering band is often more than 500 km. Gigantic vortices can separate from the main current at a sudden change in direction and can lead their own lives over weeks or months at the edge of the Gulf Stream. Vortices south of the Gulf Stream are cyclonic and cold and larger than the anticyclonic, warm vortices north of the Gulf Stream [28]. Figure 9.21a shows a loop in the Gulf Stream and a vortex "Edgar." This vortex was traced in 1950 for three weeks. In Fig. 9.21b a satellite picture reveals a large vortex about 200 km south of Cape Cod. Cyclonic vortices with a diameter of 50–400 km have existed for up to a year before they vanished in the Sargasso Sea. Fuglister and Worthington [29] estimate that five to eight vortices form about every year. They have rotational velocities of 0.5–1 m/s, travel 0.1 m/s, and extend to a depth of 2000 m (Fig. 9.22). Their higher density causes them to sink about 1 m/day. Detached vortices can sometimes recombine with the main stream, and Fig. 9.23 shows a sequence of sketches that illustrate the union process over a period of several weeks.

In addition to vortices separated from the Gulf Stream, smaller vortices of either direction are found. Many of them preserve the water which they were made of, together with its heat, salt, and other dissolved material. This preservation of their identity gives clues to where the vortices originated. Some clearly are formed behind seamounts [30]; others are believed to come from as far away as the Strait of Gibraltar, consisting of the typical, quite undiluted Mediterranean water [31].

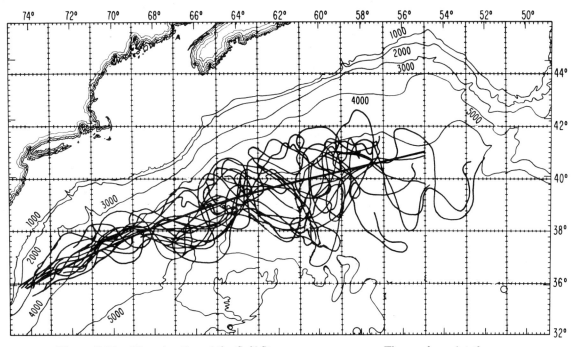

Figure 9.20. The migration of the Gulf Stream over many years. The numbers give the ocean depths in meters (from G. F. Carrier [27]. Reprinted with permission from *SIAM Review*, 1970, p. 175. Copyright 1970 by Society for Industrial and Applied Mathematics).

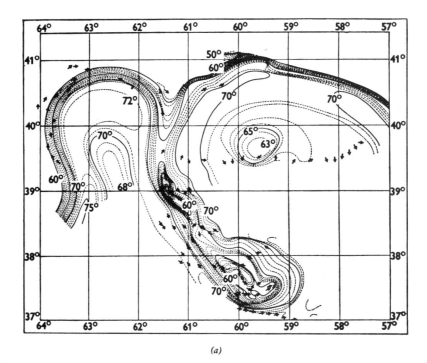

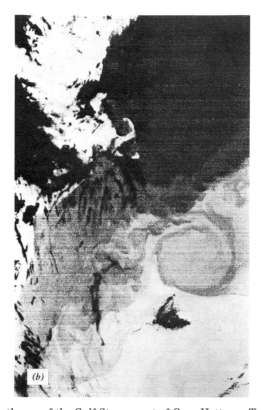

Figure 9.21. (a) Isotherms of the Gulf Stream east of Cape Hatteras. Temperature in degrees Fahrenheit (from F. C. Fuglister and L. V. Worthington [29]). (b) Photograph of a Gulf Stream vortex from a satellite, digitally enhanced (from H. M. Byrne, Pacific Marine Environmental Laboratory, Seattle, Washington).

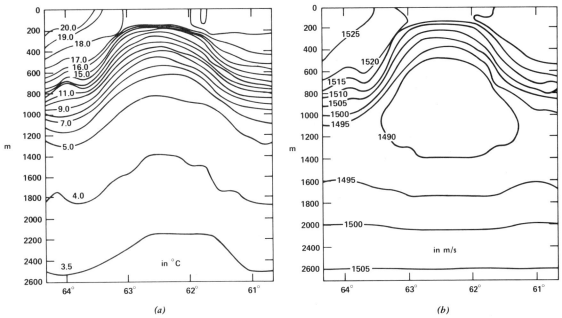

Figure 9.22. (a) Vertical temperature distribution in a cold Gulf Stream vortex; (b) lines of equal velocity of sound in m/s (from Reference 32).

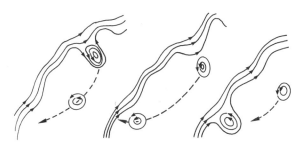

Figure 9.23 Detached vortices in the Gulf Stream can recombine with the main current (adapted from Reference 32).

Knowledge of the properties of oceanic vortices is important in acoustic surveillance of submarines in the western North Atlantic. The radiation of underwater sound is influenced by cold vortices, that can refract sound waves and form "silent zones," which sound cannot penetrate [32, 33]. Figure 9.22b shows the bell-like arrangement of lines of equal sound velocity in a cold (cyclonic) Gulf Stream vortex.

Cold spots in the Gulf Stream were discovered in 1810 by the ship *Eliza* and interpreted as molten icebergs. About 1860 scientists believed in a bandlike structure of alternating cold and warm currents. Today there are various theories for the meandering movement, one saying that the Gulf Stream becomes unstable after separation from Cape Hatteras. Another attributes the meandering to the irregular sea bottom, particularly after the Gulf Stream leaves the continental shelf. Finally, there is a jet theory, similar to the one for the atmospheric jet stream [18].

As already mentioned, the Kuroshio off the Japanese coast is very similar to the Gulf Stream. In two properties, however, the two currents are different. The Kuroshio forms meandering loops before Cape Nojima, the separation point proper. During this process vortex separation and subsequent reunion with the main current have been observed (as in Fig. 9.23). The second typical characteristic of the Kuroshio is its twofold band structure. The meandering flow patterns after separation are thus even more complicated than those of the Gulf Stream [34, 35].

9.4. LABORATORY EXPERIMENTS AND COMPUTER CALCULATIONS

We will hope ... that the introduction of the experiment into meteorology will be accepted not as a mere children's play but as a subject of highest importance.

W. Köppen, 1884

Laboratory experiments in wind tunnels and water tanks have been important aids in the development of fluid dynamics. It is therefore understandable that meteorologists and oceanographers have looked for similar possibilities. Unfortunately, their situation is much more complicated than that of the fluid dynamicists. Large-scale motions of the atmosphere and the oceans are difficult to simulate because the phenomena are so complex and because laboratory experiment restricts the geometric modelling and the choice of the flow parameters[5] [36]. The same is true for many local events like thunderstorms. On the other hand, laboratory experiments have the advantage of controlling the happening through systematic variation of the parameters.

Simple experiments to clarify the main patterns of the general circulation were made relatively late in the history of science. Probably Vettin in 1857 made the first experiments [37]. He placed a flat dish on a rotating table, packed ice in the center of the dish, and observed the meridional circulation of the air (Fig. 9.24).

In 1923 Exner [38] repeated Vettin's experiments with water instead of air. With the aid of ink, the development of vortices at the bottom of the container was clearly visible.

Systematic laboratory experiments started after World War II at the University of Chicago under Fultz and at Cambridge University under Hide [Reference 28, Chapter 1]. Here too, the "rotating dishpan" was used to study the atmosphere.

Despite the obvious shortcomings of these experiments basic phenomena of meteorological importance were observed. According to Hide one can distinguish the following types of motion (Fig. 9.25):

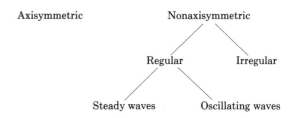

The essential properties of axisymmetric flow were mentioned in Section 8.6. With faster rotation th axisymmetric motion changes to a nonaxisymmetric one, which according to Hide's scheme can be divided into regular and irregular flow. For a steady regular flow, that is, for a time-independent flow in a reference frame rotating with the dish, spatial waves occur whose number depends on outside parameters like angular velocity, temperature difference, and geometric dimensions. This state can oscillate under certain conditions, in which case the waves are not steady but change their form and velocity periodically. In the irregular state, which occurs with further increase in the angular velocity, the temporal and spatial formation of waves becomes irregular. This irregular state and the oscil-

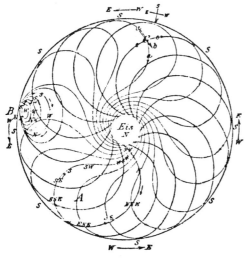

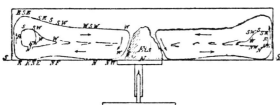

Figure 9.24. Vettin's experiment of 1857.

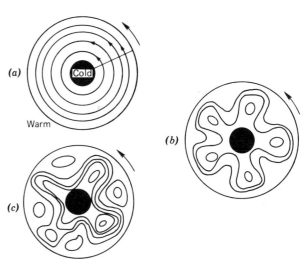

Figure 9.25. Streak lines at the surface of a rotating, annular dish: (a) axisymmetric, (b) steady waves, and (c) irregular.

lating state have similarities with the large-scale motions of the atmosphere. The existence of the jet stream and of the migrating vortex system is here demonstrated (see also Reference 39).

For ocean circulations the experiment must be essentially different from that of atmospheric circulation. The shear stress of the wind must be simulated. Baker [40] has described an experimental set-up that consists of a spherical shell obliquely placed on a rotating table. This model can answer questions on the influence of inertial forces in relation to frictional forces. A survey of laboratory experiments for the study of ocean circulations is given in Reference 36.

With computers the general circulation can be simulated numerically much better than with experimental methods. While in laboratory experiments only a limited number of flow parameters can be chosen to agree with those in nature, and, hence, only partial aspects of reality can be investigated, computer experiments permit full simulation, provided the capacity of the computer and the mathematical model are adequate. Here, too, restrictions and compromises must be made. But before they are discussed, some historical data are given.

The first attempt to describe atmospheric motion through the solution of mathematical equations goes back to L. F. Richardson in 1922. Without a computer and without knowledge of modern numerical methods his attempt had to fail. A milestone in the direction of a mathematical description was Rossby's discussion of the vorticity equation in 1939. After World War II, with the arrival of electronic computers, the mathematician John von Neumann recognized immediately that numerical weather prediction was the first practical field in which the computer could prove itself. In 1949 the first computations were made on the ENIAC computer by Charney and Fjørtoft in collaboration with von Neumann [41]. At that time a 24-hour prediction needed 24 hours of computer time. Von Neumann's student Smagorinsky headed the first laboratory for the numerical modelling of the general circulation, first in Washington, D.C. and later in Princeton, New Jersey. A survey on the state of the art at the time of the first numerical experiments can be found in the Proceedings of the Princeton Conference of 1955 [Reference 41, Chapter 8]. In the following years centers for numerical weather prediction were established in various countries. The National Center for Atmospheric research (NCAR) in Boulder, Colorado; the European Centre for Medium Range Weather forecasts in Reading, England; and the Deutsche Wetterdienst in Offenbach, West Germany may be mentioned.

Parallel to these endeavors to describe atmospheric motions numerically were treatments of the problem of oceanic circulations by Charney, Sverdrup, Munk, Stommel, Morgan, Bryan, and others [18].

The basic idea of numerical calculations of meteorological and oceanic processes has been given in Section 4.3 for numerical fluid dynamics: A system of partial differential equations must be solved for a certain initial state in a finite network, that is, in space and time. This is weather prediction with the aid of a computer. One does not need, however, be a meteorologist to realize the difficulties of such a computer program. The total surface of the earth with the layer of air above it must be used as "integration space." Since only a certain number of grid points is available (see Section 4.3), the quality of the calculations depends on the density of the net. The distance between grid points directly controls the phenomena that can be described and the local events that are filtered out. A grid with a horizontal distance of 200 km and 10 layers in height (which at present is frequently used) cannot describe a local thunderstorm. Computer programs for global air circulations can thus describe only large-scale events. This is also the main reason for the temporal limitation on weather prediction. Local disturbances, which in the course of days or weeks bring about large-scale changes, are often not included in the initial conditions. Further difficulty is caused by the shortcomings of the basic mathematical model. Size of net and kind of model determine the computer time and the effort required for data storage. A collection of numerical models has been published by the World Meteorological Organization in Reference 42. The short life and the variety of these programs make it impractical to discuss them in detail [43–45].

9.5. AN EXCURSION INTO THE EARTH'S HISTORY

And when the sea here retracts, there extends, and so forth, then one recognizes that not always from the whole earth this part is sea and that part continent but that everything changes in time.
    Aristotle, Meteorologica

The earth is not a solid body, as can be assumed for meteorological and oceanic studies without great restriction. In reality the earth has elastic, plastic, and liquid properties. Matter inside the earth moves and circulates. Because of the extremely high viscosity of the earth's material, the velocity of the circulation is many orders of magnitude smaller than that of the air and water masses on the earth's surface. Still little is known about the kind and extent of the circulations inside the earth, and the processes described here are only one version of many theories.

It is conjectured that the earth as a member of the family of planets was created about 4.5 billion years ago from a rotating gas or dust cloud, either through condensation of gases or through accretion of solid material particles (Section 12.2). This process probably took place so quickly that the earth was a quite homogeneous mixture of various ingredients. Through radioactive heating the iron and nickel components melted and flowed to the center. According to a theory proposed by Vening Meinesz [46] this motion caused a convection of the vortex-ring type, which transported silicates to the outer layer of the earth and after half a revolution formed a solid plate there. Through this bilateral asymmetry a land and an ocean mass were formed. When the bottom of the primeval ocean cooled off, the earth's crust became unstable and convection cells developed which broke the primeval continent. As uncertain as this conjecture may be, it is considered today as assured that this primeval continent and ocean were not the supercontinent (Pangaea) and the world ocean (Panthalassa) from which the present state developed. The paleogeography of the last 570 million years reveals that Pangaea was assembled 250 million years ago from clusters of continents that have no relation to the continents of today [47]. About 200 million years ago Pangaea split into two parts (Gondwana and Laurasia) and then into several parts from which today's continents emerged [48].

There is still no certainty about the kind of convection in the earth's mantle. Runcorn [49] imagined a Bénard convection, Morgan [50] a jetlike upward flow only a few hundred kilometers wide, and Elder [51] the idea of turbulent cell flows. The size of the cells in the interior of the earth is also disputed. Runcorn and Schubert and Turcotte [52] assume that the circulations extend over the whole mantle depth; Gordon [53] believes that they are restricted to the layer immediately below the earth's crust. These theories depend on how the change of viscosity with depth is conceived. For this, however, still no reliable data exist. The problem is compounded by the fact that the mantle is not a liquid heated only from below, that is, from the earth's interior, but one which also heats itself through radioactive decay. The Rayleigh number is, according to McKenzie and Richter [54], of the order of 10^6. In this range convection is three dimensional and "spokelike" This version appears to be the most plausible one.

The break up of Pangaea and the subsequent continental drift can be proven by a number of independent methods, among them geological and biological conformities in various continents and paleomagnetic measurements. Since the continental drift plays an important role in the evidence of circulations inside the earth, it is taken up in more detail.

In 1620 the English natural philosopher and statesman Francis Bacon discussed the possibility that North and South America were once united with Europe and Africa. The German scientist Alexander von Humboldt noted the conformity of the east coast of South America and the west coast of Africa (Fig. 9.26). In 1858 Antonio Snider's book *La Création et Ses Mystères Dévoilés* was published, in which he pointed to the similarity between American and European fossil plants and in which he expressed the belief, as Francis Bacon did, that once all continents were a single land mass. But it was only at the beginning of this century, in 1912, that Alfred Wegener interpreted the separation of the

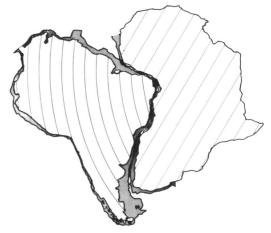

Figure 9.26. The continents can be put together as a puzzle. The best fit is obtained when the edges of the continental shelves and not the coasts are chosen.

continents as a drift. In careful and detailed studies he developed a theory that has become known as "Wegener's continental drift." According to this theory, 250 million years ago the present continents formed a single land mass, which Wegener named "Pangaea." This supercontinent then broke apart, and the individual pieces drifted away on the hot liquid earth's mantle like ice floes. Wegener's theory was vehemently disputed and rejected because Wegener could not explain where the force for the drift originated. After his death in the Greenland ice in 1930 discussion of his theory subsided, but after World War II it was revived by a number of discoveries: new finds of fossils on both sides of the Atlantic Ocean, rock magnetism, and, the observation based on this, that the bottom of the Atlantic spreads.[6]

Continental plates have drifted apart where the mantle circulations have an upward flow toward the earth's surface and new material from the mantle reaches the earth crust. Such an area is the mid-Atlantic ridge (Fig. 9.27). South America and Africa draw apart with a velocity of 1–5 cm/year. Where the circulation of the mantle is downward and surface material is drawn into the interior of the earth, a trench forms and the edge of the continent is raised, which leads to the generation of mountain ridges. In this way, the Andes were created.

Continental plates not only collide or separate, they can also slide along each other. Moreover, continents or parts of them can rotate about axes normal to the earth's surface. Figure 9.26 shows clearly that South America and Africa have rotated in relation to each other. Smaller areas can also be affected by rotations. For instance, Italy has rotated clockwise the last 40 million years by 50°, and a similar rotation of the Iberian peninsula has taken place. In areas of stagnation, even intake vortices in the earth's mantle circulation may appear. Wunderlich [55] believed that the arc of the Westalps was formed by such a vortex. This "alpine vortex" would have its center near the city of Torino.

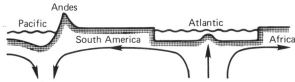

Figure 9.27. Schematic representation of the circulation of the earth mantle near the surface and the continental drift caused by it. The cross section is along a latitude south of the equator.

Remarks on Chapter 9

1. An interesting problem, which contributes to the understanding of climatic changes, has been studied by Hunt [56]. What would happen if the solar radiation suddenly stopped? Apparently the atmosphere would cool off quickly and its kinetic energy would decrease. The single processes, however, are quite complicated, and the decay is slower than expected. Hunt found with the aid of a computer program that the decrease of the general circulation is dictated by the decay of the turbulent eddies (including cyclones). The falling temperature interrupts the exchange between the zonal flow and the vortices. The latter dissipate quickly, while the zonal winds diminish only slowly. The consequence is that within 50 days the total kinetic energy decreases to only 25% of the initial value, but the energy of the vortices decreases to 3%. Over the same time span the average temperature of the atmosphere would drop from 261 K to 207 K.

2. The longest waves observed in the atmosphere at the 500 mb level may be interpreted as Rossby waves, which are a solution of the drastically truncated vorticity equation. Atmospheric waves, however, are much more complicated in their structure. They interact with waves of smaller lengths (nonlinear coupling), change in the presence of vertical flows, and influence the stability of the westerlies. Disturbances of this complicated system (baroclinic instability) give rise to jet streams, cyclones, and fronts. The Rossby waves themselves also owe their existence to baroclinic instability. The significance of baroclinic instability for the development of cyclones in the midlatitudes was first recognized by Charney [57] in 1947 (see also Eady [58]).

3. Streamlines at the ocean surface behave like those on a solid surface. Singular and regular separation points can occur (Figs. 5.56 and 5.58), but the nonslip condition at solid bodies is now replaced by the slip condition on the water surface. Figure 9.28 shows surface vortices in the current around South Africa. Maps for surface currents must be distinguished from those in which streamlines are averaged over the depth (for instance Fig. 9.16, which, by the way, is a very general schematic representation of the circulation).

4. The statement that an area of the ocean is at rest must be made with caution. As has been explained in Section 9.3, large-scale vortices of 50–400 km diameter separate from the Gulf Stream and migrate into the Sargasso Sea. Some even unite again with the Gulf Stream off the coast of Florida. Such vortices are also generated in other

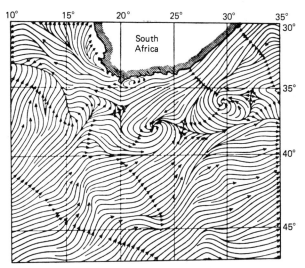

Figure 9.28. Surface streamlines near South Africa (from A. Merz [59]).

currents as in the Kuroshio and the circumpolar current, and off the coast of California. Also in the Baltic Sea and the English Channel, vortices have been observed with diameters of 50–60 km [60]. In addition to these large-scale vortices, which determine the "weather" in the ocean, there are smaller fluctuations near the ocean surface that form the "microstructure" of the ocean. These irregular oscillations of the order of centimeters indicate that the sea is turbulent in the upper layer of about 30 m. These oscillations are caused by winds and tides [61].

5. The Reynolds numbers in laboratory experiments are smaller than 10^4 compared to the Reynolds numbers of the order of 10^8 in nature. However, if the kinematic viscosity in the geophysical Reynolds number is replaced by the eddy viscosity (which is admittedly a crude turbulence model) with typical values of 10–100 m^2/s for horizontal motion, then the two Reynolds numbers of the experiment and the flow in nature are of the same magnitude. A similar argument can be made for the turbulent Prandtl and Schmidt numbers for the transport of heat and salinity, respectively. The true test for the usefulness of laboratory experiments is the anwer to the question of whether the most important parameters, the Rossby and internal Froude numbers, can be matched for laboratory and geophysical flows. This answer is affirmative [36].

6. It is not without irony that Wegener in his first paper in 1912 proposed sea-floor spreading as the cause of continental drift but that he abandoned this idea 3 years later in his famous book *Die Entstehung der Kontinente und Ozeane.*

10. Single Vortices in Atmosphere and Ocean

10.1. PREFERRED FREQUENCY RANGES IN THE SPECTRUM OF VORTICES

The expositions in Chapter 9 have shown that the heat exchange between equator and poles by means of the general circulation and the ocean currents is not a steady process. In addition to the daily, annual, and other periodic variations, there are aperiodic events caused by instabilities of the circulation systems. In midlatitudes cyclones and anticyclones are the most important mechanisms for maintaining the balance of energy and angular momentum. Also sporadically occurring disturbances like hurricanes contribute to the transport of energy and angular momentum. In oceanic circulations the fast western boundary currents (like the Gulf Stream and Kuroshio) transport heat to colder latitudes. In those currents water vortices of large size can form, which influence the heat environment (and also the weather of the surrounding areas).

In the train of these large-scale fluctuations smaller disturbances are observed, which often have singular characteristics in the sense that they occur only in a sporadic and irregular way. These individualists in the family of vortices are particularly fascinating. Chapters 10 and 11 are devoted to them.

Atmospheric and oceanic motions can be divided into three ranges depending on the period of the fluctuations (or for single disturbances on their lifetime) [1]:

	Micro-range	Meso-range	Synoptic range
Period, lifetime	1 hour	1–48 h	48 h
Wave length, diameter	20 km	20–500 km	500 km

In the ocean the lifetime in the mesorange and synoptic range can be much longer. If one considers the total spectrum of atmospheric and oceanic vortices, of which the most important are collected in the following list, one notes that vortices favor the microrange and synoptic range. To the mesorange, which is only sparsely represented, belong sea breezes and atmospheric and oceanic vortex streets behind islands. This means that the distribution of kinetic energy is not uniform in the atmosphere and ocean but has a minimum in the mesorange [1].

Vortices, Waves	Order of magnitude of Diameter Wavelength	Lifetime Period
Atmosphere		
Cyclones and anticyclones	1000–2000 km	days
Hurricanes	500–1000 km	days
Sea breezes	200 km	hours
Vortex streets behind islands[1]	30 km	hours
Lee waves	10 km	hours
Thunderstorms, cumulus clouds	10 km	hours
Tornadoes and waterspouts	500 m	minutes
Turbulent eddies	200 m	minutes
Dust devils	50 m	minutes
Ocean		
Ocean circulations	2000 km	quasisteady
Fluctuations of currents	500 km	months
Vortices of the Gulf Stream	50–400 km	months
Vortex streets behind islands	30 km	days
Langmuir vortices	200 m	hours
Tidal vortices (whirlpools)	10 m	minutes
Turbulent eddies	200 m	minutes

Vortices in the synoptic range are exclusively of the disk type because of the relative thinness of the atmospheric and oceanic layers (Fig. 2.4a). These vortices are generated in horizontal temperature and wind gradients essentially through barotropic and baroclinic instabilities.[2] Their order of magnitude is comparable or smaller than that of the Rossby waves. Although most ocean circulations are limited by the continents, their extents are of the same order of magnitude.

Vortices in the microrange have the same dimensions in the horizontal and vertical directions. They owe their origin to Helmholtz and Rayleigh–Taylor instabilities in vertical shear flows and vertical temperature gradients.

10.2. COLD FRONT AND SQUALL LINE AS ORIGINS OF LOCAL VORTICES

Cyclones and anticyclones are typical atmospheric phenomena in the midlatitudes of the earth. Their significance for the general circulation has been explained in Section 9.1. Cyclones (lows) and anticyclones (highs) play a decisive role in the daily weather and will be more closely studied in the context of the formation of local vortices [Reference 17, Chapter 9].

The development of a low and a high is presented in Fig. 10.1 by means of isobars and isotherms in a horizontal cross section for events both near the surface and at higher altitudes. Disturbances in the 500 mb level travel as Rossby waves from west to east (Fig. 7.9). The local vorticity increases from crest to trough and decreases from trough to crest. The absolute vorticity, which is assumed to be constant, can change, however, if one allows vertical flows. It follows from Fig. 7.4 that in an upwind, that is, at a horizontal convergence, cyclonic circulation increases; in a downwind, the anticyclonic circulation increases. The change in the absolute vorticity also changes the length and the amplitude of the Rossby waves. The wavelength shrinks from 5000–8000 km

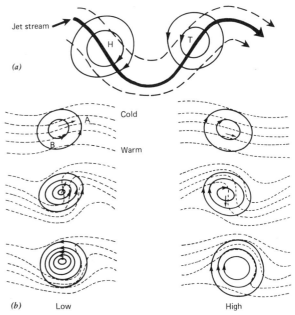

Figure 10.1. Development of highs and lows. (*a*) Isobars near the ground (solid) and streamlines in the midtroposphere on the 500 mb level (dashed). The jet stream is on the 200 mb level. (*b*) Isotherms (dashed) and isobars (solid) near the earth.

for constant absolute vorticity to 1600–3500 km for vorticity with a vertical flow.

The change in the circulation caused by vertical motion is clearly visible near the surface, where the westerlies are weak. There, cyclones develop at horizontal convergences, anticyclones at divergences. Under certain conditions, for instance, when a flow is disturbed by mountains (see also remark 2, Chapter 9), a horizontal convergence may develop between warm and cold air. If the incoming air is pulled upward by the sucking effect of the jet stream, the low-pressure area deepens to a cyclone. The center is then at the inflection point of the upper tropospheric wave. This means that the air (converging and being cyclonic) on the surface, is pulled up through divergence in high altitudes in an anticyclonic manner (Fig. 8.6). For a high on the surface, the situation is reversed.

The change of the isotherms in the course of cyclone and anticyclone formation will now be described. Suppose cold air is situated to the north, warm air to the south (Fig. 10.1b). In the subsequent development of associated vortices, the transition from cold to warm air takes place in an increasingly narrow area, until finally a sort of jump occurs. Such a discontinuity line in the temperature distribution is called a "front." When cold air is replaced at a certain place by warm air, one talks about a "warm front," which is designated on the weather map by half circles. The sudden change from warm to cold indicates a "cold front," marked on the weather map by triangles (Fig. 10.3). The most important mechanisms for the development of a front are:

1. The horizontal deformation, which is sketched in Fig. 10.2a and which occurs in Fig. 10.1 at the point A.
2. The horizontal shearing, which is indicated in Fig. 10.2b and which occurs in Fig. 10.1 at the point B [2].

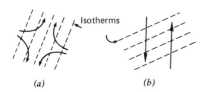

Figure 10.2. (a) Horizontal deformation and (b) horizontal shearing, which lead to the formation of a front (from B. J. Hoskins and F. P. Bretherton [2]).

Historically, the frontal theory was developed in the famous Norwegian school of Bergen about 1920. The names of V. Bjerknes, J. Bjerknes, and T. Bergeron are connected with this school. Their theory, however, did not yet include the reason for the development of cyclones, particularly the suction effect of the jet stream. In spite of its merely descriptive nature the frontal theory has become the basis for the meteorology of today. The classical theory of the Bergen School, which is sketched in Fig. 10.3 is well adapted to the dynamical theory of cyclone generation. The essential difference becomes clear in the vertical cross-section (Fig. 10.4). It may be mentioned, however, that a satisfying theory still does not exist [Reference 43, Chapter 9; 3].

Cold fronts not only move faster than warm fronts (Fig. 10.3), they can also generate an unstable state of the atmosphere when humid cold air is pushed upward. This instability causes violent storms in cold fronts, which can produce thunderstorms, hail, and tornadoes. Such storms occur, in particular, in intense lows, called "low-pressure

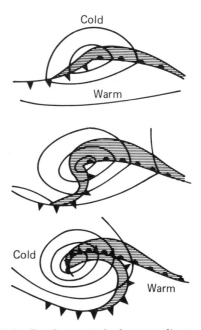

Figure 10.3. Development of a low according to Bjerknes. Areas of rain are shaded.

Figure 10.4. Vertical cross section through a warm front according to Bjerknes. The modern version is seen in Fig. 9.12.

190 Single Vortices in Atmosphere and Ocean

Figure 10.5. Low-pressure storm 1900 km north of Hawaii, seen from the Apollo 9 spaceship (NASA—photograph).

storms." Figure 10.5 shows the spiral cloud bands of such a storm.

Squall lines can coincide with cold fronts or can be precursors of cold fronts [Reference 17, Chapter 9]. Squall lines consist of series of single thunderstorms that form a chain up to several hundred kilometers long. They are more violent and last much longer than thermal thunderstorms (Section 8.2). Figure 10.6 shows a "storm cell" in a squall line, which is simultaneously a cold front. Hail and tornadoes can develop in such squall lines.

It may be mentioned that wavy disturbances in the trade winds can also roll up into vortices under certain atmospheric conditions. These tropical whirlwinds, which are much larger than tornadoes, will be described in Chapter 11.

10.3. TORNADOES AND WATERSPOUTS

Voices hoarse and shrill and sounds of blows, all intermingled, raised tumult and pandemonium that still whirls on the air forever dirty with it as if a whirlwind sucked at sand.

Dante, The Inferno

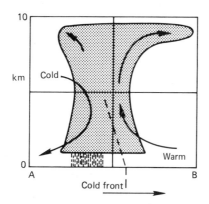 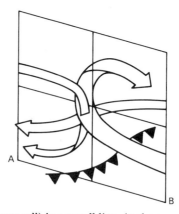

Figure 10.6. Sketch of a single thunderstorm (storm cell) in a squall line. At the cross point of the three air currents, hail and tornadoes may develop.

The highest local wind velocities on earth are observed in tornadoes. These columnar whirlwinds, with a diameter up to 500 m and a height of about 3 km, reach rotational velocities up to 200 m/s. If one thinks of the corresponding stagnation and lowest pressures occurring on an obstacle, one understands the enormous violence of these atmospheric vortices. Buildings collapse like houses of cards, trees are uprooted, people and beasts are lifted up (Fig. 10.7). Some tales about tornadoes border on the grotesque, but despite such exaggerations, the remaining facts speak for themselves. One of the best known tornadoes occurred on March 18, 1925 in the United States. It killed 695 people, injured over 2000, and touched the surface for 3 hours. This tornado traveled 350 km with a velocity of 95 km/h [4]. The 47 Palm Sunday tornadoes of April 11, 1965 killed 257 and injured 3000 and caused damage in six states—Illinois, Indiana, Ohio, Iowa, Michigan, and Wisconsin—all in one day [5].

Tornadoes are observed on all continents, but they are rare except in the United States and Australia [6]. In France they are called "trombe," in Germany "Trombe" or "Windhose," in the Soviet Union "Smerch." Since the most frequent and violent whirlwinds of this kind occur in the United States, the name "tornado" has been chosen for this species. The states of Texas and Oklahoma in the midwest have the most tornadoes. Texas leads with more than 130 a year. Wegener [7], who developed the theory of continental drift (Section 9.5), reported on the occurrence of tornadoes in Europe. If tornadoes appear over water, they are called "waterspouts," but they are no less dangerous than tornadoes on land. The well-known French pilot Jean Mermoz died in December 1936 in a waterspout.

The atmospheric vorticity required by tornadoes and waterspouts for their generation and maintenance is in most cases derived from cyclones [8]. Most tornadoes in the United States are found in cold fronts and squall lines near the jet stream, particularly in spring in the afternoon hours. Tornadoes are also observed in hurricanes, usually north of the center. A rare spectacle of nature is the formation of tornadoes during the eruption of a volcano [9]. Since vorticity in the mesorange and synoptic range is mostly cyclonic, tornadoes correspondingly rotate counterclockwise, although in general no influence of the Coriolis force can be expected because of its relatively small size. But there are also anticyclonic tornadoes, and sometimes two or more can be observed simultaneously.

Tornadoes are recognized by the funnel-shaped cloud, which broadens with hight and blends into the parent cloud. The funnel is about 800–1500 m high. Figure 10.8 shows four types of tornado with photographic examples in Figs. 10.9 through 10.11. In all of them the border of the funnel may possibly represent an area in which the vertical velocity is zero (Fig. 10.12).[3] Inside and outside the funnel a secondary flow is superposed on the horizontal rotation. Combined with the primary rotation the flow is spiral (Fig. 2.4b). Near the ground there is an Ekman layer in which the air flows in radially to the center [11]. Uprooted trees give evidence of the spiral flow near the ground (Fig. 10.13, see also the spiral water waves in Fig.10.11).[4]

A meridional flow different from that of Fig. 10.12 was found by Barcilon [13] and Maxworthy [14] and is depicted in Fig. 10.14. The phenomenon near the axis is labeled "boundary-layer eruption" and may be interpreted as vortex breakdown (Sections 5.4 and 7.5). In fact, vortex breakdown above

Figure 10.7. Tornado in the midwest of the United States (NOAA—painting).

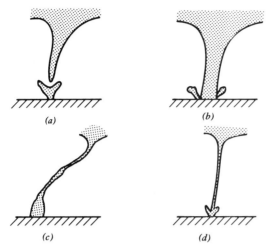

Figure 10.8. Four different types of funnel-shaped clouds of tornadoes. (a) The funnel does not quite reach the ground where a separate part of dust and debris is formed. This type is quite unsteady and jumps over the ground. (b) The broad and stable hose (sometimes called "elephant trunk") denotes a strong tornado (Fig. 10.7). (c) In the state of decay, the columnar cloud looks often ropelike (Fig. 10.9). (d) Waterspouts have mostly smooth, narrow cloud tubes (Fig. 10.11) (adapted from B. R. Morton [10], with permission from Pergamon Press, Ltd., Copyright 1966).

the ground can develop in tornadoes, as was confirmed in drawings from the 18th century [7] and in a recent one by Reber [15] (Fig. 10.15). The appearence may indicate a weakened or decaying vortex. The abrupt enlargement of the funnel with its convex shape can pass into the parent cloud. In Fig. 10.16b the typical paraboloid form, which was recorded by Bruzek in a photograph [16] and by numerous drawings from earlier centuries [7, 17], is compared with the usual concave shaped funnel (Fig. 10.16a).

Exact data on pressure, velocity, and temperature distribution are scarce, and those available are inaccurate because of the difficulty of obtaining measurements. Most data were obtained indirectly. Moreover, they vary considerably among individual tornadoes. In the following text some reference data are given for tornadoes in the United States; European tornadoes are generally weaker.

Velocities in the boundary layer at ground level have been estimated from tree damage to be up to 150 m/s. Just above the boundary layer, at about 40 m, rotational velocities of 200 m/s and vertical velocities of 70 m/s can occur. The profile of the ro-

Figure 10.9. Tornado at four different times (NOAA—photograph).

Tornadoes and Waterspouts   193

Figure 10.10.   Tornado (NOAA—photograph).

Figure 10.11. Waterspout (photograph by J. Golden, NOAA).

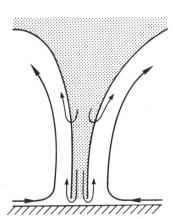

Figure 10.12.   Meridional flow in a tornado.

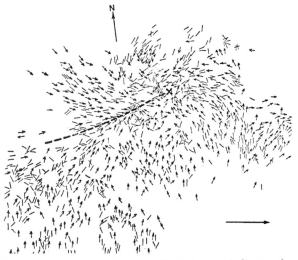

Figure 10.13.   The direction of felled trees indicates the flow pattern of a tornado near the ground (from L. J. Budney [12], *Weatherwise*, Vol. 18 (1965), p. 74. A publication of the Helen Dwight Reid Foundation).

194 Single Vortices in Atmosphere and Ocean

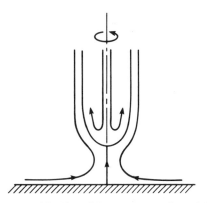

Figure 10.14. Meridional flow in a tornado with boundary layer eruption (adapted from T. Maxworthy [14], with permission from Pergamon Press, Ltd., Copyright 1972).

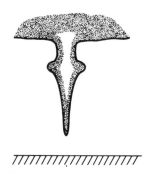

Figure 10.15. Vortex breakdown in a tornado. Drawing by C. M. Reber [15].

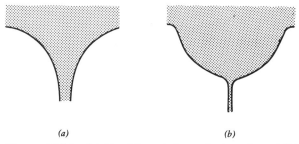

Figure 10.16. (a) Usual funnel shape of a tornado and (b) convex form, which indicates vortex breakdown near the parent cloud.

tational velocity corresponds roughly to the Rankine vortex (Fig. 3.10), in which the flow is strongly turbulent. Values for pressure drop of 20–200 mb have been recorded. The diameter of the vortex extends from 5 m up to 500 m. So far, not much is known about the temperature field. Supposedly the vortex center is about 10°C colder than its surroundings.

The path of destruction of a tornado is on the average about 200 m wide and 7 km long.[4] The travel speed averages 50 km/h. A tornado, however, can also remain almost stationary. Extremal values, which go far beyond the average, are possible as the examples in the beginning of this section show.

Tornadoes are audible over many kilometers. Their strong roaring can resemble the noise of a passing freight train, or they can hum like a swarm of bees [7]. The validity of these comparisons was confirmed in Reference 18 by means of acoustic spectral analysis. The noise is caused partly by the destruction on the ground and partly by the vibration of the vortex. In addition, tornadoes are often accompanied by electric discharges. The heating of the vortex by lightning was thought by some scientists like Vonnegut [19] to cause tornadoes, but electric discharges should have little to do with their origin.

Outside the vortex center secondary vortices have been observed, which are responsible for suction marks on the ground [20]. Not only can secondary vortices, which are smaller than the primary vortices, occur in tornadoes; several vortices of about equal strengths can develop from the same parent cloud. The twin vortices of Elkhart, Indiana, on Palm Sunday of 1965 became well known. A theory on the splitting of a tornado vortex was given by Rotunno [21].

The condition for the creation of tornadoes is the existence of an unstable shear flow with a sufficiently strong background vorticity in the mesorange. As has been already mentioned, such conditions are found in cold fronts and squall lines, hurricanes, and volcanic eruptions. In a squall line winds from opposite directions meet and are twisted by the Coriolis force. With the vertical upward motion, a mesorange circulation is generated which is 5–15 km across at a height of 5 km. This vortical motion is called a "tornado cyclone" and is visible in its lower part as the "parent cloud" of the tornado. Unstable cold air in the parent cloud sinks downward (Figs. 10.6 and 10.17) and concentrates to a tornado. This process resembles the formation of a bathtub vortex (Section 3.4), which is caused by a sink with background vorticity. Near the surface, then, an Ekman layer forms, which generates a radial influx with strong uplift near the vortex core, supported by the release of latent heat (Figs. 10.8 and 10.17). The ground effect strengthens the vortex and draws it completely to the earth through the supply of additional vorticity in the Ekman layer.

The kind of shear flow, with the upward wind and background vorticity that generate the tornado

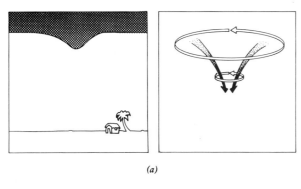

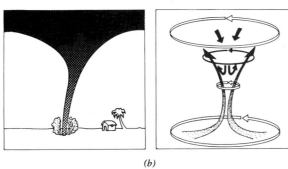

Figure 10.17. Cloud pattern and flow direction (*a*) in the initial phase and (*b*) in the mature state of a tornado.

cyclone, is decisive for the development of a tornado. In the so-called "Wokingham model" [22] it is assumed that tornadoes form in the turbulent mixing zone of three air currents: two warm horizontal currents of low heights compared with the height of a thunderstorm and a cold downward motion in the "anvil" of the thunderstorm (Fig. 10.6). In the Wokingham model it is furthermore assumed that the angle between the two currents of warm air is a critical parameter for tornado generation. This limitation to a certain angular range is probably one of the reasons that tornadoes are few compared to the number of violent thunderstorms [23, 24]. The model of the shear currents of warm air also explains why sometimes—despite the cyclonic background—anticyclonic tornadoes occur. The role of the jet stream, however, does not seem to be properly appreciated in the Wokingham model. As in extratropical lows, the jet stream also acts in the local field of a tornado in the form of a suction pump. The devastating Palm Sunday tornadoes of 1965 occurred along a jet stream.

Tornadoes that form over land and travel out to sea are called waterspouts. Not all waterspouts, however, are tornadoes. Whirlwinds that form over water are weaker than tornadoes and more closely resemble dust devils in size and development (Section 10.4). They are observed in sultry weather when the temperature of the ocean surface is relatively high. Instability of the lower air layers causes local convection. Here, also, background vorticity is necessary for the creation of a columnar vortex. However, surface winds with velocities greater than 10 km/h prevent the generation of waterspouts [25].

Waterspouts, whose tubes are much thinner than those of tornadoes, are observed in the Mediterranean, in the Gulf of Mexico, in the Bahamas, in the Bay of Bengal, and in the South China Sea (Fig. 10.11). They appear often in groups of up to 15 vortices (Fig. 10.18) [10].

Numerous laboratory experiments have been performed since the days of Wilcke (Fig. 1.11) to simulate tornadoes. Although it is relatively easy to produce concentrated vortices in the laboratory, none of the experiments described in the literature contains all aspects of a real tornado.

Nevertheless, partial aspects have been successfully studied, and the most important ones are listed:

1. The interaction of a columnar vortex with the ground was investigated by Wan and Chang [26], Davies–Jones [27], and by Hsu and Fattahi [28].

2. The simulation of the upper part of a tornado in the spirit of the Wokingham model was attempted by Gillies, Withnell, and Glass [23].

3. Secondary vortices were created in the laboratory by Weske [Reference 15, Chapter 6] (Fig. 10.19), Ward [29], and Church, Snow and Agee [30].

4. Vortex breakdown and multiple vortices in a single big vortex were modelled by Ward [29], Church, Snow, and Agee [30]. Depending on the strength of the swirl, vortex breakdown occurs in the center part of the vortex (Fig. 10.15). A stronger swirl moves the location of vortex breakdown toward the ground (Fig. 10.14), and with further increase of the swirl multiple vortices occur.

The most difficult features of tornado modelling are obviously the generation of vorticity in the parent cloud together with the vertical updraft and the influence of latent heat. Their realistic simulation remains a challenge to experimentalists.

Figure 10.18. Two waterspouts (NOAA—photograph).

10.4. DUST DEVILS

··· all the land is level, and the wind travels freely and wreaks the fury of Aeolia all over the desert. There is no rain in the cloud of whirling dust which it drives furiously in circles; most of the land is lifted up by it and is suspended in the air, as the eddying motion is continuous.
                                        Lucanus, Pharsalia

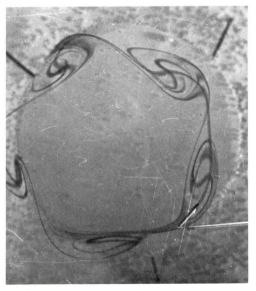

Figure 10.19. Steady secondary vortices at the edge of the boundary layer of a primary vortex (photograph by J. R. Weske, University of Maryland).

Dust devils have a certain similarity to tornadoes. They develop through unstable stratification of the air, receive their rotational energy from the vorticity of the surrounding shear flow, and are turbulent, locally restricted columnar vortices with strong uplift near the ground. Dust devils, however, do not attain the intensity of tornadoes and differ in the cause of the unstable stratification of the air. Whereas tornadoes are generated by the sinking of heavier air from a rotating parent cloud, dust devils develop from thin, hot air layers on the ground [7]. On clear summer days solar radiation in the desert can produce temperatures of 70°C near the surface (that is, within a layer 2 m thick). There, air becomes superadiabatic and unstable. As long as the unstable air layer is nearly horizontal, it cannot move up as a whole, and shifting takes place through local convection. If the ground is not flat, such convection does not occur, which explains why dust devils are not observed in the mountains. Dust devils receive their angular momentum from weak shear flows of superposed winds around local obstacles. It has been said that even animals can cause dust devils [31]. Thus, they are generated by a combination of Helmholtz and Rayleigh–Taylor instabilities in a vertical shear layer. The wind velocity, however, must not exceed 7–10 m/s, because higher velocities would hinder local convection. The high rotational speed is generated as in tornadoes by radial inflow (for dust devils in the boundary layer of the ground, for tornadoes in the parent cloud and, subsequently, also in the surface boundary layer) [32, 33].

The direction of rotation depends on the shear flow and the kind of obstacle. Therefore, both directions are found with equal frequency. Dust devils become visible only through sand, dust, and other flying particles, but they do not contain water droplets. As a curiosity it may be mentioned that small animals have been observed lifted up in dust devils [34].

Because of the way they are formed, dust devils are observed frequently in desert regions (Arizona, Mojave Desert, Sudan, Egypt, and Iraq), but they have also been seen in colder areas as in Iceland [35]. In Germany dust devils appear on hot days over the Autobahn and in places with high temperature differences. The largest dust devils are created in the sand storms of the desert. However, they cannot be distinguished there from tornadoes, because the way they are formed is difficult to observe [36].

The literature provides various quantitative data on size and intensity of dust devils. The following list gives some estimates and the most frequently observed values in parentheses.

Diameter near the ground	1–50 m (10 m)
Height	1.5–1500 m (100 m)
Duration	20 min (2 min)
Tangential velocity, maximum	15 m/s
Vertical velocity	10 m/s
Radial velocity near the ground	5 m/s
Pressure drop	2 mb
Temperature increase in the center	4°C
Velocity of migration	4–7 m/s

The tangential velocity distribution is similar to that of the Rankine vortex [37].

10.5. LANGMUIR VORTICES

Vortex rolls, known in the atmosphere in unstable air layers (Fig. 8.21) and generated by heat convection, can also occur in water. Here, however, they have in most cases a different origin. If a sufficiently strong wind blows over the ocean surface, stripes can occur in the direction of the wind. These "Windrows" are visible through the accumulation of foam, debris, seaweed, or ice on the sea or in a lake.

Windrows are caused by vortex rolls beneath the surface (Fig. 10.20). The spacing of these rolls, which play an important part in the mixing process of the ocean layer close to the surface, probably depends on the depth of the thermocline and the size of the surface waves.

In 1938 Langmuir investigated the phenomenon of windrows for the first time, and his name is associated with this type of vortex [38]. He reported on 500-m-long ribbons of seaweed in the ocean, each ribbon 2–6 m wide, and spaced 100–200 m apart. When the wind turned 90°, the ribbons formed anew within 20 minutes. Usually the distance between the ribbons is smaller, about 10 m.

The explanations for the occurrence of Langmuir vortices are controversial. However, it appears certain that they are not caused by heat convection. Faller [39] proposed that an instability of the Ekman layer may be the reason[5] (Section 7.6). Today the formation of Langmuir vortices is usually attributed to the interaction of surface waves with a shear flow (drift current) according to Leibovich [40] and Craik [41]. Comprehensive critical reviews have been written by Faller [42], Pollard [43], and Leibovich [44].

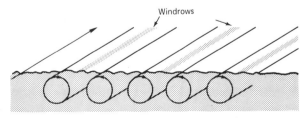

Figure 10.20. Windrows at the water surface are caused by vortex rolls.

10.6. TIDAL VORTICES

*And it boils and it roars, and it hisses and seethes,
As when water and fire first blend;
To the sky spurts the foam in steam-laden wreaths,
And wave presses hard upon wave without end.
And the ocean will never exhausted be,
As if striving to bring forth another sea.*

*But at length the wild tumult seems pacified,
And blackly amid the white swell
A gaping chasm its jaws open wide,
As if leading down to the depths of hell:
And the howling billows are seen by each eye
Down the whirling funnel all madly to fly.*

F. Schiller, The Diver

No vortex has been described more often in myths and legends than the tidal whirl (Section 1.1). It develops in currents caused by the tides, especially when the incoming tide hits the retreating waters of the previous tide. In straits with high flow velocities, very strong whirlpools can form. In his story *Descent into the Maelstrom* the poet Edgar Allan Poe paints a gloomy and morbid picture of the dangers of this vortex. The yarn which seamen spun about it was no doubt exaggerated, but the fact remains that tidal vortices are dangerous for fishing boats and swimmers. Everywhere on earth, wherever strong tidal currents occur because of rugged coasts and offshore banks or islands, one finds local tales about tidal vortices. The most famous one is no doubt Charybdis of the *Odyssey*. Personified as female monsters, Charybdis and the rock Scylla decimated the Homeric heroes in the Straits of Messina. The word Charybdis was sometimes used as a synonym for tidal vortex (Fig. 10.21) but was replaced by the word "maelstrom." Maelstrom was originally the name for tidal vortices in the Lofoten (Norway) and surfaced for the first time in Mercator's atlas of 1595. Olaus Magnus in his Carta Marina called it "horrenda caribdis" (Fig. 10.21, 16th century), and Athanasius Kircher talked about the "Norvegianus Vortex" or "gurges mirabilis" in his map "Mundus Subterraneus" from 1665.

Other known whirlpools are the Swilkie of Pentland Firth between Scotland and the Orkney Islands and the whirlpools of St. Malo in the English Channel (Fig. 10.22). The largest whirlpools are probably encountered in the Naruto Straits in the Inland Sea of Japan. The legend goes that a monster (this time a male) named Kappa threatens the fishermen [45]. Another male vortex monster, Kaegyihl Depguesk, supposedly devoured many a brave warrior of the Indian tribes of the coast of British Columbia [46].

Smaller tidal vortices exist in the shallows. They form in rills and cause indentations called "kolk" in the bottom of the sea (Section 7.5).

One may distinguish vortices that are locally fixed and vortices that are carried away with the currents. Locally fixed vortices are in most cases slowly rotating waters without noteworthy downward flow. They develop from flow separation behind obstacles or from widenings of the flow and are thus wake vortices. Figure 10.23 depicts tidal vortices in the sound of Tromsoe, Norway. Their locations depend on the flow direction. The northbound current is pushed toward the east by the Coriolis force, and vortices are generated at the western bank. When the flow direction changes, the situation reverses.

Rapidly rotating vortices of the intake type (Section 3.4) with strong downward motion (whirlpools) develop in shear flows when tidal waves bounce at each other, and they are formed behind obstacles. Here density differences in the tidal currents can increase the strength of the vortices. Boundary-layer boiling at the bottom of the sea can also generate vortices [Reference 65, Chapter 7] (Section 7.5). Most vortices of this kind are carried away with the current, but there are locally fixed concentrated vortices, such as those in kolks, which have secured a place by washing out the bottom (Fig. 7.37). Tidal vortices with horizontal axes can form behind shallow sills at the mouth of fjords [48], and dangerous whirlpools are also created in maritime caves and passages in the rock, such as the "Blue caves" in the limestone cliffs and underwater grottos of the Bahamas. In these caves and channels the tide generates strong currents that form columnar vortices which reach the surface of the ocean (Fig. 4.21). These tidal vortices are thus generated in the open sea without land nearby. The strength of such vortices is indicated by the discovery of the wreckage of a fishing boat at the entrance of a blue cave in water 25 m deep [49].

Only a few data on the size and intensity of tidal vortices are available. A report on tidal vortices in the Naruto Straits mentions that the diameter of the bell-shaped water surface can be 15 m or more [45]. The current in the straits reaches velocities of 4–5 m/s.

Figure 10.21. The Carta Marina of Olaus Magnus (16th century) shows the Maelstrom in the Lofoten Islands, called here "horrenda caribdis" (from the former Tall Tree Library).

The tidal current in the Straits of Messina is of interest not only for its Homeric description of Scylla and Charybdis but also for the oceanic peculiarity of this current [50]. Figure 10.24 shows that the greatest vertical constriction of the straits coincides with the narrowest horizontal constriction off Ganzirri–Punta Pezzo.[6] At this location the highest flow velocities are to be expected. In addition, the tides of the Ionian and Tyrrhenian Seas are exactly opposite. That is, when the Ionian Sea has a high tide at the Straits of Messina, about 3 hours after the meridional passage of the moon (MM), the Tyrrhenian Sea has a low tide. Nine hours after MM the situation is reversed. Maximal flow velocities of about 2 m/s occur off Ganzirri–Punta Pezzo, that is, northward 4.7 hours after MM, southward 10.7 hours after MM. The current does not reverse simultaneously throughout the whole straits, but changes earlier in the north than in the south. About 2.5 and 7 hours after MM such flow reversals occur at the narrowest passage. They take place in form of turbulent layers of opposing currents accompanied by vertical flows, so-called "flow convergences" (Fig. 10.25) [51]. Since the water of the Io-

Figure 10.22. Tidal vortex near St. Malo in the English Channel (photograph by George Silk, *Life Magazine*, © 1969 Time Inc.).

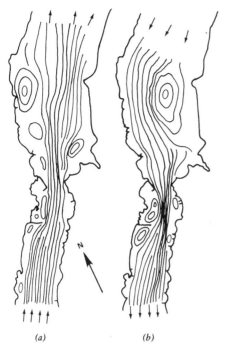

Figure 10.23. Surface streamlines in the straits near Tromsoe, Norway: (a) northbound tidal current and (b) southbound tide (adapted from H. Mosby [47]).

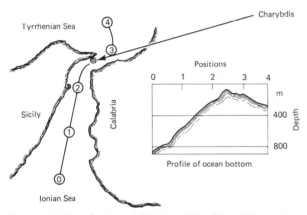

Figure 10.24. Straits of Messina. The Charybdis vortices form at the narrowest place where the highest flow velocities occur (adapted from A. Defant [50]).

nian Sea is colder and saltier than that of the Tyrrhenian Sea, making Ionian water heavier than Tyrrhenian water, stratified shear flows occur, in which instabilities cause strong vertical and horizontal vortices [52] (Section 8.2). Stratified shear flows are generated horizontally when countercurrents with different densities occur near the coast and vertically through superposition of the lighter water over the heavier one. The particular geography of the coast then favors the development of

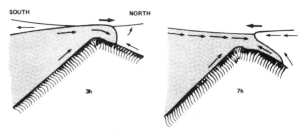

Figure 10.25. Sketch of flow reversals in the Straits of Messina. The occurrence of "flow convergences" causes unstable stratified shear layers and the formation of tidal vortices (adapted from A. Defant [50]).

dangerous vertical vortices. About 1 km off Cape Peloro the whirlpools of Charybdis form, off the port of Messina the whirlpools of Punta San Ranieri.

The arrival of the two periods of flow convergence is visible from the coast through the appearance of tidal bores (Fig. 10.25). The connection between bores and whirls is indicated already in Dante's seventh song of Inferno:

*Just as the surge Charybdis hurls to sea
crashes and breaks upon its countersurge, ...*

Remarkes on Chapter 10

1. Vortex streets behind islands have been detected on photographs from weather satellites. According to Chopra and Hubert [53] vortices behind the Canary Islands had diameters from 20 to 40 km. The length and width of the vortex street were 400–500 km and 40–50 km, respectively. The pattern lasted 18–30 hours. A survey is given in Reference 54.

2. The difference between barotropic and baroclinic instability is described in the following way: Consider in a barotropic fluid a geostrophic horizontal main flow, which does not change in the vertical direction. A disturbance of this main flow, which increases (and can grow only in the horizontal plane) represents a barotropic instability. In a baroclinic fluid, however, the main flow can change (as well as the superposed disturbance) with height. If this flow becomes unstable, one talks about baroclinic instability. The difference between these types of instability is important in understanding energy transport. Barotropic waves transfer the energy of zonal westwinds to vortices, whereas baroclinic waves cause the transition from available potential energy of zonal westwinds to kinetic energy of the vortices [55, 56].

3. The study of columnar vortices in the laboratory has raised the question of whether artificial tornadoes can be exploited for energy production. This idea was probably first expressed by A. Spilhaus, who suggested building a huge funnel-shaped tower. The temperature difference between the sun-heated ground and the colder air on top of the tower would cause the colder air to descend inside the tower. Through suitable guide vanes a tornado could be generated, which would propel a turbine. More realistic is Yen's proposal [57], to let air flow tangentially into a 60-m-high tower of 20 m diameter. Through rotation a low pressure is generated in the center of rotation. At the bottom of the tower air is sucked in through an opening from outside, which flows upward. The sucked-in air could also propel a turbine.

4. In Scandinavian mythology (and still in today's tales of Swedish farmers), the visible path of destruction through woodland caused by tornadoes is called "Asgard's way," which means the road to the residence (Asgard) of the gods [7].

5. Faller [58] has pointed out that in the atmosphere, too, not all vortex rows need be explained by heat convection. Rather, instability of a shear layer can be the reason for it.

6. Charybdis is said to have been relocated and weakened since antiquity, and the whirls of Scylla have become insignificant. Defant [50] believes that the straits between Sicily and Calabria were narrower 3000 years ago and that, therefore, the tidal currents were stronger than they are today. It may be recalled that in geological times Sicily and southern Italy were not separated by the sea.

11. Hurricanes

In June   —It's too soon.
In July   —Stand by.
In August   —Look out you must.
In September—Remember.
In October —They're all over.
Rhyme from the West Indies

Under certain conditions atmospheric disturbances can originate near the equator and develop within a few days into vortices 500–1000 km in diameter. These almost circular tropical storms can generate very high wind velocities near the center of rotation. If the tangential velocity exceeds 120 km/h, the storm is called a hurricane. Maximum wind velocities may reach 250 km/h on land, and up to 300 km/h over open seas. Torrential rains, high seas, and inundations of coastal areas are also typical of such powerful storm systems. Hurricanes move with a forward speed of 20–30 km/h. They usually migrate with the trade winds to the west and then turn eastward at higher latitudes (Fig. 11.1).

A hurricane is known by different terms in various parts of the world. In the Atlantic and along the west coast of Central and North America it is called "hurricane," in the Pacific "typhoon," and in the Indian Ocean "cyclone." Local names are "cardonazo" on the west coast of Mexico, "baguio" in the Philippines, and "willy-willy" in Australia.

11.1. HISTORICAL NOTES

The first records in human history of the fury of tropical storms are as obscure as the periods from which they originate. They are found in the Old Testament and in the cuneiform texts of the Babylonians as the story of the Deluge, which inundated part of the then known world. In 1929 the archeologist Wolley discovered in Chaldea a 2-m-thick layer of clay, which must have resulted from a huge inundation. He dated this tidal catastrophe to 4000 BC [1]. Meteorologists know that tropical storms occur in the Persian Gulf and that they can cause great inundations and destruction of life and property. But has this layer of clay anything to do with the Deluge, and was it caused by a cyclone? Nobody knows, but even Leonardo as long ago as the Renaissance argued that the Deluge must have been a local event and not a worldwide catastrophe [2].

Reports on tropical storms from later times are more reliable. In the 13th century one of the most capable successors of Genghis Khan, his grandson Kublai Khan, prepared an expedition of ships to conquer Japan. This campaign came to a premature end in 1281 when the whole fleet of Kublai Khan was lost in a huge typhoon in the Hakata Bay of Japan. An army of 100,000 Mongolian, Chinese, and Korean soldiers was said to have drowned. The Japanese called this saving storm "Kamikaze" (divine wind)[1] [3].

It was not until two centuries later that a steady source of information on tropical storms became available with the discovery of America. On October 12, 1492 Christopher Columbus landed on one of the Bahama islands after a 71-day voyage. He did not recognize the historical importance of that moment nor was he aware of the danger of tropical storms. He crossed the Atlantic at a time, and then disembarked on a shore, when and where hurricanes most frequently occur without encountering a single storm. His incredible luck did not desert him even on his return, which was, however, jeopardized by

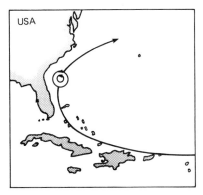

Figure 11.1. Typical path of a hurricane in the Atlantic. Since the direction of the vortical rotation is determined by the Coriolis force, it is always counterclockwise in the northern hemisphere.

the storms of the Azores. On his second trip to Central America Columbus did hear of the Indian name "Furacane" from which today's word "hurricane" originated. But only on his fourth voyage did Columbus experience the full fury of a hurricane south of Haiti. He barely escaped. Earlier, Columbus had warned the governor of Haiti about this storm, since a Spanish flotilla was ready to sail in Santo Domingo. The governor, unfriendly toward Columbus, ignored the warning. Of 30 ships laden with treasures and slaves, only four escaped the storm [4].

Such catastrophes recurred many times in the ensuing centuries of exploration of America. The history of discoveries on the American continent in the 16th and 17th centuries has, in fact, also become a chronicle of hurricanes. Shakespeare devoted the play *The Tempest* to this subject. It is surprising that not until 1821 did a storekeeper by the name of Redfield in Connecticut find out accidentally that a hurricane is a whirlwind: A passing hurricane had uprooted trees in opposite directions.

In human history tropical storms have caused enormous natural catastrophes. As in the biblical Deluge, it has been not the high wind velocities but the storm-intensified tidal waves that have brought about the greatest disasters. For example, on October 7, 1737 the 12-m-high tidal wave of a cyclone reached the Bay of Bengal and destroyed 20,000 ships and fishing boats. More than 300,000 people were said to have drowned. The same number of fatalities was caused by a typhoon in Indochina in 1881. A year later 100,000 people died in Bombay. In the Galveston hurricane of 1900, 6000 people lost their lives, and in 1906 Hongkong reported 10,000 dead from the impact of a typhoon.

11.2. HURRICANES IN THE 20TH CENTURY

Modern technology has not yet been able to lessen the fury of hurricanes. Of course, modern communication systems with the aid of weather satellites, reconnaissance flights, and seismic apparatus provide advance storm warnings. These techniques have enabled people in urban and rural areas to be moved to safe locations in time. There are computer programs with which position predictions of modest accuracy can be made [5]. Even influencing hurricanes with dry ice has been tried on the U.S. east coast.

Nonetheless, tropical storms still create great anxiety and wreak havoc. On September 19, 1945 the typhoon "Makurazaki" swept over Japan. More than 4000 dead and almost 56,000 completely destroyed houses were counted [6]. In this century the largest losses have resulted from cyclones in the Bay of Bengal, where the rectangular shape of the flat coast of Bangladesh and the width of the continental shelf increase the height of the tidal waves (Fig. 11.2). The large population density at the coast and the flatness of the islands turn every cyclone-caused inundation into a catastrophe. The statistics of the annual dead are terrible. On November 13, 1970 a cyclone with 240 km/h tangential velocity and a 6-m-high tidal wave hit the Bengalian coast at the Ganges delta (Fig. 11.2). The number of dead was 300,000 [7]. On Christmas Day, 1974, the city of Darwin in northern Australia was 90% destroyed by a cyclone.

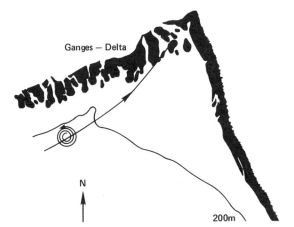

Figure 11.2. Gulf of Bengal at the Ganges delta. The 200-m-depth line shows the widespread area of the continental shelf. The path of the cyclone of November 13, 1970 is drawn (adapted from G. R. Flierl and A. R. Robinson [7]).

The significance and impact of tropical storms in the strategies of World War II is revealed by the following true account. On December 18, 1944 the Third U.S. Fleet under Admiral Halsey crossed the path of a typhoon in the waters east of the Philippines. Without knowing of the possibility of a typhoon, the fleet had the task of supporting the impending landing on the island of Luzon, which was occupied by the Japanese, and gaining air control over Luzon. The fuel reserves of many ships were so low that refueling from tankers became a necessity. Tankers were ordered to the task and the refueling started on the morning of December 17th. The weather at that time was already very bad, but nobody thought of a typhoon. The refueling during such weather caused tremendous difficulties and had to be postponed several times. In the meantime a low-pressure storm was reported which, on the morning of December 18th, was recognized as a typhoon. Unfortunately, the meteorologist of the flagship predicted the wrong position of the typhoon, so Admiral Halsey ordered a change of course, which led the fleet directly into the path of the typhoon. This happened about noon of December 18th. Three destroyers with half-empty tanks sank in the fury of the storm and 28 ships were damaged, 7 of them heavily. Almost 800 men lost their lives, and about 150 airplanes crashed on the aircraft carriers or were swept from the decks. The planned attacks on Luzon could not be carried out on time [8].

Another more recent tragedy, although less costly in human terms, will help the reader become more familiar with the occurrence and migration of hurricanes. In early September 1957 a hurricane was reported in the developing stage south of the Cape Verde Islands. This hurricane, named "Carrie" by the meteorologists, was tracked by reconnaissance planes, and its intensity was measured. Its migration is one of the longest ever recorded (Fig. 11.3), the path being about 10,000 km. On September 21, 1957, south of the Azores, Carrie met the German sailing vessel *Pamir* which sank. Of the 86 men on board only 6 were rescued from a lifeboat [9].

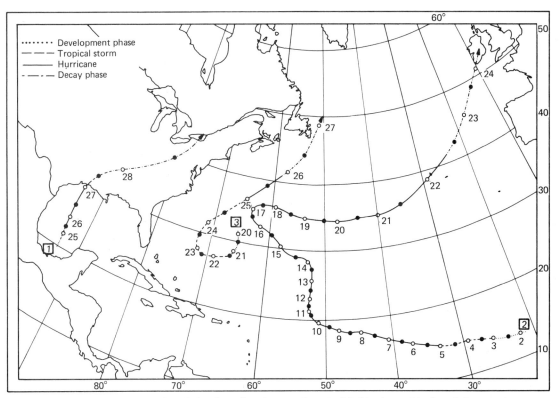

Figure 11.3. The paths of the three hurricanes of 1957: (1) Hurricane "Audrey," June 25th through 28th, (2) hurricane "Carrie," September 2nd through 24th, and (3) hurricane "Frieda," September 20th through 27th. ○, Position at 7:00 Eastern Standard Time, ●, Position at 19:00 (adapted from Monthly Weather Review [9]).

11.3. CONDITIONS FOR THE DEVELOPMENT OF HURRICANES

From the knowledge of where and when tropical storms occur most frequently, the first conclusions can be drawn on their causes and sources of energy. The world map in Fig. 11.4 shows the main routes of tropical storms. At the start, their paths are almost all parallel to the equator and move westward without, however, coming to the equator or crossing it. As these storms continue their journey, they move farther and farther from the equator and in most cases turn to the east, where they slowly die in colder regions. The exact path of an individual storm depends on the overall weather conditions at the time, in particular, on the position of the jet stream and the locations of highs and lows nearby. In 1976 the jet stream over the United States shifted unusually far to the south. As a consequence, the paths of the hurricanes changed in such a way that the storms in the western part of the North Atlantic occurred only east of the Bermudas and thus spared the east coast of the United States. West of Mexico, however, a hurricane was deflected to Baja California, where it caused great damage.

The most favorable season for the development of tropical storms in the northern hemisphere is late summer; in the southern hemisphere, it is during the spring. According to Colón [10] the following average monthly occurrence has been observed for hurricanes in the Atlantic and in the Pacific west of Mexico:

May	June	July	August	September	October	November
0.1	0.4	0.5	1.6	2.4	1.9	0.4

This distribution is reflected in the rhyme from the West Indies cited at the beginning of this chapter. Hence, on the average, 7.3 hurricanes occur per year. The average number of typhoons in the middle and western parts of the North Pacific is 20.4 per year.

The direction of rotation of all tropical storms is without exception cyclonic, that is, they rotate counterclockwise in the northern hemisphere and clockwise in the southern hemisphere. From this it can be concluded that the Coriolis force determines the direction of rotation.

Evaluation of the information thus far results in the following conditions for the development and maintenance of tropical storms (after Palmén and Riehl [11, 12]).

1. The Coriolis force must have a certain minimum value. Since this force decreases toward the equator and vanishes there, this condition for the generation of such vortices does

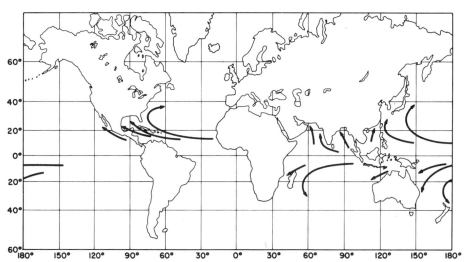

Figure 11.4. The main paths of tropical storms on earth. Arrows indicate the direction of motion.

not occur in a belt of 5°–8° on either side of the equator (Fig. 11.4).

2. Because the intensity of the vortex diminishes in colder waters and over land, a sufficiently large ocean surface with a minimum temperature of 27°C must exist for the generation and maintenance of tropical storms. This minimum temperature requirement and the condition that the Coriolis force must have a certain magnitude restrict the development of tropical storms to the warmer months.[2] During this time the doldrums, the belt in which the northern and southern Hadley cells meet (Fig. 9.7), are farthest away from the equator.

3. Tropical storms presume certain atmospheric disturbances. Compared to the frequency of such disturbances in the tropics, however, the development of a tropical storm is a rare event.[3] Thus, a number of additional factors must come together to trigger a tropical storm.

Although all factors are not yet completely known, the following approximate picture of the development of a tropical storm can be drawn: A low-pressure trough in the trade winds can become the starting mechanism for the development of a storm. Observations have shown that hurricanes can form at a latitude of about 15° from such low-pressure regions, whose nuclei often lie over West Africa [13]. In a low-pressure region air is sucked in from its surroundings near the surface. If at larger heights over the low a horizontal air current exists (in particular, a jet stream), it will draw air out of the low. The vertical flow increases, and the low becomes more intense. Heavy rain favors this process because it increases the humidity. The air releases part of this humidity in the updraft of the low through condensation and heats up through the release of latent heat (heat which becomes available during condensation). This heating increases the buoyancy of the air, and a tropical depression is thus born. The motion of the wind is now circular due to the influence of the Coriolis force, and closed isobars show up on the weather maps. At this stage the fate of the tropical storm is decided. Up to this time the updraft at the center is still so weak that disturbances in the surrounding air can suppress the upward motion. This actually occurs in most cases. However, if the depression is located over a warm ocean surface, and if other conditions (such as the presence of a jet stream) are favorable, then the air flowing over the sea surface is further enriched with humidity, which condenses at the center and releases latent heat. The pressure at the center decreases further, and the tangential velocity increases. In less than a day the tropical depression can reach the strength of a tropical storm or even that of a hurricane. Above a certain wind speed, the Ekman layer becomes detached near the center and forms a separated region, the "eye" of the storm.

The World Meteorological Organization has agreed on the following classification for tropical low-pressure disturbances, according to their intensity [14]:

1. Tropical disturbance: Slight rotary circulation, no closed isobars on the weather map.
2. Tropical depression: Some closed isobars, a little rotation near the earth's surface. Highest wind velocities: 63 km/h.
3. Tropical storm: Closed isobars, clear rotation of the wind with velocities between 63 and 120 km/h.
4. Hurricane: Wind velocities above 120 km/h.

As soon as a tropical depression in the Atlantic has reached the strength of a tropical storm, it receives a name. The temporal sequence in each season is designated alphabetically by a given name (formerly female names only, but now alternately male and female). From Fig. 11.3, for instance, it can be seen that hurricane "Audrey" was the first storm of the 1957 season.. The second storm "Bertha" did not reach the strength of a hurricane and is not shown in Fig. 11.3. The third storm of the season was hurricane "Carrie."

11.4. THE HURRICANE AS A HEAT ENGINE

Although the way in which tropical storms develop is not entirely clear, fairly reliable statements can be made about the structure and mechanisms of a mature hurricane. Observations made from airplanes and satellites above such storms, measurements made inside the storm by reconnaissance flights, and theoretical considerations result in a picture of a meridional flow as schematically presented in Fig. 11.5. One may distinguish two regions: The vortex ring about the central core in which all the action of the storm takes place, and the center itself, the eye of the storm, which is quite calm [15].

Immediately above the sea surface the vortex ring consists of the Ekman layer, in which the air flows almost radially toward the center. The Ekman layer is about 1.5 km thick. The strong radial wind pulls the water upward toward the center,[4] which is the reason for the inundation of coastal areas. If this rise of the surface at the center coincides with the tide, catastrophic floods can occur (Sections 11.1 and 11.2). The Ekman layer is deflected upward immediately outside the center and forms a vertical cloud wall around the eye. In this updraft, which has an average velocity of about 8 m/s, the humid air condenses, and torrential rainfall results. At a height of 10 km the air turns away from the center, descends at the outer part of the storm, and closes the cycle.

Superposed on this vertical circulation is a horizontal rotation generated by the action of the Coriolis force (as in the upward flow described in Section 8.2). The rotation is cyclonic near the surface up to a height of 10 km, and is anticyclonic above (and farther away from the center).

The azimuthal velocity profiles from numerous observations are plotted in Fig. 11.6. Despite the differences in individual storms, the following average characteristic properties are obtained [16]: If V denotes the azimuthal velocity (velocity of rotation) and r is the distance from the center, then the following relation holds for the lower troposphere (900 to 500 mb):

$$Vr^x = \text{const}, \quad \begin{aligned} x &= -1 \text{ for } 0 \leq r \leq r_{\max} \\ &= 0.5 \text{ for } \quad r > r_{\max}, \end{aligned}$$

where $r_{\max}$ is the location of the greatest velocity. Although the standard deviations are quite large, it is obvious that below $r_{\max}$ hurricanes rotate like solid bodies, but above $r_{\max}$ they deviate from the potential vortex by $x = 0.5$. Of course, the flow inside the storm is turbulent and has radial and axial velocity components.

The rotation is in most cases anticyclonic above a height of 10 km [17]. In contrast to the cyclonic motion near the surface this rotation is quite asymmetric, because the superposed stratospheric winds act as suction pumps on the ascending air of the vortex [18, 19]. A slight asymmetry near the surface

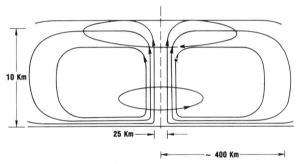

Figure 11.5. Schematic cross section of a hurricane. The rotation is cyclonic in the lower part, anticyclonic above. The vertical scale is stretched.

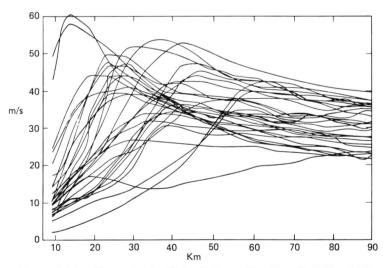

Figure 11.6. Observed azimuthal velocity profiles (from D. J. Shea [16]).

is visible in Fig. 11.7. The rule (important for the sailor) is that the highest wind velocities occur to the immediate right of the storm center in the direction of movement (in the northern hemisphere). If the translational velocity is superposed on the rotational velocity (which gives the velocity measured on the surface), the asymmetry is even more pronounced (Fig. 11.8).

Thermodynamically, a tropical storm may be likened to a heat engine. On the open sea the air takes up humidity from the ocean surface through evaporation. During this process the top layer of the ocean over which the storm passes cools off by about 2°C. Energy is thus transmitted to the air in the form of latent heat. Near the center, the humid air is deflected upward and cools through adiabatic expansion. The humidity condenses and returns to the sea in the form of torrential rainfall. Latent heat is thereby freed and warms the air. The generated low pressure sucks air along the Ekman layer, while the heated air flows upward through buoyancy. During this process heat energy is transformed into kinetic energy. At the storm's center on the open sea, pressure differences of 90 mb relative to the surrounding atmosphere have been measured.[5] The lower the pressure, the more intense is the storm. In Fig. 11.9 the average maximum velocity of rotation is plotted against the pressure at the center. Large fluctuations of the data, however, occur. The figure also

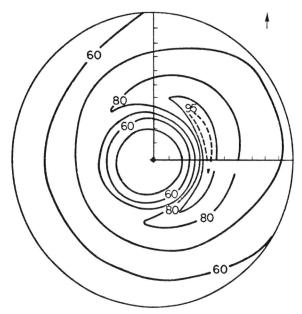

Figure 11.8. Azimuthal velocity with superposed translational speed near the surface (900 mb) (from D. J. Shea [16]). (Data in knots; 1 knot = 1.852 km/h.)

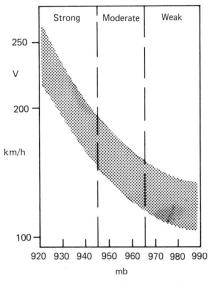

Figure 11.9. Approximate data for the maximum azimuthal velocity versus the measured pressure at the center of the storm. The real data, however, deviate considerably from these average values (adapted from D. J. Shea [16]).

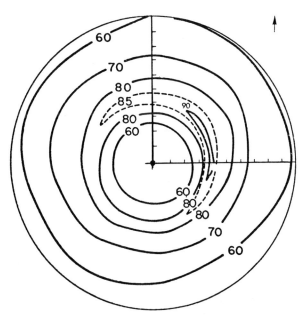

Figure 11.7. Azimuthal velocity relative to the center of the storm near the surface (900 mb) (from D. J. Shea [16]). (Data in knots; 1 knot = 1.852 km/h.)

shows the pressure regions in which a storm is considered weak, moderate, or strong. At heights above 10 km, the air flows away from the center, partly back into lower layers where the cycle ends, partly carried away by the currents of the upper atmosphere and the stratosphere. The thermodynamic process resembles a heat engine. Nature, however, does not work very economically, because the efficiency is only about 3%. The seawater is the actual heat donor, and one realizes immediately that the power of the storm must diminish as soon as it moves over colder waters or onto land [20]. By that time, the hurricane has brought rain to arid regions and has carried away heat that builds up in the tropics. A hurricane, after all, does perform a useful task.

It has been suggested that the delicate interplay among friction caused by air transport in the Ekman layer, released latent heat in the updraft (whose quantity depends on the sea's temperature), and the suction in the anticyclonic high-altitude current are the deciding factors in the survival of a storm. If less air is sucked away to the upper layers than is brought in near the surface, the center fills up, and the storm dies. If, however, more air is sucked away than flows in, the intensity of the storm increases until an equilibrium is reached.

Radar has shown for the first time that the clouds around the eye of the storm are spirally arranged, with cloudless intervals of 50–80 km (Fig. 11.10). This phenomenon, as well as the almost cloudless eye itself, has made it possible to track hurricanes with satellites and to estimate their strength [22] (Figs. 11.11 and 11.12). In April 1960 the weather satellite "Tiros I" made meteorological history when it discovered 1300 km off Australia a typhoon, whose existence until then had not been known. The spiral formation of clouds can be explained with the initial condition of the storm: The warm humid, cloud-covered tongue in the tropical trough

Figure 11.11. Typhoon "Opal," September 4, 1967, seen from a weather satellite (ESSA—photograph).

Figure 11.12. Two typhoons in the South China Sea, September 4, 1968, seen from a satellite (ESSA—photograph).

Figure 11.10. Composite radar picture of hurricane "Daisy," August 27, 1958, taken from a height of 4000 m (from B. Ackerman [21]).

rolls up during the development of the storm (Fig. 11.13). The amount of precipitation during an average storm is about 20 cm/day near the center. It sometimes happens that a storm stays in an area for several days before it moves on. Then, the tremendous amount of rain leads to large inundations.[6]

The eye of the storm is a strange phenomenon. It is created through separation of the radial Ekman

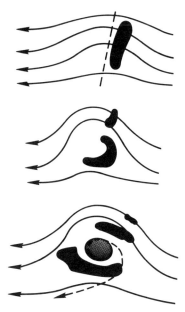

Figure 11.13. The formation of spiral cloud bands. In some tropical storms the sky is completely covered with clouds except for the eye (adapted from R. W. Fett [23]).

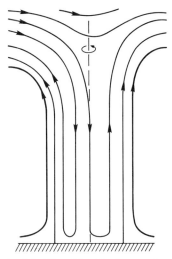

Figure 11.14. Meridional flow in the eye of a tropical storm, parallel to the stratospheric suction flow (sketch).

Remarks on Chapter 11

1. Japanese pilots in World War II, who sacrificed themselves, were called "Kamikaze" after the typhoon of 1281.

2. In the South Atlantic no tropical storms occur because the ocean is relatively cold. The doldrums are almost always located in the northern hemisphere.

3. In the trade winds of the North Atlantic about 100 low-pressure disturbances are generated annually. From these, however, only 4.8 hurricanes develop on the average.

4. On the sea east of Bombay, a wave height of over 15 m (including tide) was measured at an oil rig [24].

5. Data vary on the lowest barometric pressure ever measured at the center of a tropical storm. Tannehill [10] reports a low pressure of 889 mb measured on board a ship in a typhoon.

6. In October 1963 hurricane "Flora" stayed for four days over the eastern part of Cuba. The rainfall destroyed half of the sugar crop of Cuba and 90% of the coffee crop.

layer flow immediately next to the center. Within the eye a weak downflow of dry adiabatic air prevents cloud formation (Fig. 11.14). Behind this sober explanation is hidden a unique drama of nature. After the furious rage of the storm passes over, the spectator suddenly encounters a weird stillness: the eye of the storm. The pressure drop at the center is considerable, and the sultriness is almost unbearable. Joseph Conrad has vividly described such an experience in his novel *Typhoon*. The calm of the center, however, is deceptive, because soon afterward the storm bursts forth again, this time with winds from the opposite direction.

The air circulation inside the eye is not yet fully understood, since the flow velocity is so small. This is particularly true for the upper part of the eye. Figure 11.14 shows schematically a cross section of the eye parallel to the stratospheric suction flow.

12. Extraterrestrial Vortices

12.1. PERIODICITY AS AN ORDERING PRINCIPLE

Since prehistoric times humans have observed the stars and been aware of their orderly movements as proof of order in the universe. The essence of this order lies in the predictability of the returning stars, their periodicity. The ancient Greeks considered this world order perfect in the mathematical sense. Accordingly, the orbits of the stars could only be circular and their periodicity eternally unchanging (Aristotle). In modern times the geometric periodicity of the stars has been replaced by the dynamic one of classical mechanics. Periodicity as an inherent property of rotation remains an expression of an ordering principle, but it is subject to evolution. Stars, clusters of stars, and galaxies are rotating systems, gigantic vortices. However, the description of their orbits no longer suffices. How have they developed? Where does their angular momentum come from?

Before the cause of the rotation of stars and galaxies is investigated, the question must be asked why rotary motion is predominant in the universe. The answer reveals a deeper meaning of the ordering principle than the mere predictability of cyclic events.

In previous chapters rotation and vortical motion have been viewed in a continuum. Section 2.2 outlined the fundamental concept that every movement in a finite space must be rotary, a consequence of the law of conservation of mass. The continuum concept, however, must be abandoned for explaining the orbits of stars. In relation to the immensity of space, in which the stars move, the volume of the stars is vanishingly small.[1] The permanency of circling stars and clusters of stars is explained by the conserving interplay of gravitation and rotation (except for the slow course of evolution). They do not enlarge the entropy of the system. Thermodynamics says that all forms of energy change finally to the lowest one, that of heat energy. A measure for this change is entropy. In a closed system entropy increases continuously until the final state of heat death. Rotation delays the process toward heat death.

A simple example will illustrate this concept (Fig. 12.1). A large and small body meet in space. If they hit each other head on, gravitational energy will be transformed immediately into heat energy. However, if they do not meet each other directly, they will be deflected in such a way that they circle each other. Gravity and centrifugal force are in equilibrium, and entropy does not increase.

Rotation is one factor among others responsible for the existence of the universe for about 16 billion years and for the fact that it has not already perished in a "gravitational collapse." Other mechanisms also delay this process, the most important one being the simple fact that space is so thinly filled with matter. The time that a body requires to strike another one is already very large without rotational delay, and the expansion of the universe supports this form of delay [1].

In the following sections astronomical data, theories, and hypotheses on rotational properties, and the sources of angular momentum of stars and larger systems are briefly summarized.

12.2. THE SOLAR SYSTEM

Compared to material rotations on earth, which have been described in previous chapters, planetary

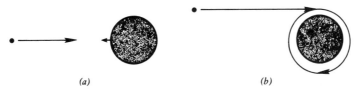

Figure 12.1. Transformation of gravitational energy (a) into heat with an increase in entropy and (b) into rotational energy without increase in entropy.

motions about the sun appear quite simple. The law of gravitation alone is sufficient to compute the orbits of the planets around the sun. According to Kepler's third law, the average velocity V of the planets decreases with distance r from the sun, $Vr^{0.5}$ = const. In constrast, the evolution of the solar system is extremely difficult to explain, first because the initial state is unknown, and second because almost all physical and chemical processes are involved in the development process. Exactly as was done for the complicated atmospheric and oceanic circulations, it is useful to start by collecting observational data and then drawing conclusions from them. The solar system has the following characteristics:

1. Almost all matter in the solar system rotates in the plane of the ecliptic.
2. All planets rotate in the same direction around the sun.
3. The distances of the planets and those of their satellites are determind roughly by the Titius–Bode rule.
4. The solar system has a nonzero angular momentum.
5. The planets have almost all the angular momentum of the solar system (98%), the sun only 2%. The ratio of the masses of the planets to that of the sun, however, is only 0.1 to 99.9.
6. The group of inner planets (from Mercury to Mars) consists of small planets with relatively high densities and only a few moons. The outer planets are much larger and have relatively low densities and many satellites [2].

These facts about the solar system leave ample room for the question on how the solar system was created. Were the planets and the sun part of a rotating primeval nebula? According to Reeves [3] the many existing theories may be arranged into four groups: (1) Sun and planets formed simultaneously out of a rotating primeval nebula. This hypothesis was first put forward by Kant in 1755 and Laplace in 1796 (although an indication of this idea can be found in Descartes' work of 1644). (2) Sun and planets formed at the same time but the planets assembled after the young sun had attracted interstellar matter or ejected gaseous material. This planetary matter had stellar temperatures ("hot" initial state). (3) The sun was once part of a binary system. The companion star disintegrated for some reason and was transformed into planets. (4) The last group includes theories based on a collision event, in which the sun drew matter from a passing star. Such a "catastrophe theory" was conceived by Buffon in 1745 and by Chamberlin and Moulton in 1906, among others.

Although no theory is universally accepted, the most probable or "best fit" theory is today considered to be that of Kant and Laplace. Among the modern versions built on this theory, the one by von Weizsäcker [4] from 1943 will be discussed in more detail.

Von Weizsäcker assumed a turbulent motion of the primeval gas cloud and the gradual condensation of matter to small bodies, called "planetesimals," which revolved around the developing sun in Keplerian orbits. If the reference frame of an observer is fixed to the average rotation of the planetesimals around the sun, the individual planetesimals moved in elliptic paths for small eccentricity of the Keplerian orbits, on "bean-shaped" paths for larger eccentricity. Assemblies of planetesimals formed vortices in configurations like those shown in Fig. 12.2, and the planets developed through collision of planetesimals at the border of two vortices and accretion to larger bodies. The way in which such an accretion process could have taken place was simulated on a computer by Cox of the Massachusetts Institute of Technology [5].

The Titius–Bode rule raises questions about what happened after the planets formed. In 1772 Titius from Wittenberg, Germany, found that the distances of the planets from the sun can be expressed in a simple sequence of numbers. Four years

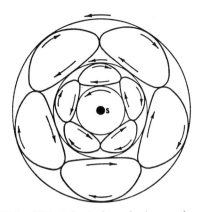

Figure 12.2. Weizsäcker's hypothesis on the origin of planets from vortex rings. The gaseous disk rotates in the direction of the arrow. At the border between two vortex rings, matter collides and condenses to planets on the circles drawn in the picture. The sketch is designed for three planets. S = sun (adapted from C. F. von Weizsäcker [4]).

later Bode published this result, and since then this "rule" is known as the Bode or Titius–Bode rule. If the distance from the ith planet to the sun is designated r_i, and i assumes the values from 1 to 10, then the rule is

$$r_i = 0.4 + 0.3 \cdot 2^{(i-2)},$$

where earth's distance from the sun is chosen to be unity. The following data result for the individual planets:

i	Planet	Titius–Bode	Actual value
1	Mercury	0.55	0.39
2	Venus	0.7	0.72
3	Earth	1	1
4	Mars	1.6	1.52
5	Asteroids	2.8	2.77
6	Jupiter	5.2	5.20
7	Saturn	10	9.54
8	Uranus	19.6	19.2
9	Neptune	38.8	30.1
10	Pluto	77.2	39.5

Similar data are obtained for the distances of the satellites from the planets Jupiter, Saturn, and Uranus. Has this number game a physical explanation? Von Weizsäcker used the radii of the circles on which the vortices meet (Fig. 12.2) to compute the distances of the planets, and he found, at least for the outer planets next to earth, the following remarkably good values:

Planet	r_i
Mars	1.45
Jupiter	5.20
Saturn	9.86
Uranus	18.67

More recent arguments are based on the assumption that the initial states of the systems of planets and satellites were probably different from their present states. They could not have been formed directly during the process of creation of the solar system. Rather it suggests itself that for a (quite) arbitrary initial state of the multibody system a certain stable end state results. Computer calculations by Hills [6] seem to confirm such a conclusion. According to Ovenden [7] and Lecar [8] the Titius–Bode rule is probably based on the principle of least interaction of bodies; in other words, the planets do not want to come too close to each other.

The planets not only circle around the sun (and the moons around the respective planets), they also rotate about their own axes as the earth does. This eigenrotation, too, is subject to evolution. It is known that over geological time spans the earth has transferred angular momentum to the moon through tidal friction. This means that the earth rotated faster in earlier times than it does today, and that the moon was closer to the earth. Quantitative data depend on the origin of the moon, about which there is no single preferred hypothesis [9].

The sun also rotates around its axis, which is almost perpendicular to the ecliptic. However, it does not rotate like a solid body, but rotates faster at the equator than at the poles. At the equator the sun's surface rotates once in 25 days; at 60° northern and southern latitudes, once in 31 days. The interpretation of this "differential rotation" is controversial. Starr and Gilman [10] assume that through turbulent transport of angular momentum to the equator a higher velocity is reached. Another theory is based on the spin-down effect (Fig. 7.19). According to Dicke [Reference 56, Chapter 1] the core of the sun rotates with the large velocity of one revolution per day, while the rotation of its shell is slowed down by magnetic forces. Instability produces meridional convections, which explain the differential rotation.

The surface of the sun (photosphere) is covered

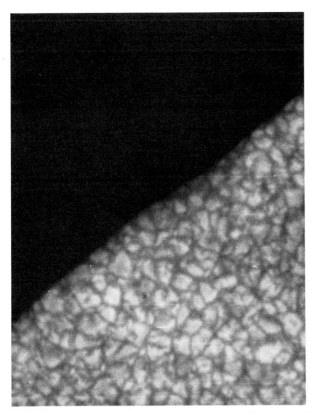

Figure 12.3. Granulation of the solar surface (Courtesy of W. Mattig, Kiepenheuer-Institut für Sonnenphysik, Freiburg, Breisgau).

with convection cells of the Bénard type (Fig. 12.3). This structure was first observed in 1877 by Jansen, who called it "granulation" [11]. The cells have a diameter of about 1000 km and a life span of several minutes. The cell convection is accompanied by the appearance of locally strong magnetic fields whose duration coincides with that of the cells [12]. There is also a "supergranulation," which consists of Bénard cells of the order of 20,000 km. These large cells are said to be the cause of solar oscillation with a period of 2 hours and 40 minutes [13].

12.3. CIRCULATIONS AND VORTICES IN PLANETARY ATMOSPHERES

Most planets are either enveloped by gaseous layers or consist largely of gas with a small solid or liquid core. In this gaseous environment meteorological events in the form of general circulation systems with temporal weather patterns and long-lived disturbances can occur. However, it cannot be expected that they are all similar to those on earth because of the diversity of the physical and chemical parameters involved. Rate of rotation, density, stratification, balance between solar radiation and planetary reflection, chemical composition, greenhouse effect, and depths of the atmospheres with solid or liquid bottoms are important factors. They differ for each planet and moon.

Atmospheric motions are determined essentially by the way angular momentum and heat exchange between equator and poles are balanced. The change of solar radiation with latitude, as depicted in Fig. 9.2 for the earth, is found in other planets too (as in Mars and Venus), although with different intensity and duration. One should, therefore, expect a certain similarity of the flux of heat and angular momentum over the planetary spheres with that on earth, but this is not necessarily the case. The greenhouse effect and heat from the interior of the planets may change the kind of balance. For this, visible evidence of incredible beauty has been brought to earth from Jupiter in the photographs of the spacecraft Voyagers 1 and 2, which reveal organized zonal bands and arrays of vortices alien to the earth's atmosphere. Figures 12.4 and 12.5 show such impressive pictures from those spacecraft missions. What makes Jupiter's weather systems so different from those on earth?

The answer lies in the following observations. The season on Jupiter barely changes, because the axis of rotation is almost parallel to the axis of the orbit around the sun. Jupiter, the largest planet in the solar system, consists of dense gas (except for a small liquid or solid core) and rotates 2.4 times faster than the earth. It emits nearly twice the energy it receives from the sun. The balance is derived from the inner heat of the planet, which remains from the primordial energy left after the gravitational contraction. In contrast to the earth, the effective temperature difference between the equator and the poles is no more than 3°C. Consequently, there is almost no energy transport between equator and poles [Reference 58, Chapter 1; 14–16].

All these factors cause a crucial difference in the atmospheric motions of the two planets. Whereas the earth's atmosphere is on the overall scale baroclinic (Section 9.1), Jupiter's atmosphere is barotropic with far-reaching consequences: The surface is divided into convection bands of high and low pressure (called "zones" and "belts," respectively) parallel to the Jovian equator (Fig. 12.6). These bands are gaseous jets with opposite directions in the zones and belts. The winds can reach 500 km/h

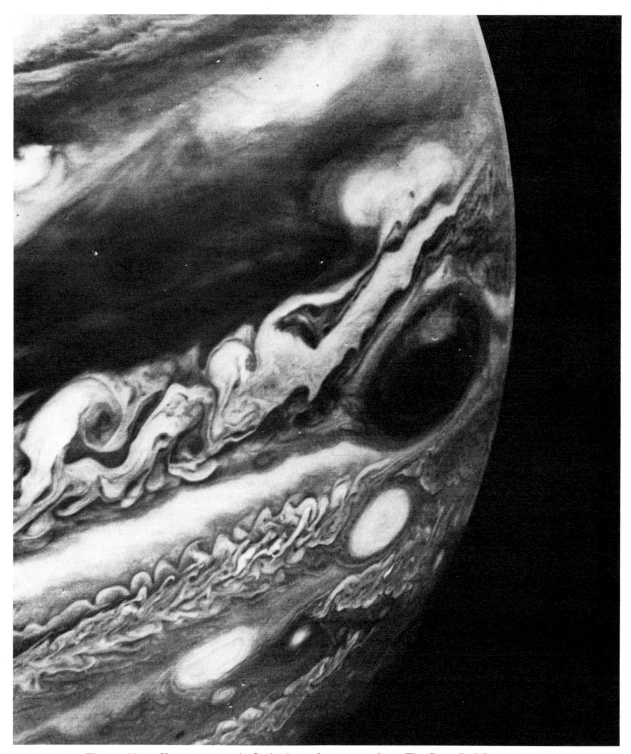

Figure 12.4. Vortex patterns in Jupiter's southern atmosphere. The Great Red Spot is to the right of the center, 22° below the equator. The wake of the Great Red Spot consists of large cloud whirls. To the lower left of the Great Red Spot is a white oval with vortical cloud arrays parallel to the latitudes (NASA—photograph).

Circulations and Vortices in Planetary Atmospheres   217

Figure 12.5. Saturn (NASA—photograph).

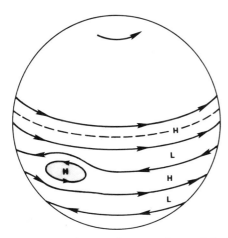

Figure 12.6. The barotropic atmosphere of Jupiter. At lower left is the Great Red Spot.

relative to the planet. Obstacles and other disturbances can cause violent hurricanes in these highly turbulent wind bands. Even so, the band configuration sketched in Fig. 12.6 is quite orderly for the equatorial zone and adjacent bands, as can be seen in photographs. Beyond $\pm 45°$ latitudes, the bands gradually break down due to the prevalence of internal heating. The anticyclonic disturbances appearing in the form of ovals in the lower latitudes are long-lasting phenomena compared to terrestrial hurricanes. This is essentially a scale effect because of the much larger masses involved. Additional factors in the longevity of these disturbances are the low temperature of the gas (and thus the small loss of energy) and the absence of an appreciable liquid or solid surface that could break up the disturbances. The ovals or spots are high-pressure (anticyclonic) vortices, with vorticity of the order of 3×10^{-5} s^{-1} and Ro ≈ 0.2 (compare these values with those of the vortices on earth in Section 7.3). They form in the shear flow of the zones and belts and may be interpreted as "solitary waves" in a rotating, stratified fluid according to Maxworthy and Redekopp [17] (Other explanations are given by Ingersoll and Busse [18].) The spots may be a mechanism for exchanging heat between the interior and the outside. They move relative to each other with speeds much smaller than the winds of the bands. For instance, the "Great Red Spot" in Fig. 12.4 moves to the west with a velocity of 11 km/h, whereas the white oval just below the Great Red Spot travels eastward at 14 km/h. These phenomena are obstacles in the jet streams, which cause turbulent wakes visible as vortical cloud patterns.

Among the spots and trains of waves the Great Red Spot deserves special attention. It is certainly the largest spot, 22,000 by 11,000 km, and is distinguished by its deep orange color. (The various shades of color, ranging from white to reddish, are due to complex chemical reactions in the atmosphere.) Clouds from the east pass by the Great Red Spot to be partially deflected from the southeastern border or sucked into the churning wake at the northwestern boundary. In this inferno the Great Red Spot is remarkably stable and persistent. Observations show only a slow decay; the Great Red Spot was twice as large 100 years ago.

Saturn's atmosphere reveals band structures, spots, and cloud whirls similar to those of Jupiter's, although less dramatic and less colorful. These features are humbled by the flat rings around Saturn, which are vortices of the most abstract and bizarre appearance in the universe (Fig. 12.5). Although they are not phenomena inside the atmosphere, they are mentioned here since they are vortices. The flatness of the myriad rings, their almost geometric perfection (except for temporal radial "spokes" and "braids"), and, above all, their permanence in time seem to contradict the laws of nature. To try to explain them raises more questions than can be given definite answers. One possible explanation is that the formation of moons was not completed and that the planetesimals in the form of ice particles remained at their present size when the planetary gas disk was gone [19]. Reference 19 gives a brief history of the rings since their discovery by Galileo and a survey on the theories about them.

Jupiter, by the way, has only one ring, so thin it was not discovered until Voyager flew by. Also Uranus' (at least) nine rings remained unknown until only a few years earlier because of their darkness [20].

The inner planets are summarily described because so little is known about their atmospheres. Viking Orbiter 1 has photographed extratropical cyclones on Mars, which indicate the existence of baroclinic waves owing their occurrence to the temperature difference between the freezing polar gas and the warmer midlatitude region. These features are similar to the midlatitude storms on earth (Section 9.1) [21].

Although Venus has nearly the same mass and size as the earth and its radiation balance is similar to that of the earth, the atmospheric motions are different, mainly due to the extremely slow rotation of the planet (one rotation in 243 days), to the different atmospheric temperature profile, and to a lesser extent to the lack of seasons. The Venusian atmosphere consists of 96% carbon dioxide, which with other constituents absorbs the infrared solar radiation and produces a greenhouse effect: the lower atmosphere is hot, the upper atmosphere cold—the reverse of the earth's situation. The heat flux from the equator to the poles, which is smaller than that on earth, causes a Hadley cell (Fig. 9.7) as on earth, but it extends to the poles because of the slow rotation of Venus (hence there is no difference between the Hadley cell and the Hadley circulation, Figs. 9.4 and 9.7). Moreover, this meridional circulation is located higher in the Venusian atmosphere because of the stable ground layers, and it is conjectured that as many as three cells may exist, one on top of the other. In addition, and surprisingly, a very strong westward wind has been observed (the sun moves toward the east) in the higher altitudes. This wind circles Venus with velocities up to 100 m/s [22].

12.4. ROTATING STARS

Like the sun, other stars can rotate about their axes. Which stellar properties are affected by the rotation or are greatly influenced by it? Because the stars are gaseous and are held together by their gravity, the shape of a star depends on its rotary velocity (and also to a certain extent on the meriodional circulation). A nonrotating star is spherical, a rotating one flattened at the poles. The temperature of a star also depends on its rotation. Since the centrifugal force counteracts gravity, the rotating star is less compressed and hence is heated less. Furthermore, since the development of a star depends on its temperature, rotation also influences the life process of a star [23].

Stars are designated by their spectral class and their luminosity. The spectral class is directly related to the surface temperature of the star, luminosity to its mass. If luminosity is plotted versus spectral class, one obtains the "Hertzsprung–Russel diagram," in which 99% of all observed stars lie on a curve, the "main sequence," among them the sun (Fig. 12.7). The larger the mass of a star, the greater is the compression of gravitation and the higher is the temperature. This curve, however, says nothing about the evolution of a star except that stars with large masses burn faster and hence do not remain as long on the curve as those with smaller masses. This will be explained in more detail in the following description of the actual life process of a star [24].

Stars are continually forming and decaying. In the process their lives pass through the following phases:

1. Gas and dust in interstellar space accumulate and aggregate through changes in their gravitational fields, but it is unlikely that these masses of gas and dust will contract to completely spherical shapes. Rather, they flatten through rotation or magnetization and form a material disk. The beginning of such an assemblage of interstellar matter to form a star can be triggered by shock waves. It is conjectured that massive stars, which are much heavier and hotter than the sun, are generated by shock waves from exploding supernovas or by spiral density waves in galaxies. The most favorable place for such a development might be the spiral arms of galaxies [25, 26]. Stars of the magnitude of the sun and smaller probably originate from dark, cold gas clouds through contraction (e.g., from "Bok-globules" [27]).

2. Shrinking matter heats up through compression. The source of energy is the gravitational field. As soon as the temperature of the young star is high enough to trigger a nuclear reaction of the hydrogen, a period of almost constant radiation of nuclear energy begins. The star is now on the main sequence of the Hertzsprung–Russel diagram. Compared to the first phase and to later states, this period encompasses most of the life of the star. This is why 99% of all stars lie on the main sequence. Their duration of stay depends on their mass. The larger the star, the hotter it is and the faster it burns.

3. After most of the hydrogen has been converted to helium, the star expands and cools.

4. The star is now a "red giant."

5. When the last hydrogen has been changed to helium and the mass of the core has reached a maximum, the core contracts and again heats up.

6. Other nuclear reactions follow, producing materials of higher atomic weights, until the most stable form, iron, has been reached.

7. The star continues to shrink until it has arrived at the last period of its life. Its nuclear energy is now consumed, but its remaining gravitational energy, which is released during subsequent shrinking, is sufficient (because of the small surface of this "white dwarf") to make the star continue to radiate for a long time. This period is longer than the present age of galaxies (Section 12.6). In a white dwarf matter is so compressed that the density is between 10^4 and 10^8 g/cm^3. The radius of the star is only about 5000 km. Matter is in a "degenerate" state: the electrons are densely packed. However, there are still enough fast electrons to produce the pressure necessary to resist gravity [28].

Whether or not a star reaches the state of a white dwarf depends on its mass. Chandrasekhar found that white dwarfs cannot exist if the stellar mass is larger than 1.44 times that of the sun (Chandrasekhar limit). Such a massive star can explode during contraction if in the outer layers of the star nuclear reactions are still possible. These explosions are known as supernovae. Simultaneously, a mist of gas, called a planetary nebula, can be expelled. Some have the form of rings and are reminiscent of vortex rings[2] (Fig. 12.8). A supernova explosion is not the only possible way a star beyond the Chandrasekhar limit can die. Other ways will be discussed in Section 12.5.

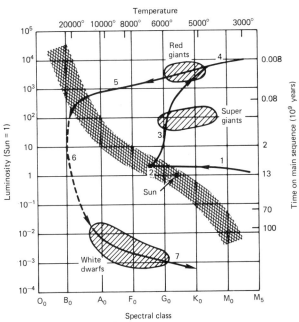

Figure 12.7. Hertzsprung–Russel diagram; 99% of all stars lie on the "main sequence." The lines 1 through 7 represent the single phases of the lifetime of a star.

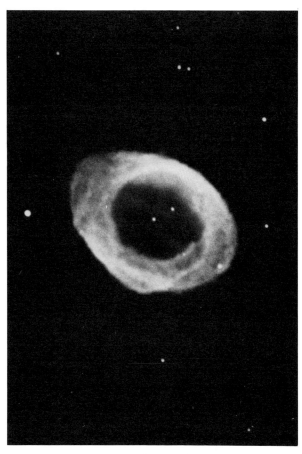

Figure 12.8. Planetary nebula NGC 6720 in the constellation *Lyra* (Messier 57) (photograph from Hale Observatories).

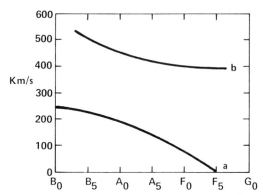

Figure 12.9. Average rotational velocities of stars on the main sequence of the Hertzsprung–Russel diagram (*a*). The small or vanishing rotation beyond F_5 suggests that the stars have planets (from Reference 24). The highest possible rotation is plotted in curve (*b*) (from Reference 56, Chapter 1).

What role does rotation play in the life of a star? Observations have shown that the rotational velocity of the stars on the main sequence decreases with diminishing temperature. The order of magnitude of this decrease is seen in Fig. 12.9. It may be mentioned, however, that the deviation from this curve is quite large [Reference 56, Chapter 1]. Schatzman [29] has tried to explain the lower rotary velocities of stars with a decrease in temperature. During the contraction of the star in phase 1 matter can be expelled locally on the stellar surface. In the magnetic field of the star matter follows the magnetic field lines and rotates spirally outward until it leaves the gravitational field of the star. This matter takes its angular momentum with it. Thus, in a magnetic field a relatively small amount of expelled mass can transport a considerable portion of angular momentum away from the star. The expelled mass depends on the surface activity of the star. Those with large convection zones are more active than those with small ones. Stars in spectral classes *A* and *B* have small or no convection. Stars in the *F* class have large convection. The longer a star stays in phase 1 in the range of large convection zones, the more mass with angular momentum it loses. Huang [30] put forward the hypothesis that the loss of angular momentum in F_5 stars could be explained by the generation of planets.

Since rotating stars are cooler than nonrotating ones because of the smaller compression, the main sequence in Fig. 12.7 shifts to the right with increasing rate of rotation [23]. The critical velocity is reached when a further increase in rotation would tear the star apart. Then, binary stars or clusters of stars are generated. A quarter of all stars are binary stars. The magnitude of the critical velocity is shown in Fig. 12.9 for stars on the main sequence.

During the development of a star to a red giant in phase 3 the rotational velocity decreases, provided that the total angular momentum of the star remains constant. Little is known about the rotary properties of stars in the other phases. White dwarfs with their extremely high densities can rotate very rapidly, about one revolution in a quarter of a second.

12.5. EXOTIC VORTEX STARS

The black hole is the Charybdis of energy–matter, from which there is no escape.

Stars end their lives in various ways. Some explode and others contract. White dwarfs are the final state if the stellar mass is smaller than 1.44 solar masses. Above this Chandrasekhar limit, stars contract through gravitation to neutron stars or black holes [31].

A white dwarf has ceased to produce nuclear energy and is, therefore, cold. However, it still has fast electrons which exert a certain pressure. For neutron stars the collapse is so strong that the electrons of the atomic shell are pressed into the atomic nucleus and combine with the protons to form neutrons. The stellar matter now consists of neutrons with a density of atomic nuclei of the order of 10^{11}–10^{15} g/cm^3. This incredible density squeezes a star with the mass of the sun to a sphere of 10 km radius. The angular velocity can then reach 1000 s^{-1}. Only the extremely high gravitational force can resist the centrifugal force of such a vortex.

Stars which emit X rays or light or radio waves of high frequency are called "pulsars." They are related to neutron stars [28]. It is conjectured on the basis of estimates that a rotating magnetic neutron star is the source of energy for pulsars.

Neutron stars were predicted theoretically in 1934 by Baade and Zwicky. In 1967 Hewish and his collaborators discovered pulsars, which in 1968 were identified by Gold as neutron stars. The neutron stars so far found are partners in binary stars. Examples are Centaurus X-3 and Hercules X-1 [32].

Astrophysics states on the basis of general relativity that, in a star with mass larger than three suns, the nuclear forces can no longer prevent a gravitational collapse. The space–time coordinates are so curved near and inside the matter of such heavenly bodies that at a certain distance from the material center the escape velocity is equal to the velocity of light (Fig. 12.10). If energy or matter passes this "event-horizon," there is no escape. A light ray cannot get out; it is caught in the "black hole" [Reference 57, Chapter 1].

How can a black hole be observed and identified?

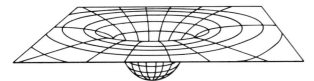

Figure 12.10. To envision a local space curvature one may consider a two-dimensional space represented by a plane with orthogonal coordinates. Through bending of such two-dimensional space due to local strong gravitational fields a "bump" develops. With increasing gravitation the bump gets pointed downward and finally ruptures. A black hole is created. The corresponding curvature in three-dimensional space, of course, cannot be envisioned but only mathematically described.

This is possible only through the influence it exerts on its surroundings. For instance, a black hole can be the partner of a binary star, like the neutron stars found so far. Although the existence of black holes was predicted theoretically in 1939 by Oppenheimer and Volkov, no black hole so far has been identified with certainty. It is conjectured that the source of X rays in the constellation *Swan*, called Cygnus X-1, is a black hole [33]. The visible partner of the binary star is the supergiant HDE 226 868. From this star the black hole attracts gas, which circles spirally around the black hole, like water in a bathtub vortex, and then vanishes behind the event-horizon (Fig. 12.11). In this process the gas becomes hot through friction. A large part of the released energy is emitted in the form of X rays.

Although almost all information on black holes is hidden behind the veil of the event—horizon, three physical quantities remain, which uniquely determine the structure of the black hole: mass, charge, and angular momentum. If neither charge nor angular momentum is present (which is unlikely), the black hole is "dead" (Schwarzschild structure). This means that no energy can be taken out of it. However, if charge and angular momentum are there (Kerr–Newman structure), these parts of the energy can be taken out of the black hole. It "lives" [Reference 57, Chapter 1].

When a particle falls through the event-horizon into a black hole, it approaches the center of the black hole. This special case occurs only if the black hole is dead. In a rotating black hole a particle does not reach the center but travels spirally in the direction of the center, reverses direction, and again approaches the event-horizon. However, it will not return to the original space–time world, but passes into another. A rotating black hole is thus the bridge to another space–time world (Fig. 12.12). This idea is so fantastic that it outstrips the content of any science fiction. It must be remembered, however, that this is only a physical model whose usefulness

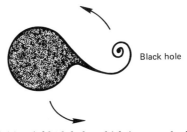

Figure 12.11. A black hole, which is part of a binary star, sucks matter from its partner; this matter is attracted spirally.

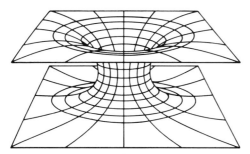

Figure 12.12. In a rotating black hole two worlds are connected with each other (in the figure simplified by two two-dimensional spaces)

and limitations are at present the subject of intense research. In fact, voices have been raised [34] in doubt that with today's models such "world bridges" can be demonstrated. This criticism, however, applies only to the presently existing mathematical models and does not claim that such bridges could not exist.[3]

The theory of black holes has been improved by Hawking through consideration of quantum-mechanical effects [35]. According to him the black hole is not completely black but can emit particles by means of a "tunnel effect" and can vanish in an explosion. It is possible that the big bang at the birth of the universe (Section 12.7) was the explosion of a black hole.

12.6. GALACTIC VORTICES

The solar system belongs to a large assembly of stars that forms an isolated material island in the vastness of space. It is called the Milky Way, and a portion of it is visible in the night sky as a broad, light ribbon. The Milky Way consists of about 100 billion individual stars. Its form is disklike with a bulge at the center (Fig. 12.13). The rest of the sphere formed by the radius of the disk is, however, not completely devoid of stars, but contains clusters of stars in irregular distribution. The diameter of the Milky Way in the plane of the disk is about 75,000 light years; perpendicular to the core region the diameter is about 15,000 light years. The sun is 30,000 light years away from the center of the Milky Way.

The flat form of the Milky Way indicates that it rotates in space, and this has been confirmed by measurements. The sun requires about 250 million years at a velocity of 250 km/s for one revolution about the center of the Milky Way. The orbit is almost circular. The stars farther away from the center move more slowly, those closer to it faster than the sun. Hence, the Milky Way does not rotate like a solid body but has a differential rotation with a velocity distribution as shown in Fig. 12.14. Another property of the Milky Way is the spiral arrangement of the stars in the plane of the disk.

The Milky Way is not the only island of stars in the universe. There are billions of them; they are the galaxies.

Kant had already interpreted stellar nebulae like that in the constellation *Andromeda* as a material island, but the discussion on whether spiral nebulae are outside our Milky Way or members of it became more intense at the beginning of this century and culminated in the "great debate" of 1920 between Shapley and Curtis. The latter advocated the hypothesis that spiral nebulae are stellar assemblies outside the Milky Way [36]. This idea, however, found a definite recognition only in 1924 when Hubble determined the distances of spiral nebulae and, subsequently, interpreted the observed red shift in the spectrum of galaxies as "flight velocities": the universe expands. However, stars inside of galaxies and groups of galaxies do not participate in this expansion, since gravity prevents them from doing so. Hubble distinguished the types of galaxies shown in Fig. 12.15.

About 75% of all galaxies belong to the *S* class, 20% to the *E* class, and the rest are irregular. The

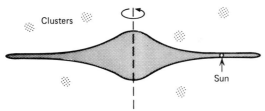

Figure 12.13. Side view of the Milky Way.

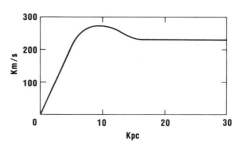

Figure 12.14. Velocity distribution of the *Andromeda* nebula. This galaxy is greatly similar to the Milky Way. (1 kpc = 3263.3 light years.)

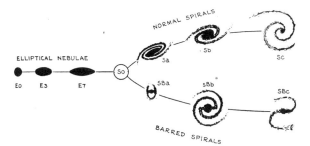

Figure 12.15. Hubble's classification of galaxies.

masses of the galaxies are between 10^8 and 10^{12} solar masses.

Hubble's classification has nothing to do with the evolution of galaxies. Their shapes probably depend on the magnitude of the angular momentum of the protogalaxy in the nascent state. A protogalaxy with a small angular momentum would not flatten and would assume a spheroidal shape. With increasing rotation, ellipsoidal galaxies would develop and finally become disklike. The flattening would then take place so fast that there would not be enough time for the gas and dust of the protogalaxy to be used completely for the creation of stars. This idea is supported by the fact that in the row $S_a \to S_c$ the proportion of gas and young stars increases.

All galaxies probably rotate. It has been observed that even galaxies of chaotic appearance rotate. Most measured data on the velocity distribution in a galaxy are available for spiral structures. A typical example is given in Fig. 12.14. The data for parts far away from the center are, however, quite uncertain. There are values (from different observers) that show a decrease in the velocity, some even show an increase [37].

The center of a galaxy is marked by a high stellar density (Figs. 12.16 and 12.17), and this region is generally also very active in the sense that electromagnetic waves are radiated and new stars are formed. Such activity is observed in the Milky Way [38]. In comparison, however, some galaxies have cores that radiate unusually high energy and that reach luminosities of the order of billions of suns. Among them are quasars and Seyfert galaxies, which also reveal a tremendous red shift. They may be the most distant objects in the universe and could thus give information on the early history of the cosmos. Sources of energy may be either huge explosions of massive giant stars or gigantic black holes, in whose region of attraction up to a million stars are whirled around and sucked in [39].

The spiral structure of galaxies has been an

Figure 12.16. Spiral galaxy (Messier 81) in constellation *Ursa Major* (photograph from Hale Observatories).

enigma to astronomers for a long time and is still not understood today (1983). Since galaxies do not rotate like solid bodies but turn faster in the inside than on the outside, any material collection would have been completed after a few revolutions. In 1925 Lindblad suggested that the spiral arms are quasisteady density waves that migrate outward while rotating. C. C. Lin and his group at the Massachusetts Institute of Technology have developed this idea into a theory [40]. Many questions, however, remain unanswered, in particular the question of what causes the density waves. Another hypothesis on the spiral structure was offered by Larson [41]. According to him intergalactic matter is continuously drawn into the galactic disk and of necessity forms elongated spirals. More recent computer simulations point to still another possibility: Exploding supernovae can cause a redistribution of matter in galaxies. The stars created by the explosion form long chains, which wind up in spirals around the galaxy [42].

Figure 12.17. Whirlpool galaxy (Messier 51) with satellite galaxy in constellation *Canes venatici* (photograph from Hale Observatories).

Galaxies may also come so close that they interact, resulting in galaxies of sometimes bizarre forms, called "peculiar galaxies." They indicate that gravitation can act even among galaxies. Long spiralling arms of galaxies may have originated from such close encounters [43].

Galaxies occur in clusters, and these clusters are again combined into larger groups. The hierarchical order of these systems of galaxies gives information on the distribution of matter in the universe and its evolution (Section 12.7) [44]. For instance, the Milky Way belongs to a small family of about 20 galaxies, of which the *Andromeda* nebula is the best known. A cluster, however, can have up to 1000 galaxies. Among these so-called "large clusters" are those that have an unusually high density of galaxies at the center. They are called "rich clusters" and their centers are extremely active.

Whereas the galaxies in ordinary clusters circle around a common center like planets around the sun, the number of galaxies in rich clusters is so large that galaxies collide and exchange energy. In this situation the massive galaxies cannot maintain their orbits and fall spirally toward the center, losing their galactic structure. Their stars either form giant elliptical central galaxies or generate a halo around an already existing central galaxy. These central giant galaxies are called "cD galaxies" [45].

Rich clusters of galaxies are thus the largest vortices in the universe. Entire galaxies are swallowed in this maelstrom of gravitation and lose their separate identity.

12.7. THE LIMITS OF HUMAN PERCEPTION

In the beginning was the logos ...
The gospel according to St. John

Galaxies and clusters of galaxies rotate. Where did these largest material structures in the universe get their angular momentum? The discussion of this question requires knowledge about the origin of the universe.

Discoveries in the last few decades support the view that the universe was created in a big bang. This hypothesis was put forward by Lemaitre in 1931 and by Gamov in 1948 and is substantiated today by two phenomena: the red shift and the cosmic background radiation. The red shift of the stellar spectra indicates a Doppler effect. Stars, galaxies, and clusters of galaxies not bound to other systems by gravitation drift apart; the universe expands. The cosmic background radiation is a remnant of the primeval fireball after the big bang, as will be explained in the next paragraph.

The age of the universe has been estimated by independent methods to be about 16 billion years. The three major epochs of this existence (a division sufficient for the purpose of this book) are given on the logarithmic scales of a density–time diagram in Fig. 12.18 [46]. Two curves for the densities of matter and radiation show the decrease of density (and thus a decrease of temperature) in an expanding universe.

1. The time period up to 1 second after the big bang comprises the "quark, hadron, and lepton eras" of the very early universe. These eras are characterized by the forces that dominated and controlled the interaction between elementary particles and antiparticles, a process of annihilation in which

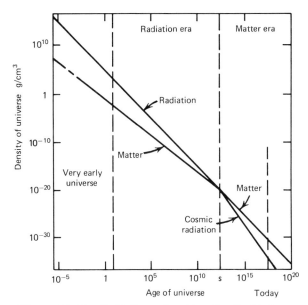

Figure 12.18. Radiation and matter after the big bang.

energy was released and new matter and antimatter were created. With falling temperature first the quarks ceased to be recreated, then the heavy particles (hadrons) like protons and neutrons, on which the "strong interaction forces" acted; and finally the light particles (leptons) like electrons and positrons, whose fate was determined by "weak interaction forces." Photons remained and their dominance signalled the beginning of the radiation era.

2. The radiation era lasted from 1 second to about 1 million years after the big bang. Radiation and matter interacted until the universe had expanded so much that radiation and matter decoupled. The period of the fireball was over.

3. Matter was now denser than radiation and cooled more quickly. The new era of "materialization" started with the reaction of electrons and ionized gas and, independently, the radiation of energy. This radiation during decoupling has persisted over the course of time and exists today in the microwave range as black-body radiation at a temperature of 2.8 K. Since it is present throughout the universe, it is called "cosmic background radiation" [47].

Because the cosmic background radiation is uniform in all directions, it is assumed that space is essentially homogeneous and isotropic. This means that space is everywhere equally structured and has the same properties in all directions. This basic assumption is called the "cosmological principle." However, it holds only in the large, that is, on the scale of the universe, for if the fireball was completely symmetric, how could matter have been created and assembled? In the asymmetry "in the small," therefore, must be sought the origin of material collections (galaxies) and their angular momentum.

There are two basic hypotheses on the asymmetry after the big bang. The first assumes that the initial state was chaotic or turbulent in the sense that deviations from isotropy and homogeneity were large. From turbulent eddies in the fireball after the decoupling of radiation and matter, protoclusters of galaxies developed. Their dimensions were determined at a time when the radiation density was equal to that of matter. Through later break-up of the clusters, then, galaxies were generated, and angular momentum was transferred directly from the turbulent eddies to the galaxies. Proponents of this "chaotic cosmology" are von Weizsäcker (1951), Gamov (1952), and, more recently, Ozernoi and Chernin (1968), J. Oort (1969), and others [48, 49].

The second hypothesis assumes that the irregularities in the initial state were small ("quiescent cosmology"). This means that there were only slight disturbances in density in an otherwise homogenous fireball; these disturbances then increased through instability and developed to protogalaxies. Clusters of galaxies formed later. According to this hypothesis the angular momentum was not directly transferred but was carried from density fluctuations to the protogalaxies through gravitational interaction. Advocates of this hypothesis are Gamov and Teller (1939), Hoyle (1949), Sciama (1955), and Peebles (1965) [50]. More recent evidence suggests that the irregularities of the fireball before decoupling, estimated quantitatively as 1 part in 1000, have been that way since 10^{-35} second after the big bang [51]. The fraction of time between 10^{-35} second and the "cosmological singularity" at time zero is determined by the laws of quantum cosmology, a field of the physics of superdense energy–matter.

Whichever idea one prefers, the chaotic or the quiescent cosmology, the following statement can be made: At the beginning of the universe irregularities in the homogeneity and isotropy of the energy–matter sphere must have already existed, and they provide the angular momentum for the present local assemblages. These irregularities in the initial singularity are the "primeval vortices" of the cosmos. They were local vortices, not the mythological, all-embracing cosmic whirl of antiquity (Section 1.1).[4]

The search for the source of angular momentum

of matter has now reached the very beginning of the universe at time zero. Conjectures on the initial state of the big bang are stated in the form of "principles." An interesting summary was made by Ellis and Harrison [52]. Their "uniformity principle" says that the physical laws have been valid at all times everywhere in the universe. Such a principle is necessary to allow to extrapolate to the initial state. The problem is thus reduced to a search for suitable initial conditions that can theoretically predict the essential structures of the universe, that is, isotropy and homogeneity of material distribution in the large, and the existence of material concentrations like galaxies in the small. Can such initial conditions be expressed in the general form of a chaotic primeval state, which means that the present situation is quite independent of the special type of initial conditions [53, 54]? Or must the initial conditions be prescribed in the form of small perturbations as in the hypothesis of instability [55]?

Both models actually require that the initial conditions be specified more finely than by the existence of galaxies in the small. They must guarantee the preconditions for the existence of intelligent life. This requirement is expressed in the "anthropic principle" (from the Greek "anthropos," man). Its utility and even legitimacy are being debated today by philosophers of science [56]. In this connection, the principle of minimum initial conditions should also be considered [52]. This principle requires that the set of initial conditions be self-consistent and that it not be extended arbitrarily. A nice historical example of the wrong application of initial conditions has been cited by Harrison in Reference 52: In the 17th century Bishop Ussher declared that the universe was completely created in all details in 4004 BC. Later, when fossils were found that were much older, it was declared that the fossils, too, were created with the universe.

Finally, it may be mentioned that the early state of the big bang was similar to a black hole, which is completely characterized by only three physical quantities: mass, charge, and angular momentum. Thus this theoretical model contains angular momentum in the primeval state too.

These discussions on the initial conditions of the universe are within the realm of physics. The answer to the question of what triggered the big bang, or what preceded it, lies beyond any scientific perception. Such an answer can be sought only in the metaphysical concept of "logos."[5]

Remarks on Chapter 12

1. The concept of the continuum need be abandoned not only for the motion of stars and planets but also for events in the microcosm. In relation to the space in which electrons move, the volume of the atomic nucleus is vanishingly small. The number and positions of the electrons around the atomic nuclei reveal an order, which has found expression in the "periodic system" of the elements. Although the concept of the continuum is not applicable to the assembly of stars and planets, it still has a place in astrophysics. The gas of interstellar space is admittedly highly rarefied, but the characteristic distances are so large that the mean free paths of the gases are small in relation to these dimensions. Thus, the interstellar gas can be considered as a continuum. The average Reynolds number is about 10^9 (Oort, [Reference 55, Chapter 1]), hence the motion is turbulent. This is an important statement for understanding stellar evolution. For instance, the rotational axes of stars in a galaxy are not aligned perpendicular to the plane of the galactic disk but are statistically distributed, a consequence of the turbulence. Protogalaxies were probably generated from turbulent vortices (Section 12.7).

2. There are still other reasons for the development of planetary nebulae. For example, ionized gas can be expelled from the mother star when it is a red giant [57, 58].

3. Another possibility is to bend the two planes in Fig. 12.12 at one side and to join them. Then, a rotating black hole would be the bridge to another, completely different part of the universe [Reference 57, Chapter 1].

4. It is also of interest to speculate about the further development of the universe into the future. A brief scenario is given in Reference 46. The role of vortices and vorticity in a universe expanding forever (if it does) is discussed by Barrow and Tipler [59] in an article with the provocative title "Eternity is unstable" pointing to a new conjecture about the way entropy increases, different from the classical heat death theory.

5. Goethe's Dr. Faustus had difficulties in translating the Greek concept "logos" at the beginning of the gospel according to St. John. How pale is the translation "word" in the Bible! Logos is the act of creation in the sense of the Old Testament: The creation out of nothing in contrast to the Greek philosophy, in which matter is assumed eternal, and the demiurge has only to order matter. In general, logos will here be interpreted as creation of a structure, of an order in nature, either out of nothing or out of chaos.

References

Chapter 1

1. D. A. Mackenzie, *The Migration of Symbols and Their Relations to Beliefs and Customs.* A. A. Knopf, New York, 1926.
2. H. Kühn, *Der Aufstieg der Menschheit.* Fischer Bücherei Nr. 82, 1955.
3. C. Schuchhardt, *Alteuropa,* 4th ed. Berlin, 1941, W. de Gruyter, p. 205.
4. H. J. Lugt, Wirbelstürme im Alten Testament. *Biblische Zeitschrift,* **19** (1975), 195.
5. E. Neumann, *The Great Mother.* Princeton University Press, Second Printing, 1974.
6. G. de Santillana and H. von Dechend, *Hamlet's Mill.* Gambit, Ipswich, Mass., 1969.
7. A. Speiser, *Die mathematische Denkweise.* Verlag Birkhäuser, Basel, 1945.
8. H. J. Lugt, *The Vortex Concept in the History of Science.* Deutsche Luft- und Raumfahrt DLR-FB 77-16, 1977, p. 163.
9. J. M. Robinson, *An Introduction to Early Greek Philosophy.* Houghton Mifflin, Boston, 1968.
10. G. S. Kirk and J. E. Raven, *The Presocratic Philosophers.* Cambridge University Press, New York, 1971.
11. S. Tigner, "Empedocles' Twirled Ladle and the Vortex-Supported Earth." *Isis* **65** (1974), 433.
12. W. Windelband, *History of Ancient Philosophy.* Dover, New York, 1956.
13. Aristotle, *Physics.* Translated by R. Hope. University Nebraska Press, Lincoln, 1961.
14. W. Yourgrau and S. Mandelstam, *Variational Principles in Dynamics and Quantum Theory.* Pitman, New York, 1962.
15. A. Maier, "Die Anfänge des physikalischen Denkens im 14. Jahrhundert." *Philosophia Naturalis* **1** (1950), 7.
16. M. Fierz, "Vorlesungen zur Entwicklungsgeschichte der Mechanik." *Lecture Notes in Physics,* Vol. 15. Springer-Verlag, New York, 1972.
17. E. Grant (ed.), *A Source Book in Medieval Science.* Harvard University Press, Cambridge, Mass., 1974.
18. R. Dugas, *A History of Mechanics.* Central Book Co., New York, 1955.
19. E. Grant (ed. and translator), *Nicole Oresme and the Kinematics of Circular Motion.* University Wisconsin Press, Madison, Wisc., 1971.
20. O. Flachsbart, *Handbuch der Experimentalphysik.* **4,** Pt. 2 (1932), 8.
21. Leonardo da Vinci, *Del moto e misura dell' acqua.* E. Carusi and A. Favaro (eds.), Bologna, 1923.
22. Leonardo da Vinci, *Quaderni d' Anatomica* II 9, 1513.
23. B. J. Bellhouse and L. Talbot, "The Fluid Mechanics of the Aortic Valve." *J. Fluid Mech.* **35** (1969), 721.
24. K. Clark, *Leonardo da Vinci.* Penguin Books A 430, 1967.
25. E. J. Aiton, *The Vortex Theory of Planetary Motions.* American Elsevier, New York, 1972.
26. E. Whittaker, *A History of the Theories of Aether and Electricity,* Vol. 1. Harper Torchbooks, 1960.
27. A. Koyré, *From the Closed World to the Infinite Universe.* The Johns Hopkins Press, Baltimore, Md., 1957.
28. E. N. Lorenz, *The Nature and Theory of the General Circulation of the Atmosphere.* World Meteorological Organization, 1967.
29. A. Wegener, *Wind- und Wasserhosen.* Vieweg, Braunschweig, 1917.
30. D. Fultz, "Experimental Analogies to Atmospheric Motions." *Comp. Meteorology.* Waverly Press, Baltimore, 1951, 1235.
31. S. Goldstein, *Modern Developments in Fluid Dynamics,* Vol. II. Dover, New York, 1965.
32. H. J. Lugt and J. Schot, "A Review of Slip Flow in Continuum Physics." Proc. Sec. Symposium Fluid-Solid Surface Interactions. DTNSRDC, H. J. Lugt (ed.), 1974, 101.
33. R. Chevray, "Giuseppe Venturoli, Pioneer in Channel Studies." *J. Hydraulics Div.* **HY 10** (Oct. 1976), 1533.
34. C. Truesdell, *Essays in the History of Mechanics.* Springer-Verlag, New York, 1968.
35. H. Schlichting, *Boundary Layer Theory.* McGraw-Hill, New York, 1979, 7th ed.
36. W. Thomson (Lord Kelvin), *Mathematical and Physical Papers,* Vol. IV. Cambridge University Press, 1910.
37. R. H. Silliman, "William Thomson: Smoke Rings and Nineteenth-Century Atomism." *Isis* **54** (1963), 461.
38. P. J. Pauly, "Vortices and Vibrations: The Rise and Fall

of a Scientific Research Program." M. A. Thesis, University of Maryland, 1975.
39. J. Serrin, *Handbuch der Physik.* Vol. VIII/1, Springer-Verlag, New York, 1959, p. 248.
40. H. P. Greenspan, *The Theory of Rotating Fluids.* Cambridge University Press, Cambridge, 1968.
41. L. Rosenhead (ed.), *Laminar Boundary Layers.* Oxford University Press, New York, 1963.
42. J. O. Hinze, *Turbulence,* 2nd ed. McGraw-Hill, New York, 1975.
43. A. Roshko, "Structure of Turbulent Shear Flows: A New Look." *AIAA J.* **14** (1976), 1349.
44. I. Prigogine, *From Being to Becoming.* Freeman, San Francisco, 1980.
45. R. Giacomelli and E. Pistolesi, in *Aerodynamic Theory,* Vol. 1. W. F. Durand (ed.). Dover, New York, 1963.
46. S. Goldstein, "Fluid Mechanics in the First Half of this Century." *Ann. Rev. Fluid Mech.* **1** (1969), 1.
47. D. P. Riabouchinsky, "Thirty Years of Theoretical and Experimental Research in Fluid Mechanics." *J. Roy. Aeronaut. Soc.* **39** (1935), 282.
48. K. P. Cohen, *The Theory of Isotope Separation as Applied to the Large Scale Production of U-235.* McGraw-Hill, New York, 1951.
49. D. G. Lilley, "Swirl Flows in Combustion: A Review." *AIAA J.* **15** (1977), 1063.
50. C. DuP. Donaldson and R. D. Sullivan, "Examination of the Solutions of the Navier–Stokes Equations for a Class of Three-Dimensional Vortices." Aer. Res. Ass., Princeton, AFOSR TN 60-1227, 1960.
51. P. Roache, *Computational Fluid Dynamics.* Hermosa, Albuquerque, New Mexico, 1976.
52. M. J. Lighthill, "On Sound Generated Aerodynamically." *Proc. Roy. Soc. Lond. Ser. A* **211** (1952), 564.
53. A. Powell, "Theory of Vortex Sound." *J. Acoust. Soc. Am.* **36** (1964), 177.
54. T. Sarpkaya, "Vortex-Induced Oscillations." *J. Appl. Mech.* **46** (1979), 241.
55. *Problems of Cosmical Aerodynamics.* Proceedings of the Symposium on the Motion of Gaseous Masses of Cosmical Dimensions. Central Air Documents Office, 1951.
56. A. Slettebak (ed.), *Stellar Rotation.* Gordon and Breach, New York, 1970.
57. C. W. Misner, K. S. Thorne, and J. A. Wheeler, *Gravitation.* Freeman, San Francisco, 1973.
58. "Voyager 1 Encounter with the Jovian System." *Science* **204** (1979), 945. "Voyager 2." *Science* **206** (1979), 925.
59. R. P. Feynman, "Atomic Theory of the Two-Fluid Model of Liquid Helium." *Phys. Rev.* **94** (1954), 262.
60. F. Lund and T. Regge, "Unified Approach to Strings and Vortices with Soliton Solutions." *Phys. Rev. D* **14** (1976), 1524.
61. J. Campbell, *The Masks of God: Primitive Mythology.* Penguin Books, New York, 1978.
62. T. A. Cook, *The Curves of Life.* Dover, New York, 1979.
63. J. E. Cirlot, *A Dictionary of Symbols.* Routledge & Kegan Paul, London, 1967.
64. V. G. Childe, "Rotatory Motion." In *A History of Technology,* Oxford University Press, New York, 1954, Vol. I.
65. M. L. West, *Early Greek Philosophy and the Orient.* Oxford University Press, New York, 1971.
66. A. V. Shubnikov and V. A. Koptsik, *Symmetry in Science and Art.* Plenum, New York, 1974.
67. T. Schwenk, *Sensitive Chaos.* Schocken Books, New York, 1976.
68. C. F. Krafft, "Life a Vortex Phenomenon." *Biodynamica* No. 18 (Dec. 1936).

Chapter 2

1. H. J. Lugt, "The Dilemma of Defining a Vortex." In *Recent Developments in Theoretical and Experimental Fluid Mechanics.* Springer-Verlag, New York, 1979, p. 309.
2. T. von Kármán, *Aerodynamics: Selected Topics in the Light of their Historical Development.* Cornell University Press, Ithaca, N.Y., 1954.
3. R. J. Bethke and H. Viets, *Data Analysis to Identify Coherent Flow Structures.* AIAA Atmospheric Flight Mech. Conf., AIAA-80-1561-CP, Aug. 1980, p. 42.
4. H. J. Lugt, *Recent Advances in the Numerical Treatment of the Navier–Stokes Equations.* AGARD Lecture Series No. 86, April 1977.
5. W. Merzkirch, *Flow Visualization.* Academic, New York, 1974.
6. L. Prandtl and O. G. Tietjens, *Fundamentals of Hydro- and Aeromechanics.* Dover, New York, 1957.
7. P. A. Lagerstrom, "Solutions of the Navier–Stokes Equation at Large Reynolds Number." *SIAM J. Appl. Math.* **28** (1975), 202.
8. P. G. Saffman and G. R. Baker, "Vortex Interactions." *Ann. Rev. Fluid Mech.* **11** (1979), 95.
9. U. B. Mehta and Z. Lavan, "Starting Vortex, Separation Bubbles, and Stall: A Numerical Study of Laminar Unsteady Flow Around an Airfoil." *J. Fluid Mech.* **67** (1975), 227.
10. P. V. Danckwerts and R. A. M. Wilson, "Flow Visualization by Means of a Time-Reaction." *J. Fluid Mech.* **16** (1963), 412.

Chapter 3

1. K. Oswatitsch, *Handbuch der Physik,* Vol. VII/1. Springer-Verlag, New York, 1959.
2. G. I. Taylor, "Low Reynolds Number Flow." Encyclopedia Britannica Educational Corp., Chicago.
3. C. W. McCutchen, "Ghost Wakes Caused by Aerial Vortices." *Weather* (Jan. 1972).
4. E. R. Hoffmann and P. N. Joubert, "Turbulent Line Vortices." *J. Fluid Mech.* **16** (1963), 395.
5. S. P. Govindaraju and P. G. Saffman, "Flow in a Turbulent Trailing Vortex." *Phys. Fluids* **14** (1971), 2074.
6. L. D. Landau and E. M. Lifshitz, *Fluid Mechanics.* Pergamon, London, 1959.
7. H. O. Anwar, "Flow in a Free Vortex." *Water Power* (Apr. 1965), 153.
8. A. K. Jain, K. G. Ranga Raju, and R. J. Garde, "Vortex Formation at Vertical Pipe Intakes." *J. Hydr. Div., ASCE* (Oct. 1978), 1429.

9. H. O. Anwar, J. A. Weller, and M. B. Amphlett, "Similarity of Free-Vortex at Horizontal Intake." *J. Hydraul. Res.* **16** (1978), 95.
10. E. N. da C. Andrade, "Whirlpools, Vortices, and Bathtubs." *New Scientist* **7** (1963), 302.
11. H. Y. Rajagopal, "Vortex Formation at Vertical Pipe Intakes" (Discussion of Reference 8). *J. Hydr. Div., ASCE* (Jan. 1980), 211.
12. A. M. Binnie and J. F. Davidson, "The Flow under Gravity of a Swirling Liquid through an Orifice-Plate." *Proc. Roy. Soc. Lond. Ser. A* **199** (1949), 443.
13. R. Fearn and R. P. Weston, Vorticity Associated with a Jet in a Cross Flow. *AIAA J.* **12** (1974), 1666.
14. C. L. Stong, *Sci. Am.* (Oct. 1971), 110.
15. H. Klein, *Small Scale Tests on Jet Engine Pebble Aspiration*. Douglas Aircraft Co. Rep. SM 14895, Aug. 1953.
16. D. E. Glenny and N. Pyestock, *Ingestion of Debris into Intakes by Vortex Action*. Aer. Res. Council C.P. No. 1114, 1970.
17. E. Levi, "Experiments on Unstable Vortices." *J. Hydr. Div., ASCE* (June 1972), 539.
18. M. C. Quick, "Efficiency of Air-Entraining Vortex Formation at Water Intake." *J. Hydr. Div., ASCE* **96** (1970), 1403.
19. O. Turmlitz, "Ein neuer physikalischer Beweis der Achsendrehung der Erde." *S.B. Akad. Wiss. Wien*, Abt. IIa, 117, 1908, 819.
20. A. H. Shapiro, "Bath-Tub Vortex." *Nature* **196** (1962), 1080.
21. G. K. Batchelor, *An Introduction to Fluid Dynamics*. Cambridge University Press, New York, 1967.
22. J. H. Olsen, A. Goldburg, and M. Rogers (eds.), *Aircraft Wake Turbulence and its Detection*. Plenum, New York, 1971.
23. N. F. O'Connor, "The Fujiwara Effect." *Weatherwise* (Oct. 1964), 232.
24. L. Rosenhead, "Formation of Vortices from a Surface of Discontinuity." *Proc. Roy. Soc. Lond. Ser. A* **134** (1931), 170.
25. K. V. Roberts and J. P. Christiansen, "Topics in Computational Fluid Dynamics." In *The Impact of Computers on Physics*. G. R. Macleod (ed). North Holland, Amsterdam, 1972.
26. S. E. Widnall, "The Structure and Dynamics of Vortex Filaments." *Ann. Rev. Fluid Mech.* **7** (1975), 141.
27. J. H. B. Smith, "Theoretical Work on the Formation of Vortex Sheets." *Progress in Aeronautical Sciences*, Vol. 7. Pergamon, New York, 1966, p. 35.
28. A. E. H. Love, "Stabilität von Ringwirbeln." *Encycl. Math. Wiss.* Teubner Verlag, Leipzig, 1901, Bd.4, Teil 3, 86.
29. Lord Kelvin, *Phil. Mag.* **10** (1880), 155.
30. H. Hasimoto, "A Soliton on a Vortex Filament." *J. Fluid Mech.* **51** (1972), 477.
31. R. A. Granger, "Observations of a Surge in a Forced Vortex Flow." *J. Ship Res.* **18** (March 1974), 12.
32. P. H. Roberts and R. J. Donnelly, "Superfluid Mechanics." *Ann. Rev. Fluid Mech.* **6** (1974), 179.
33. E. J. Yarmchuk, M. J. V. Gordon, and R. E. Packard, "Observation of Stationary Vortex Arrays in Rotating Superfluid Helium." *Phys. Rev. Lett.* **43** (1979), 214.
34. R. A. Ashton and W. I. Glaberson, "Vortex Waves in Superfluid ^{4}He." *Phys. Rev. Lett.* **42** (1979), 1062.
35. K. W. Schwarz, "Theory of Turbulence in Superfluid ^{4}He." *Phys. Rev. Lett.* **38** (1977), 551.
36. R. M. Linsley, "Shell Form and the Evolution of Gastropods." *Am. Sci.* **66** (1978), 432.
37. M. Gardner, *Sci. Am.* (Apr. 1970), 108.
38. M. Gold, "Who Pulled the Plug on Lake Peigneur? *Science* **81** (Nov. 1981), 56.
39. P. T. Fink and W. K. Soh, "A New Approach to Roll-Up Calculations of Vortex Sheets. *Proc. Roy. Soc. Lond. Ser. A* **362** (1978), 195.

Chapter 4

1. C. Truesdell, *The Kinematics of Vorticity*. Indiana University Press, Bloomington, Ind., 1954.
2. M. J. Lighthill, in *Laminar Boundary Layers*, L. Rosenhead (ed.). Oxford University Press, New York, 1963.
3. F. H. Harlow and J. E. Fromm, "Computer Experiments in Fluid Dynamics." *Sci. Am* (March 1965), 104.
4. "Proceedings of the International Conferences on Numerical Methods in Fluid Dynamics." *Lecture Notes in Physics* No. 8, 18, 35, etc., Springer, New York.
5. *Numerical Methods in Fluid Dynamics*, C. A. Brebbia and J. J. Connor (eds.). Pentech Press, London, 1974.
6. T. J. Chung, *Finite Element Analysis in Fluid Dynamics*. McGraw-Hill, New York, 1978.
7. *Numerical Methods in Applied Fluid Dynamics*, B. Hunt (ed). Academic Press, New York, 1980.
8. R. B. Payne, "Calculation of Unsteady Viscous Flow Past a Circular Cylinder." *J. Fluid Mech.* **4** (1958), 81.
9. V. G. Jenson, "Viscous Flow Round a Sphere at Low Reynolds Numbers (<40)." *Proc. Roy. Soc. Lond. Ser. A* **249** (1959), 346.
10. J. Happel and H. Brenner, *Low Reynolds Number Hydrodynamics*. Prentice-Hall, Englewood Cliffs, N.J., 1965.
11. W. Jacobs, *Fliegen, Schwimmen, Schweben*. Springer-Verlag, New York, 1954.
12. J. M. Bourot, "On the Numerical Computation of the Optimum Profile in Stokes Flow. *J. Fluid Mech.* **65** (1974), 513.
13. H. Weyl, *Symmetry*. Princeton University Press, Princeton, N.J., 1952.
14. T. Y. Wu, C. J. Brokaw, and C. Brennen (eds.), *Swimming and Flying in Nature*. Plenum, New York, 1975.
15. M. J. Lighthill, *Mathematical Biofluiddynamics*. SIAM, Philadelphia, 1975.
16. H. Hertel, *Structure—Form—Movement*. Reinhold, New York, 1966.
17. T. L. Jahn and J. J. Votta, "Locomotion of Protozoa." *Ann. Rev. Fluid Mech.* **4** (1972), 93.
18. H. C. Berg, "How Bacteria Swim." *Sci. Am.* **233** (Aug. 1975), 36.
19. L. Prandtl, "Über Flüssigkeitsbewegung bei sehr kleiner Reibung." *Verh. III. Intern. Math. Kongress*, Heidelberg, 1904.

20. H. von Helmholtz, "Über Integrale der hydrodynamischen Gleichungen, welche den Wirbelbewegungen entsprechen." *Crelles J.* **55** (1858), 25.
21. J. Hadamard, *Leçon sur la propagation des Ondes et les équations de l' Hydrodynamique.* Hermann, Paris, 1903. See also: J. Hadamard, La formation des discontinuités dans les fluides." Proc. Second Int. Congress Appl. Mech., Zürich, 1926, p. 507.
22. F. Klein, "Über die Bildung von Wirbeln in reibungslosen Flüssigkeiten." *Z. Math. Phys.* **58** (1910), 259.
23. A. Betz, "Wirbelbildung in idealen Flüssigkeiten und Helmholtzscher Wirbelsatz." *ZAMM* **10** (1930), 413.
24. A. Betz, "Wie entsteht ein Wirbel in einer wenig zähen Flüssigkeit?" *Die Naturwissenschaften* **37** (1950), 193.
25. J. Ackeret, "Über die Bildung von Wirbeln in reibungslosen Flüssigkeiten." *ZAMM* **15** (1935), 3.
26. L. Prandtl, Über die Entstehung von Wirbeln in der idealen Flüssigkeit, mit Anwendung auf die Tragflügeltheorie und andere Aufgaben (1922). See: *Gesammelte Abhandlungen*, zweiter Teil. Springer-Verlag, 1961, p. 697.
27. L. Prandtl, "Tragflügeltheorie." *Nachr. K. Ges. Wiss. Göttingen, Math.-Phys. Kl.* (1918), 451.
28. C. F. Meerwein, *Der Mensch—sollte der nicht auch mit Fähigkeiten zum Fliegen gebohren seyn?* J. J. Thurneisen, Jünger, Basel, 1784. See also: *Badische Neueste Nachrichten*, Karlsruhe (Aug. 10, 1957), 15.
29. C. H. Gibbs-Smith, *The Invention of the Aeroplane (1799–1909).* Taplinger, 1966.
30. T. von Kármán and J. M. Burgers, "General Aerodynamic Theory—Perfect Fluids." In *Aerodynamic Theory*, Vol. II. W. F. Durand (ed). Dover, New York, 1963.
31. S. Hoerner, *Fluid-Dynamic Drag.* Published by the author.
32. H. Glauert, "Airplane Propellers." In *Aerodynamic Theory*, Vol. IV. W. F. Durand (ed.). Dover, New York, 1963.
33. A. R. S. Bramwell, *Helicopter Dynamics.* Halsted Press, New York, 1976.
34. J. M. V. Rayner, "A Vortex Theory of Animal Flight. Part 1. The Vortex Wake of a Hovering Animal." *J. Fluid Mech.* **91** (1979), 697.
35. H. J. Lugt, "Autorotation." *Ann Rev. Fluid Mech.* **15** (1983), 123.
36. G. V. Parkinson, "On the Performance of Lanchester's "Aerial Tourbillion." *J. Roy. Aeronaut. Soc.* **68** (1964), 561.
37. H. Glauert, "The Rotation of an Airfoil About a Fixed Axis." Advisory Committee for Aeronautics, Rep. and Mem., No. 595, March 1919.
38. B. M. Jones, "The Spin." In *Aerodynamic Theory,* Vol. V. W. F. Durand (ed.). Dover, New York, 1963, p. 204.
39. S. B. Anderson, "Historical Overview of Stall/Spin Characteristics of General Aviation Aircraft." *J. Aircraft* **16** (1979), 455.
40. F. Hess, "The Aerodynamics of Boomerangs." *Sci. Am.* (Nov. 1968), 124.
41. T. C. Soong, "The Dynamics of Discus Throw." *ASME J. Appl. Mech.* (Dec. 1976), 531.
42. J. Shelton, in "Frisbee: A Practitioner's Manual and Definitive Treatise." S.E.D. Johnson, Workman, 1975.
43. Z. Plaskowski, "Über ringförmige Tragflügel." Fourth International Astronautical Congress, Zürich, August 1953.
44. H. J. Lugt,"Entstehung und Ausbreitung von Wirbeln unter der 'Perfect-Slip' Bedingung." Deutsche Luft- und Raumfahrt Forschungsbericht Nr. 72-27, 1972.
45. Z. Takusagawa and I. Yoshioka, "A Study on Ro (Oriental Sweep)." *Bull Fac. Eng., Yokohama Nat. Univ.* **25** (1976), 69.

Chapter 5

1. S. C. R. Dennis and J. D. A. Walker, "Calculation of the Steady Flow Past a Sphere at Low and Moderate Reynolds Numbers. *J. Fluid Mech.* **48** (1971), 771.
2. H. R. Pruppacher, B. P. Le Clair, and A. E. Hamielec, "Some Relations Between Drag and Flow Pattern of Viscous Flow Past a Sphere and a Cylinder at Low and Intermediate Reynolds Numbers." *J. Fluid Mech.* **44** (1970), 781.
3. K. V. Beard and H. R. Pruppacher, "A Determination of the Terminal Velocity and Drag of Small Water Drops by Means of a Wind Tunnel." *J. Atm. Sci.* **26** (1969), 1066.
4. B. P. Le Clair, A. E. Hamielec, H. R. Pruppacher, and W. D. Hall, "A Theoretical and Experimental Study of the Internal Circulation in Water Drops Falling at Terminal Velocity in Air." *J. Atm. Sci.* **29** (1972), 728.
5. R. H. Magarvey and J. Hoskins, "Entrainment of Small Particles by a Large Sphere." *Nature* **218** (1968), 460.
6. H. Takami and H. B. Keller, "Steady Two-Dimensional Viscous Flow of an Incompressible Fluid past a Circular Cylinder." *Phys. Fluids,* Suppl. II (1969), II-51.
7. S. C. R. Dennis and G. Chang, "Numerical Solutions for Steady Flow Past a Circular Cylinder at Reynolds Numbers up to 100. *J. Fluid Mech.* **42** (1970), 471.
8. J. Lee and Y. Fung, "Flow in Locally Constricted Tubes at Low Reynolds numbers." *J. Appl. Mech.* (March 1970), 9.
9. E. M. Parmentier, W. A. Morton, and H. E. Petschek, "Platelet Aggregate Formation in a Region of Separated Blood." *Phys. Fluids* **20** (1977), 2012.
10. J. Ackeret, F. Feldman, and N. Rott, *Investigations of Compression Shocks and Boundary Layers in Gases Moving at High Speed.* NACA TM-1113, Jan. 1947.
11. T. C. Adamson and A. F. Messiter, "Analysis of Two-Dimensional Interactions Between Shock Waves and Boundary Layers." *Ann. Rev. Fluid Mech.* **12** (1980), 103.
12. J. M. Dorrepaal, "Stokes Flow Past a Two-Dimensional Lens." *ZAMP* **30** (1979), 405.
13. D. J. Jeffrey and J. D. Sherwood, "Streamline Patterns and Eddies in Low-Reynolds-Number Flow." *J. Fluid Mech.* **96** (1980), 315.
14. H. von Helmholtz, "Über discontinuierliche Flüssigkeitsbewegungen." Monatsberichte der königl. Akademie der Wissenschaften, Berlin, 1868, p. 215.
15. R. L. Pitter, H. R. Pruppacher, and A. E. Hamielec, "A Numerical Study of Viscous Flow Past a Thin Oblate Spheroid at Low and Intermediate Reynolds Numbers." *J. Atm. Sci.* **30** (1973), 125.
16. Y. Rimon, "Numerical Solution of the Incompressible Time-Dependent Viscous Flow Past a Thin Oblate

Spheroid." *Phys. Fluids,* Suppl. II (1969), II-65. See also *Phys. Fluids* **12** (1969), 949.

17. H. J. Lugt and H. J. Haussling, "Laminar Flow Past an Abruptly Accelerated Elliptic Cylinder at 45° Incidence." *J. Fluid Mech.* **65** (1974), 711.
18. U. B. Mehta and Z. Lavan, "Flow in a Two-Dimensional Channel With a Rectangular Cavity." *J. Appl. Mech.* (Dec. 1969), 897.
19. G. Birkhoff and E. H. Zarantonello, *Jets, Wakes, and Cavities.* Academic, New York, 1957.
20. F. O. Ringleb, "Separation Control by Trapped Vortices." In *Boundary Layer and Flow Control.* Pergamon, New York, 1961, 265.
21. O. R. Burggraf, "Analytical and Numerical Studies of the Structure of Steady Separated Flows." *J. Fluid Mech.* **24** (1966), 113.
22. U. Ghia, K. N. Ghia, and C. T. Shin, "High-Re Solutions for Incompressible Flow Using the Navier–Stokes Equations and a Multi-Grid Method. *J. Comp. Phys.* (Fall 1982).
23. S. Taneda, "Visualization Experiments on Unsteady Viscous Flows Around Cylinders and Plates." In *Recent Research On Unsteady Boundary Layers.* IUTAM Symposium 1971, E. A. Eichelbrenner (ed.). Les Presses de l' Université Laval, Québec, 1972, 1165.
24. R. Bouard and M. Coutanceau, "The Early Stage of Development of the Wake Behind an Impulsively Started Cylinder for $40 < \text{Re} < 10^4$. *J. Fluid Mech.* **101** (1980), 583.
25. Ta Phuoc Loc, "Numerical Analysis of Unsteady Secondary Vortices Generated by an Impulsively Started Circular Cylinder." *J. Fluid Mech.* **100** (1980), 111.
26. J. C. Williams, "Flow Development in the Vicinity of the Sharp Trailing Edge on Bodies Impulsively Set into Motion." *J. Fluid Mech.* **115** (1982), 27.
27. E. Wedemeyer, "Ausbildung eines Wirbelpaares an den Kanten einer Platte." *Ing. Arch.* **30** (1961), 187.
28. L. Prandtl, *Gesammelte Abhandlungen.* Springer, Berlin, 1961.
29. D. W. Moore and P. G. Saffman, "Axial Flow in Laminar Trailing Vortices." *Proc. Roy. Soc. Lond. Ser. A* **333** (1973), 491.
30. D. I. Pullin and A. E. Perry, "Some Flow Visualization Experiments on the Starting Vortex." *J. Fluid Mech.* **97** (1980), 239.
31. R. J. Tabaczynski, D. P. Hoult, and J. C. Keck, "High Reynolds Number Flow in a Moving Corner. *J. Fluid Mech.* **42** (1970), 249.
32. N. Rott, "Diffraction of a Weak Shock with Vortex Generation." *J. Fluid Mech.* **1** (1956), 111.
33. H. U. Hassenpflug, "Wirbelentstehung durch Stosswellenbeugung." *Abh. Aer. Inst. Aachen* **22** (1975), 104.
34. J. Siekmann, "On a Pulsating Jet from the End of a Tube. *J. Fluid Mech.* **15** (1963), 399.
35. D. Küchemann and J. Weber, "Vortex Motions." *ZAMM* **45** (1965), 457.
36. E. C. Polhamus, *A Concept of the Vortex Lift of Sharp-Edge Delta Wings Based on a Leading-Edge-Suction Analogy.* NASA TN D-3767, Dec. 1966.
37. J. E. Lamar, "Recent Studies of Subsonic Vortex Lift Including Parameters Affecting Stable Leading-Edge Vortex Flow." *J. Aircraft* **14** (1977), 1205.
38. D. J. Peake and M. Tobak, "Three-Dimensional Interactions and Vortical Flows With Emphasis on High Speeds." *AGARDograph* **252** (July 1980) (also published as NASA TM 81169, March 1980).
39. E. J. Cross, *Experimental and Analytical Investigation of the Expansion Flow Field Over a Delta Wing at Hypersonic Speeds.* ARL 68-0027, Feb. 1968.
40. J. Szodruch, "Zur Systematik der Leeseiten-Strömung bei Deltaflügeln." *Z. Flugwiss. Weltraumf.* **4** (1980), 72.
41. C. duP. Donaldson and A. J. Bilanin, "Vortex Wakes of Conventional Aircraft." *AGARDograph* **204** (May 1975).
42. G. Sovran, T. Morel, and W. T. Mason, *Aerodynamic Drag Mechanisms of Bluff Bodies and Road Vehicles.* Plenum, New York, 1978.
43. A. J. Scibor-Rylski, *Road Vehicle Aerodynamics.* Halsted, New York, 1975.
44. K. Wieghardt, "Messungen im Strömungsfeld an zwei Hinterschiffsmodellen." *Schiffstechnik* **4** (1957), 78.
45. T. Tagori, "Investigations on Vortices Generated at the Bilge." Proceedings of the 11th International Towing Tank Conference, Tokyo, 1967.
46. K. Oswatitsch, "Die Ablösungsbedingung von Grenzschichten." *IUTAM Symposium on Boundary Layer Research.* H. Görtler (ed.). Springer, New York, 1958, 357.
47. E. H. Hirschel and W. Kordulla, "Shear Flow in Surface-Oriented Coordinates." *Notes on Numerical Fluid Mechanics,* Vol. 4. Vieweg, Braunschweig, 1981.
48. K. C. Wang, "Separation Patterns of Boundary Layers over an Inclined Body of Revolution." *AIAA J.* **10** (1972), 1044.
49. E. A. Eichelbrenner and A. Oudart, "Methode de Calcul de la Couche Limite Tridimensionnelle." Application a un Corps Fusele Incline sur le Vent. ONERA Publ. 76, 1955, Paris.
50. H. Werlé, "Separation on Axisymmetrical Bodies at Low Speed." *Rech. Aeronaut.* 90 (1962), 3.
51. T. Han and V. C. Patel, "Flow Separation on a Spheroid at Incidence. *J. Fluid Mech.* **92** (1979), 643.
52. H. U. Meier and H. P. Kreplin, "Experimental Investigation of the Boundary Layer Transition and Separation on a Body of Revolution." *Z. Flugwiss. Raumf.* **4** (1980), 65.
53. U. Brennenstuhl and D. Hummel, "Untersuchungen über die Wirbelbildung an Flügeln mit geknickten Vorderkanten." *Z. Flugwiss. Weltraumforsch.* **5** (1981), 375.
54. H. J. Lugt, "Numerical Modelling of Vortex Flows in Ship Hydrodynamics." Proceedings of the Third International Conference on Numerical Ship Hydrodynamics. Paris, June, 1981.
55. E. C. Maskell, *Flow Separation in Three Dimensions.* Royal Aircraft Establishment, Farnborough, Rep. Aero. 2565, Nov. 1955.
56. C. J. Baker, "The Laminar Horseshoe Vortex." *J. Fluid Mech.* **95** (1979), 347.
57. C. J. Baker, "The Turbulent Horseshoe Vortex." *J. Wind Eng. Industr. Aer.* **6** (1980), 9.
58. J. C. R. Hunt, C. J. Abell, J. A. Peterka, and H. Woo, "Kinematical Studies of the Flows around Free or Surface-Mounted Obstacles; Applying Topology to Flow Visualization." *J. Fluid Mech.* **86** (1978), 179.
59. W. R. Dean, "Note on the Motion of Fluid in a Curved Pipe. *Philos. Mag.* **4** (1927), 208.

60. L. B. Leopold and W. B. Langbein, "River Meanders." *Sci. Am.* (June 1966).
61. W. H. Lyne, "Unsteady Viscous Flow in a Curved Pipe." *J. Fluid Mech.* **45** (1970), 13.
62. J. Y. Lin and J. M. Tarbell, "An Experimental and Numerical Study of Periodic Flow in a Curved Tube." *J. Fluid Mech.* **100** (1980), 623.
63. T. Mullin and C. A. Greated, "Oscillatory Flow in Curved Pipes." *J. Fluid Mech.* **98** (1980), 383.
64. R. M. Nerem and J. F. Cornhill, "The Role of Fluid Mechanics in Atherogenesis." Proceedings from a Specialists Meeting, Columbus, Ohio, 1978.
65. A. P. Rochino and Z. Lavan, *Analytical Investigation of Incompressible Turbulent Swirling Flow in Pipes*. NASA CR-1169, Sept. 1968.
66. M. P. Singh, P. C. Sinha, and M. Aggarwal, "Swirling Flow in a Straight Circular Pipe." *ZAMM* **60** (1980), 429.
67. D. H. Peckham and S. A. Atkinson, "Preliminary Results of Low Speed Wind Tunnel Tests on a Gothic Wing of Aspect Ratio 1.0." British ARC CP 508, 1957.
68. H. J. Lugt, "Einfluss der Drallströmung auf die Durchflusszahlen genormter Drosselmessgeräte." PhD Thesis, Stuttgart 1959. Partially translated by the British Hydromechanics Research Association, Rep. T 716, Feb. 1962.
69. J. K. Harvey, "Some Observations of the Vortex Breakdown Phenomenon." *J. Fluid Mech.* **14** (1962), 585.
70. M. G. Hall, "Vortex Breakdown." *Ann. Rev. Fluid Mech.* **4** (1972), 195.
71. T. Sarpkaya, "On Stationary and Travelling Vortex Breakdowns." *J. Fluid Mech.* **45** (1971), 545.
72. M. P. Escudier and N. Zehnder, "Vortex-Flow Regimes." *J. Fluid Mech.* **115** (1982), 105.
73. W. J. Grabowski and S. A. Berger, "Solutions of the Navier–Stokes Equations for Vortex Breakdown." *J. Fluid Mech.* **75** (1976), 525.
74. J. P. Narain, "Numerical Prediction of Confined Swirling Jets." *Comput. Fluids* **5** (1977), 115.
75. H. H. Bossel, "Vortex Breakdown Flowfield." *Phys. Fluids* **12** (1969), 498.
76. A. M. Skow and A. Titiriga, "A Survey of Analytical and Experimental Techniques to Predict Aircraft Dynamic Characteristics at High Angles of Attack," AGARD-CP 235, No. 19, 1978.
77. L. E. Simon, *German Research in World War II*. Wiley, New York, 1951, p. 180.
78. J. M. V. Rayner, "A Vortex Theory of Animal Flight." *J. Fluid Mech.* **91** (1979), 697.
79. T. Maxworthy, "Some Experimental Studies of Vortex Rings. *J. Fluid Mech.* **81** (1977), 465.
80. J. S. Turner, "Buoyant Vortex Rings." *Proc. Roy. Soc. Lond.* **239** (1957), 61.
81. C. Krutzsch, "Über eine experimentell beobachtete Erscheinung an Wirbelringen bei ihrer translatorischen Bewegung in wirklichen Flüssigkeiten." *Ann. Phys.* **35** (1939), 497.
82. T. Maxworthy, "The Structure and Stability of Vortex Rings." *J. Fluid Mech.* **51** (1972), 15.
83. N. Didden, "On the Formation of Vortex Rings: Rolling-up and Production of Circulation." *ZAMP* **30** (1979), 102.
84. T. Kambe and Y. Oshima, "Generation and Decay of Viscous Vortex Rings." *J. Phys. Soc. Japan* **38** (1975), 271.
85. U. Boldes and J. C. Ferreri, "Behavior of Vortex Rings in the Vicinity of a Wall. *Phys. Fluids* **16** (1973), 2005.
86. J. K. Harvey and F. J. Perry, "Flowfield Produced by Trailing Vortices in the Vicinity of the Ground." *AIAA J.* **9** (1971), 1659.
87. T. Fohl and J. S. Turner, "Colliding Vortex Rings." *Phys. Fluids* **18** (1975), 433.
88. R. K. C. Lo and L. Ting, "Studies of the Merging of Vortices. *Phys. Fluids* **19** (1976), 912.
89. M. Takamoto and K. Izumi, "Experimental Observation of Stable Arrangement of Vortex Rings. *Phys. Fluids* **24** (1981), 1582.
90. H. Viets and P. M. Sforza, "Dynamics of Bilaterally Symmetric Vortex Rings." *Phys. Fluids* **15** (1972), 230.
91. N. Rott, "Unsteady Viscous Flow in the Vicinity of a Stagnation Point." *Q. Appl. Math.* **13** (1956), 444.
92. D. P. Telionis and M. J. Werle, "Boundary-Layer Separation From Downstream Moving Boundaries." *J. Appl. Mech.* **40** (1973), 369.
93. D. T. Tsahalis, "Laminar Boundary-Layer Separation from an Upstream-Moving Wall." *AIAA J.* **15** (1977), 561.
94. J. C. Williams, On the Nature of Unsteady Three-Dimensional Laminar Boundary-Layer Separation. *J. Fluid Mech.* **88** (1978), 241.
95. R. H. Magarvey and R. L. Bishop, "Wakes in Liquid-Liquid Systems." *Phys. Fluids* **4** (1961), 800.
96. P. A. Lagerstrom, "Solutions of the Navier-Stokes Equation at Large Reynolds Number." *SIAM J. Appl. Math.* **28** (1975), 20.
97. M. Huang and C. Chow, "Trapping of a Free Vortex by Joukowski Airfoils." *AIAA J.* **20** (1982), 292.
98. R. Sedney, "A Flow Model for the Effect of a Slanted Base on Drag." Ballistic Res. Lab., Aberdeen Proving Ground, Md., Tech. Rep. ARBRL-TR-02341, July 1981.

Chapter 6

1. Lord Rayleigh, "On the Instability of Jets." *Proc. Lond. Math. Soc.* **10** (1879), 4.
2. Lord Kelvin, "On a Disturbance in Lord Rayleigh's Solution for Waves in a Plane Vortex Stratum." *Math. and Phys. Papers*, Vol. 4. Cambridge University Press, Cambridge, 1880, p. 186.
3. R. S. Brodkey, *The Phenomena of Fluid Motions*. Addison-Wesley, Reading, Mass., 1967, p. 227.
4. R. Betchov and W. O. Criminale, *Stability of Parallel Flows*. Academic, New York, 1967.
5. H. Sato, "The Stability and Transition of a Two-Dimensional Jet." *J. Fluid Mech.* **7** (1959), 53.
6. P. Freymuth, "On Transition in a Separated Laminar Boundary Layer." *J. Fluid Mech.* **25** (1966).
7. A. Michalke, "The Instability of Free Shear Layers." *Progress in Aerospace Science*. Pergamon Press, New York, 1972, p. 213.
8. K. B. M. Q. Zaman and A. K. M. F. Hussain, "Vortex Pairing in a Circular Jet under Controlled Excitation." *J. Fluid Mech.* **101** (1980), 449.
9. H. Schade and A. Michalke, "Zur Entstehung von Wir-

beln in einer freien Grenzschicht." *Z. Flugwiss.* **10** (1962), 147.
10. D. O. Rockwell and W. O. Niccols, "Natural Breakdown of Planar Jets." *J. Basic Eng.* (Dec. 1972), 720.
11. G. H. Koopmann, "The Vortex Wakes of Vibrating Cylinders at Low Reynolds Number." *J. Fluid Mech.* **28** (1967), 501.
12. F. R. Hama, "Streaklines in a Perturbed Shear Flow. *Phys. Fluids* **5** (1962), 644.
13. H. J. Lugt, "Decay of a Discontinuity Line in a Viscous Fluid." *Acta Mech.* **4** (1967), 149.
14. D. Pierce, "Photographic Evidence of the Formation and Growth of Vorticity Behind Plates Accelerated from Rest in Still Air." *J. Fluid Mech.* **11** (1961), 460.
15. J. R. Weske and T. M. Rankin, *Production of Secondary Vortices in the Field of a Primary Vortex*. Institute of Fluid Dynamics and Applied Mathematics, University of Maryland, TN BN-244, AFOSR-623, April 1961.
16. E. Mollö-Christensen, "Physics of Turbulent Flow." *AIAA J.* **9** (1971), 1217.
17. Mohamed Gad-El-Hak, R. F. Blackwelder, and J. J. Riley, "On the Growth of Turbulent Regions in Laminar Boundary Layers." *J. Fluid Mech.* **110** (1981), 73.
18. H. W. Emmons, "The Laminar-Turbulent Transition in a Boundary Layer, Part I." *J. Aer. Sci.* **18** (1951), 490.
19. A. Leonard, *Vortex Simulation of Three-Dimensional, Spotlike Disturbances in a Laminar Boundary Layer*. NASA TM 78579, May 1979.
20. T. Theodorsen, "The Structure of Turbulence." In *50 Jahre Grenzschichtforschung*. Vieweg, Braunschweig, 1955.
21. S. Taneda, "Experimental Investigation of Vortex Streets." *J. Phys. Soc. Japan* **20** (1965), 1714.
22. D. C. Thoman and A. A. Szewczyk, *Numerical Solutions of Time Dependent Two Dimensional Flow of a Viscous, Incompressible Fluid Over Stationary and Rotating Cylinders*. University of Notre Dame, Dept. Mech. Eng., TR 66-14, July 1966.
23. A. Timme, "Über die Geschwindigkeitsverteilung in Wirbeln." *Ing. Arch.* **25** (1957), 205.
24. J. W. Schaefer and S. Eskinazi, "An Analysis of the Vortex Street Generated in a Viscous Fluid." *J. Fluid Mech.* **6** (1959), 241.
25. S. K. Jordan and J. E. Fromm, "Oscillatory Drag, Lift, and Torque on a Circular Cylinder in a Uniform Flow." *Phys. Fluids* **15** (1972), 371.
26. T. Sarpkaya and R. L. Shoaff, "Inviscid Model of Two-Dimensional Vortex Shedding by a Circular Cylinder." *AIAA J.* **17** (1979), 1193.
27. A. Roshko, *On the Development of Turbulent Wakes from Vortex Streets*. NACA TN 2913, 1953.
28. L. E. Ericsson, "Kármán Vortex Shedding and the Effect of Body Motion." *AIAA J.* **18** (1980), 935.
29. M. V. Morkovin, "Flow Around Circular Cylinder—A Kaleidoscope of Challenging Fluid Phenomena." *Symposium on Fully Separated Flows*. ASME, 1964, 102.
30. L. Landweber, *Flow About a Pair of Adjacent, Parallel Cylinders Normal to a Stream*. David W. Taylor Model Basin Rep. No. 485, 1942.
31. H. Honji, "Viscous Flows past a Group of Circular Cylinders." *J. Phys. Soc. Japan* **34** (1973), 821.
32. M. Kiya, et al., "Vortex Shedding from Two Circular Cylinders in Staggered Arrangement." *J. Fluids Eng.* **102** (June, 1980), 166.
33. G. Ehrhardt, "Stabilität zweireihiger Strassen geradliniger und kreisförmiger Wirbel." *Fortschr. Ber. VDI Z.*, Reihe 7, Nr. 49.
34. M. Gaster, "Vortex Shedding from Slender Cones at Low Reynolds Number." *J. Fluid Mech.* **38** (1969), 565.
35. C. F. Chen and B. J. Mangione, "Vortex Shedding from Circular Cylinders in Sheared Flow." *AIAA J.* **7** (1969), 1211.
36. M. Kiya, et al., "Vortex Shedding from a Circular Cylinder in Moderate-Reynolds-Number Shear Flow. *J. Fluid Mech.* **101** (1980), 721.
37. S. Taneda, "Studies on Wake Vortices (I)." *Rep. Res. Inst. Appl. Mech.* **1** (1952), No.4, 131.
38. H. P. Pao and T. W. Kao, "Vortex Structure in the Wake of a Sphere." *Phys. Fluids* **20** (1977), 187.
39. E. Achenbach, "Vortex Shedding from Spheres." *J. Fluid Mech.* **62** (1974), 209.
40. H. Viets and D. A. Lee, "Motion of Freely Falling Spheres at Moderate Reynolds Numbers." *AIAA J.* **9** (1971), 2038.
41. A. E. Perry, T. T. Lim, and M. S. Chong, "The Instantaneous Velocity Fields of Coherent Structures in Coflowing Jets and Wakes." *J. Fluid Mech.* **101** (1980), 243.
42. V. Strouhal, "Über eine besondere Art der Tonerregung." *Ann. Phys. Chem.* **5** (1878), 216.
43. Lord Rayleigh, "Acoustical Observations." *Philos. Mag. Ser. 5* (7) (1879), 149.
44. H. Bénard, "Formation de centres de giration à l' arrière d' un obstacle en mouvement." *Comptes rendus* **147** (1908), 970.
45. "The Failure of the Tacoma Narrows Bridge." A Report to the Honorable John M. Carmody, Adm., Federal Works Agency, Washington, D.C., March 28, 1941.
46. A. Bertelsen, A. Svardal, and S. Tjøtta, "Nonlinear Streaming Effects Associated with Oscillating Cylinders." *J. Fluid Mech.* **59** (1973), 493.
47. A. Kundt, *Pogg. Ann. Phys. Chem.* **127** (1866), 493.
48. O. M. Griffin and S. E. Ramberg, "The Vortex-Street Wakes of Vibrating Cylinders." *J. Fluid Mech.* **66** (1974), 553.
49. O. M. Griffin and S. E. Ramberg, "Vortex Shedding From a Cylinder Vibrating in Line With an Incident Uniform Flow." *J. Fluid Mech.* **75** (1976), 257.
50. O. M. Griffin and C. W. Votaw, "The Vortex Street in the Wake of a Vibrating Cylinder." *J. Fluid Mech.* **51** (1972), 31.
51. U. B. Mehta, "Dynamic Stall of an Oscillating Airfoil." *AGARD Conf. Proc.* **227** (1977), paper no. 23.
52. A. Okajima, H. Takata, and T. Asanuma, "Viscous Flow Around a Transversally Oscillating Elliptic Cylinder." Institute of Space and Aeronautical Science, University of Tokyo, Rep. 533, 1975.
53. S. Taneda, "Visual Study of Unsteady Separated Flows Around Bodies." *Prog. Aerospace Sci.* **17** (1977), 287.
54. C. Maresca, D. Favier, and J. Rebont, "Experiments on an Aerofoil at High Angle of Incidence in Longitudinal Oscillations." *J. Fluid Mech.* **92** (1979), 671.
55. H. J. Haussling, "Boundary-Fitted Coordinates for Ac-

curate Numerical Solution of Multi-Body Flow Problems." *J. Comput. Phys.* **30** (1979), 107.

56. T. Maxworthy, "The Fluid Dynamics of Insect Flight." *Ann. Rev. Fluid Mech.* **13** (1981), 329.
57. T. Weis-Fogh, "Unusual Mechanisms for the Generation of Lift in Flying Animals." *Am. Sci.* **233** (Nov. 1975), 81.
58. A. Meier-Windhorst, *Flatterschwingungen von Zylindern im gleichmässigen Flüssigkeitsstrom.* Hydr. Inst. Mitt., Vol. 9, München, Tech. Hochschule, 1939.
59. G. V. Parkinson, "Mathematical Models of Flow-Induced Vibrations of Bluff Bodies." In *Flow-Induced Structural Vibrations*, E. Naudascher (ed.). Springer, New York, 1974, p. 81.
60. R. D. Blevins, *Flow-Induced Vibration.* Van Nostrand Reinhold, 1977.
61. R. King, "A Review of Vortex Shedding, Research and its Application." *Ocean Eng.* **4** (1977), 141.
62. T. Theodorsen, *General Theory of Aerodynamic Instability and the Mechanism of Flutter.* NACA Rep. No. 496, 1935.
63. R. M. C. So and S. D. Savkar, "Buffeting Forces on Rigid Circular Cylinders in Cross Flows." *J. Fluid Mech.* **105** (1981), 397.
64. J. C. R. Hunt, "The Effect of Single Buildings and Structures." *Philos. Trans. Roy. Soc. Lond. Ser. A* **269** (1971), 457.
65. K. J. Eaton (ed.), "Wind Effects on Buildings and Structures." Proceedings of the 4th Intern. Conf., Cambridge University Press, 1977.
66. O. M. Griffin et al., "Vortex-Excited Vibrations of Marine Cables." *J. Waterway Port Coastal and Ocean Div.* (May 1980), 183.
67. E. H. Smith, "Autorotating Wings: An Experimental Investigation." Dept. Aerospace Eng., Univ. Michigan, Ann Arbor, 01954-2-T, July 1970.
68. W. W. Willmarth, N. E. Hawk, and R. L. Harvey, "Steady and Unsteady Motions and Wakes of Freely Falling Disks." *Phys. Fluids* **7** (1964), 197.
69. D. Rockwell and E. Naudascher, "Self-Sustained Oscillations of Impinging Free Shear Layers." *Ann. Rev. Fluid Mech.* **11** (1979), 67.
70. D. Rockwell, *Oscillations of Impinging Shear Layers.* 20th Aerospace Sci. Meeting, AIAA, Jan. 1982.
71. D. Rockwell and C. Knisely, "The Organized Nature of Flow Impingement Upon a Corner." *J. Fluid Mech.* **93** (1979), 413.
72. Y. T. Tsui, "On Wake-Induced Flutter of a Circular Cylinder in the Wake of Another." *J. Appl. Mech.* (June 1977), 194.
73. A. Powell, "Jet Noise—Age 25." Noise-Con 75, National Bureau of Standards, Washington, D.C., Oct. 1975, p. 33.
74. D. Ross, *Mechanics of Underwater Noise.* Pergamon, New York, 1976.
75. E. A. Müller, "Recent Progress in Understanding Basic Processes in Flow-Acoustics." 10th International Congress Acoustics, Sydney, Australia, 1980, p. 83.
76. J. E. Ffowcs Williams, "Sound Sources in Aerodynamics—Fact and Fiction." *AIAA J.* **20** (1982), 307.
77. P. E. M. Schneider, "Experimentelle Untersuchungen über den Einfluss von Schall auf Diffusionsflammen." *Z. Flugwiss.* **19** (1971), 485.
78. H. J. Lugt and S. Ohring, "Rotating Elliptic Cylinders in a Viscous Fluid at Rest or in a Parallel Stream. *J. Fluid Mech.* **79** (1977), 127.
79. O. Sawatzki, "Das Strömungsfeld um eine rotierende Kugel." *Acta Mech.* **9** (1970), 159.
80. N. Gregory, J. T. Stuart, and W. S. Walker, "On the Stability of Three-Dimensional Boundary Layers with Application to the Flow due to a Rotating Disc. *Philos. Trans. Roy. Soc. Lond. Ser. A* **248** (1955), 155.
81. G. Magnus, *Abh. Berliner Akademie 1852* (Poggendorf Ann. 1853, 1).
82. B. Robins, *New Principles of Gunnery.* Ed. Hutton, 1805. First print: 1742.
83. H. M. Barkla and L. J. Auchterlonie, "The Magnus or Robins Effect on Rotating Spheres." *J. Fluid Mech.* **47** (1971), 437.
84. J. M. Davies, "The Aerodynamics of Golf Balls." *J. Appl. Phys.* **20** (1949), 821.
85. "A Rotor Airplane." *Sci. Am.* (June 1932), 362.
86. W. M. Swanson, "The Magnus Effect: A Summary of Investigations to Date." *J. Basic Eng.* **83** (1961), 461.
87. F. N. M. Brown, *See the Wind Blow.* University of Notre Dame, Indiana, 1971.
88. G. G. Poe and A. Acrivos, "Closed-Streamline Flows Past Rotating Single Cylinders and Spheres: Inertia Effects." *J. Fluid Mech.* **72** (1975), 605.
89. P. Vasseur and R. G. Cox, "The Lateral Migration of a Spherical Particle in Two-Dimensional Shear Flows." *J. Fluid Mech.* **78** (1976), 385.
90. H. J. Lugt, "Autorotation of an Elliptic Cylinder About an Axis Perpendicular to the Flow." *J. Fluid Mech.* **99** (1980), 817.
91. A. M. O. Smith, "On the Motion of a Tumbling Body." *J. Aer. Sci.* **20** (1953), 73.
92. "Results of the First U.S. Manned Orbital Space Flight." Feb. 20, 1962. Manned Spacecraft Center. NASA.
93. O. Reynolds, "An Experimental Investigation of the Circumstances Which Determine Whether the Motion of Water Shall be Direct or Sinuous, and of the Law of Resistance in Parallel Channels." *Philos. Trans. Roy. Soc. Lond.* **174** (1883), 935.
94. J. E. Ffowcs Williams, S. Rosenblat, and J. T. Stuart, Transition from Laminar to Turbulent Flow." *J. Fluid Mech.* **39** (1969), 547.
95. L. D. Landau, "On the Problem of Turbulence." *C.R. Acad. Sci. URSS* **44** (1944), 311.
96. H. Tennekes and J. L. Lumley, *A First Course in Turbulence.* MIT Press, Cambridge, Mass., 1972.
97. A. N. Kolmogorov, "The Local Structure of Turbulence in Incompressible Viscous Fluid for Very Large Reynolds numbers." *Dokl. Akad. Nauk SSR* **30** (1941), 301.
98. V. P. Starr, *Physics of Negative Viscosity Phenomena.* McGraw-Hill, New York, 1968.
99. J. Boussinesq, "Théorie de l' écoulement tourbillant." *Mém. prés. Acad. Sci. Paris* **XXIII** (1877), 46.
100. G. I. Taylor, "Statistical Theory of Turbulence." *Proc. Roy. Soc. Lond. Ser. A* **151** (1935), 421.

101. A. A. Townsend, *The Structure of Turbulent Shear Flow.* Cambridge University Press, New York, 1956.
102. J. Laufer, "New Trends in Experimental Turbulence Research." *Ann. Rev. Fluid Mech.* **7** (1975), 307.
103. B. J. Cantwell, "Organized Motion in Turbulent Flow." *Ann. Rev. Fluid Mech.* **13** (1981), 457.
104. H. W. Liepmann, "The Rise and Fall of Ideas in Turbulence." *Am. Sci.* **67** (1979), 221.
105. R. F. Blackwelder, "Pattern Recognition of Coherent Eddies." Proceedings of the Dynamic Flow Conference 1978, 173.
106. S. A. Orszag, *Lectures on the Statistical Theory of Turbulence. Les Houches Summer School.* Springer, New York, 1975.
107. J. H. Ferziger, "Large Eddy Numerical Simulations of Turbulent Flows." *AIAA J.* **15** (1977), 1261.
108. D. R. Chapman. "Computational Aerodynamics Development and Outlook." *AIAA J.* **17** (1979), 1293.
109. M. Macagno and E. Macagno, "Nonlinear Behavior of Line Vortices." *Phys. Fluids* **18** (1975), 1595.
110. R. H. Magarvey and C. S. MacLatchy, "The Formation and Structure of Vortex Rings." *Can. J. Phys.* **42** (1964), 678.
111. P. E. M. Schneider, *Werden, Bestehen, Instabilität, Regeneration, Vergehen eines Ringwirbels.* Max Planck Institut für Strömungsforschung, Göttingen. Bericht 17/1978.
112. T. Maxworthy, "Turbulent Vortex Rings." *J. Fluid Mech.* **64** (1974), 227.
113. P. E. M. Schneider, "Sekundärwirbelbildung bei Ringwirbeln und in Freistrahlen. *Z. Flugwiss. Weltraumf.* **4** (1980), 307.
114. F. Kreith and O. K. Sonju, "The Decay of a Turbulent Swirl in a Pipe." *J. Fluid Mech.* **22** (1965), 257.
115. J. Nikuradse, "Turbulente Strömungen in nichtkreisförmigen Rohren." *Ing. Arch.* **1** (1930), 306.
116. P. R. Kry and R. List, "Angular Motions of Freely Falling Spheroidal Hailstone Models." *Phys. Fluids* **17** (1974), 1093.
117. H. Rouse and S. Ince, *History of Hydraulics.* Institute of Hydraulic Research, State University of Iowa, 1957.
118. B. A. Toms, "Some Observations on the Flow of Linear Polymers through Straight Tubes at Large Reynolds Numbers." *Proc. First Int. Congr. Rheol., Amsterdam* **2** (1948), 135.
119. J. L. Lumley, "Drag Reduction by Additives." *Ann. Rev. Fluid Mech.* **1** (1969), 367.
120. J. W. Hoyt, "The Effect of Additives on Fluid Friction." *J. Basic Eng.* (June 1972), 258.
121. M. Rosen and N. Cornford, "Fluid Friction of the Slime of Aquatic Animals." Naval Undersea Res. and Dev. Center, NUC TP 193, Nov. 1970.

Chapter 7

1. M. Jammer, *Concepts of Mass.* Harper, New York, 1964.
2. D. W. Sciama, *The Unity of the Universe.* Cambridge University Press, New York, 1966.
3. E. Mach, *Die Mechanik in ihrer Entwicklung historisch—kritisch dargestellt.* Leipzig, 1883.
4. A. Einstein, "Prinzipielles zur allgemeinen Relativitätstheorie." *Ann. Phys.* **55** (1918), 241.
5. C. Brans and R. H. Dicke, "Mach's Principle and a Relativistic Theory of Gravitation." *Phys. Rev.* **124** (1961), 925.
6. F. Hoyle and J. V. Narlikar, "The Mass Difference of the Muon and the Electron." *Nature* **238** (1972), 86.
7. T. C. Van Flandern, "Is Gravity Getting Weaker?" *Sci. Am.* **234** (Feb. 1976), 44.
8. A. N. Strahler, "The Coriolis Effect." In *Adventures in Earth History.* P. Cloud (ed.). Freeman, San Francisco, 1970.
9. H. Riehl, *Introduction to the Atmosphere.* McGraw-Hill, New York, 1965.
10. G. Neumann, *Ocean Currents.* Elsevier, Amsterdam, 1968.
11. V. W. Ekman, "On the Influence of the Earth's Rotation on Ocean Currents." *Ark. Mat. Astron. Fys. Stockholm* **2** (1905), 1.
12. K. Stewartson, "On Almost Rigid Rotations." *J. Fluid Mech.* **3** (1957), 17.
13. C. G. Rossby and Collaborators, "Relation between Variations in the Intensity of the Zonal Circulation of the Atmosphere and the Displacements of the Semipermanent Centers of Action." *J. Marine Res.* **2** (1939), 38.
14. Lord Kelvin, "Vibrations of a Columnar Vortex." *Philos. Mag.* **10** (1880), 155.
15. H. Görtler, "Einige Bemerkungen über Strömungen in rotierenden Flüssigkeiten." *ZAMM* **24** (1944), 210.
16. H. Oser, "Experimentelle Untersuchung über harmonische Schwingungen in rotierenden Flüssigkeiten." *ZAMM* **38** (1958), 386.
17. J. Proudman, "On the Motion of Solids in Liquids Possessing Vorticity." *Proc. Roy. Soc. Lond. Ser. A* **92** (1916), 408.
18. G. I. Taylor, "Experiments with Rotating Fluids." *Proc. Roy. Soc. Lond. Ser. A* **100** (1921), 114.
19. R. Hide and A. Ibbetson, "An Experimental Study of Taylor Columns." *Icarus* **5** (1966), 279.
20. R. Hide, "Origin of Jupiter's Great Red Spot." *Nature* **190** (1961), 895.
21. G. I. Taylor, "The Motion of a Sphere in a Rotating Liquid." *Proc. Roy. Soc. Lond. Ser. A* **102** (1922), 180.
22. R. R. Long, "Steady Motion around a Symmetrical Obstacle Moving Along the Axis of a Rotating Liquid." *J. Met.* **10** (1953), 197.
23. K. Stewartson, "On the Motion of a Sphere Along the Axis of a Rotating Fluid." *Q. J. Mech. Appl. Math.* **11** (1958), 39.
24. T. Maxworthy, "The Flow Created by a Sphere Moving Along the Axis of a Rotating, Slightly-Viscous Fluid. *J. Fluid Mech.* **40** (1970), 453.
25. L. M. Hocking, D. W. Moore, and I. C. Walton, "The Drag on a Sphere Moving Axially in a Long Rotating Container." *J. Fluid Mech.* **90** (1979), 781.
26. S. N. Singh, "The Flow Past a Fixed Sphere in a Slowly Rotating Viscous Fluid." *ZAMP* **26** (1975), 415.
27. E. R. Benton and A. Clark, "Spin-Up." *Ann. Rev. Fluid Mech.* **6** (1974), 257.

28. W. R. Briley and H. A. Walls, "A Numerical Study of Time—Dependent Rotating Flow in a Cylindrical Container at Low and Moderate Reynolds Numbers." *Lecture Notes in Physics No. 8.* Springer-Verlag, New York, 1970, p. 377.
29. H. J. Lugt and H. J. Haussling, "Development of Flow Circulation in a Rotating Tank." *Acta Mech.* **18** (1973), 255.
30. H. U. Vogel, *Experimentelle Ergebnisse über die laminare Strömung in einem zylindrischen Gehäuse mit darin rotierender Scheibe.* Max Planck Institut für Strömungsforschung, Bericht 6/1968.
31. H. J. Lugt and H. J. Haussling, "Axisymmetric Vortex Breakdown in Rotating Fluid within a Container." *J. Appl. Mech.* **49** (1982), 921.
32. T. von Kármán, "Über laminare und turbulente Reibung." *ZAMM* **1** (1921), 233.
33. U. T. Bödewadt, "Die Drehströmung über festem Grund." *ZAMM* **20** (1940), 241.
34. M. A. Goldshtik, "A Paradoxical Solution of the Navier–Stokes Equations." *Prikl. Mat. Mekh.* **24** (1960), 913.
35. G. J. Kidd and G. J. Farris, "Potential Vortex Flow Adjacent to a Stationary Surface." *J. Appl. Mech.* **35** (1968), 209.
36. E. W. Schwiderski, "On the Axisymmetric Vortex Flow Over a Flat Surface." *J. Appl. Mech.* **36** (1969), 614.
37. O. R. Burggraf, K. Stewartson, and R. Belcher, "Boundary Layer Induced by a Potential Vortex." *Phys. Fluids* **14** (1971), 1821.
38. R. R. Long, *Q. J. Mech. Appl. Math.* **9** (1956), 385.
39. H. Shih and H. Pao, "Selective Withdrawal in Rotating Fluids." *J. Fluid Mech.* **49** (1971), 509.
40. C. G. Richards and W. P. Graebel, "A Numerical Study of the Flow Due to a Rotating Disk With a Center Sink." *J. Basic Eng.* (Dec. 1967), 807.
41. M. Sibulkin, "A Note on the Bathtub Vortex." *J. Fluid Mech.* **14** (1962), 21.
42. F. H. Harlow and L. R. Stein, "Structural Analysis of Tornado-Like Vortices." *J. Atm. Sci.* **31** (1974), 2081.
43. J. L. Smith, "An Experimental Study of the Vortex in the Cyclone Separator." *J. Basic Eng.* **84** (1962), 602, 609.
44. G. I. Taylor, "The Mechanics of Swirl Atomizers." *Proc. 7th Int. Congr. Appl. Mech.* **2** (1948), 280.
45. W. S. Lewellen, *A Review of Confined Vortex Flows.* NASA CR-1772, 1971.
46. M. P. Escudier and P. Merkli, "Observations of the Oscillatory Behavior of a Confined Ring Vortex." *AIAA J.* **17** (1979), 253.
47. A. J. Ter Linden, "Cyclone Dust Collectors for Boilers." *Trans. ASME* **75** (1953), 433.
48. T. Matsuda and K. Hashimoto, "Thermally, Mechanically or Externally Driven Flows in a Gas Centrifuge with Insulated Horizontal End Plates." *J. Fluid Mech.* **78** (1976), 337.
49. D. G. Avery and E. Davies, *Uranium Enrichment by Gas Centrifuge.* Mills and Boon, London, 1973.
50. R. L. Hoglund, J. Shacter, and E. Von Halle, "Diffusion Separation Methods." In *Encyclopedia of Chemical Technology,* Vol. 7, 3rd ed. Wiley, New York, 1979.
51. D. R. Olander, "Technical Basis of the Gas Centrifuge." *Adv. Nuclear Science Technology,* Vol. 6, Academic, New York, 1972.
52. D. N. Wormley and H. H. Richardson, "A Design Basis for Vortex-Type Fluid Amplifiers Operating in the Incompressible Flow Regime." *J. Basic Eng.* (June 1970), 369.
53. N. Syred and J. M. Beér, "Combustion in Swirling Flows: A Review." *Combust. Flame* **23** (1974), 143.
54. J. Jacobs, "Turbulente Mischung in Zyklonbrennkammern." PhD. Thesis, Universität Karlsruhe 1974.
55. O. Carlowitz und R. Jeschar, "Zur Entstehung, Beeinflussung und Wirkungsweise von Wirbelfäden in Zyklonbrennkammern." *VDI-Ber.* **346** (1979), 241.
56. G. P. Schivley and J. L. Dussourd, "An Analytical and Experimental Study of a Vortex Pump." *J. Basic Eng.* (Dec. 1970), 889.
57. B. Vonnegut, "A Vortex Whistle." *J. Acoust. Soc. Am.* **26** (1954), 18.
58. C. duP. Donaldson, "The Magneto-hydrodynamic Vortex Power Generator: Basic Principles and Practical Problems." In *Engineering Aspects of Magnetohydrodynamics.* C. Mannal and N. W. Mather (eds.). Columbia University Press, New York, 1962, p. 228.
59. A. J. Sellers et al., "Recent Developments in Heat Transfer and Development of the Mercury Boiler for the SNAP-8 System." Proceedings of a Conference on Applications of High Temperature Instrumentation to Liquid-Metal Experiments (ANL-7100), 1965, 573.
60. R. G. Ragsdale, "NASA Research on the Hydrodynamics of the Gaseous Vortex Reactor." NASA TN D-288, 1960.
61. K. E. Tempelmeyer and L. E. Rittenhouse, "Vortex Flow in Arc Heaters." *AIAA J.* **2** (1964), 766.
62. R. Westley, *A Bibliography and Survey of the Vortex Tube.* The College of Aeronautics Cranfield, Note 9, March 1954.
63. L. Prandtl, *Führer durch die Strömungslehre.* 6. Aufl., Vieweg, Braunschweig, 1965.
64. M. Sibulkin, "Unsteady, Viscous, Circular Flow. Part 3. Application to the Ranque-Hilsch Tube." *J. Fluid Mech.* **12** (1962), 269.
65. R. G. Jackson, Sedimentological and Fluid-Dynamical Implications of the Turbulent Bursting Phenomenon in Geophysical Flows." *J. Fluid Mech.* **77** (1976), 531.
66. Lord Rayleigh, "On the Dynamics of Revolving Fluids." *Proc. Roy. Soc. A* **93** (1917), 148.
67. G. I. Taylor, "Stability of a Viscous Liquid Contained Between Two Rotating Cylinders." *Philos. Trans. Roy. Soc. Lond. Ser. A* **223** (1923), 289.
68. H. Görtler, "Dreidimensionales zur Stabilitätstheorie laminarer Grenzschichten." *ZAMM* **35** (1955), 326.
69. C. E. Pearson, "Numerical Study of the Time-Dependent Viscous Flow between Two Rotating Spheres. *J. Fluid Mech.* **28** (1967), 323.
70. O. Sawatzki and J. Zierep, "Das Stromfeld im Spalt zwischen zwei konzentrischen Kugelflächen, von denen die innere rotiert." *Acta Mech.* **9** (1970), 13.
71. M. Wimmer, "Experiments on a Viscous Fluid Flow between Concentric Rotating Spheres." *J. Fluid Mech.* **78** (1976), 317; **103** (1981), 117.
72. K. G. Roesner, "Numerical Calculation of Hydrodynamic Stability Problems with Time-Dependent Bound-

ary Conditions." Proceedings of the Sixth International Conference on Numerical Methods in Fluid Dynamics, Tbilisi, USSR, 1978.
73. F. V. Dolzhanskii et al., *Nonlinear Hydrodynamic Systems* (in Russian), Moskow, 1974.
74. K. G. Roesner, "Instabile Strömungsformen eines Flüssigkeitskreisels." *ZAMM* **61** (1981), T 174.
75. A. Ibbetson and D. J. Tritton, "Experiments on Turbulence in a Rotating Fluid." *J. Fluid Mech.* **68** (1975), 639.
76. A. D. McEwan, "Angular Momentum Diffusion and the initiation of cyclones." *Nature* **260** (1976), 126.
77. A. Warren, "Dune trend and the Ekman spiral." *Nature* **259** (1976), 653.
78. D. L. Kelly, B. W. Martin, and E. S. Taylor, "A Further Note on the Bathtub Vortex." *J. Fluid Mech.* **19** (1964), 539.

Chapter 8

1. O. M. Phillips, *The Dynamics of the Upper Ocean.* Cambridge University Press, New York, 1966.
2. Count Rumford, "Of the Propagation of Heat in Fluids." Complete Works 1, 239. American Academy Arts and Sciences, Boston, 1870.
3. C. H. V. Ebert, "The Meterological Factor in the Hamburg Fire Storm." *Weatherwise* **16** (2) (April 1963), 73.
4. R. Kimura, "Dynamics of Steady Convections over Heat and Cool Islands." *J. Met. Soc. Japan, Ser. II* **53** (1975), 440.
5. H. J. Lugt and E. W. Schwiderski, Convective Vortex over a Horizontal Surface with Nonuniform Temperature." *J. Atm. Sci.* **23** (1966), 809.
6. K. E. Torrance and J. A. Rockett, "Numerical Study of Natural Convection." *J. Fluid Mech.* **36** (1969), 33.
7. J. S. Turner, "Buoyant Plumes and Thermals." *Ann. Rev. Fluid Mech.* **1**, (1969), 29.
8. D. J. Shlien and D. W. Thompson, "Some Experiments on the Motion of an Isolated Laminar Thermal." *J. Fluid Mech.* **72** (1975), 35.
9. C. D. Cone, "Thermal Soaring of Birds." *Am. Sci.* (March 1962), 180.
10. D. J. Shlien, "Some Laminar Thermal and Plume Experiments." *Phys. Fluids* **19** (1976), 1089.
11. J. W. Elder, "The Unstable Thermal Interface." *J. Fluid Mech.* **32** (1968), 69.
12. G. Schubert and J. Whitehead, "The Moving Flame Experiment with Liquid Mercury: Possible Implications for the Venus Atmosphere." *Science* **163** (1969), 71.
13. F. H. Busse, "On the Mean Flow Induced by a Thermal Wave." *J. Atm. Sci.* **29** (1972), 1423.
14. B. J. Daly, "Two-Fluid Rayleigh-Taylor Instability." *Phys. Fluids* **10** (1967), 297.
15. C. J. Chen and L. M. Chang, "Flow Patterns of a Circular Vortex Ring with Density Difference Under Gravity." *J. Appl. Mech.* (Dec. 1972), 869; (June 1973), 637.
16. V. O'Brien, "Why Raindrops Break Up—Vortex Instability." *J. Met.* **18** (1961), 549.
17. W. H. Bradley, "Vertical Density Currents." *Science* **150** (1965), 1423.
18. F. H. Busse, "Non-linear Properties of Thermal Convection." *Rep. Prog. Phys.* **41** (1978), 1967.
19. J. A. Whitehead, "Cellular Convection." *Am. Sci.* **59** (1971), 444.
20. E. Palm, "Nonlinear Thermal Convection." *Ann. Rev. Fluid Mech.* **7** (1975), 39.
21. E. H. Weber, "Mikroskopische Beobachtungen sehr gesetzmässiger Bewegungen, welche die Bildung von Niederschlägen harziger Körper aus Weingeist begleiten." *Ann. Phys., Leipzig* **94** (1855), 447.
22. H. Bénard, "Les tourbillons cellulaires dans une nappe liquide." *Rev. Gén. Sci. Pures Appl.* **11** (1900), 1261, 1309.
23. Lord Rayleigh, *Scientific Papers,* Vol. 6. Cambridge University Press, Cambridge, 1899–1920, p. 432.
24. Sir David Brunt, *Experimental Cloud Formation.* Comp. of Meteorology. AMS, Boston, 1951, p. 1255.
25. S. Chandrasekhar, *Hydrodynamic and Hydromagnetic Stability.* Oxford Clarendon Press, 1961.
26. R. Krishnamurti, "Some Further Studies on the Transition to Turbulent Convection." *J. Fluid Mech.* **60** (1973), 285.
27. J. Zierep, "Instabilitäten in Strömungen zäher, wärmeleitender Medien." *Z. Flugwiss. Weltraumforsch.* **2** (1978), 143.
28. H. L. Kuo, "Perturbations of Plane Couette Flow in Stratified Fluid." *Phys. Fluids* **6** (1963), 195.
29. W. V. Burt, H. Crew, and S. L. Poole, "Evidence for Roll Vortices Associated with a Land Breeze." *J. Mar. Res.* **33** (Suppl.).
30. Y. Yen, "Onset of Convection." *Phys. Fluids* **11** (1968), 1266.
31. C. H. Chun, "Marangoni Convection in a Floating Zone Under Reduced Gravity." *J. Crystal Growth* **48** (1980), 600.
32. J. Zierep, "Das Leewellenproblem der Meteorologie." *Die Naturwissenschaften* **45** (1958), 197.
33. J. W. Miles, "Waves and Wave Drag in Stratified Flows." Proceedings of the 12th International Congress on Applied Mechanics, Stanford, 1968.
34. *The Airflow over Mountains,* prepared by Queney, Corby, Gerbier, Koschmieder, Zierep. World Meteorological Organization, Tech. Note 34, 1960.
35. D. K. Lilly snd E. J. Zipser, "The Front Range Windstorm of 11 January 1972, a Meteorological Narrative." *Weatherwise* **25** (1972), No.2, 56.
36. W. Georgii, "Segelflug in grossen Höhen." *Aerokurier* (1967), 20.
37. C. S. Yih, "Stratified Flows." *Ann. Rev. Fluid Mech.* **1** (1969), 73.
38. H. J. Haussling, "Viscous Flows of Stably Stratified Fluids Over Barriers." *J. Atm. Sci.* **34** (1977), 589.
39. H. A. Panofsky, "Air Pollution Meteorology." *Am. Sci.* (Summer 1969), 269.
40. G. Veronis, "The Analogy between Rotating and Stratified Fluids." *Ann. Rev. Fluid Mech.* **2** (1970), 37.
41. D. Fultz, "Experimental Models of Rotating Fluids and Possible Avenues for Future Research." In *Dynamics of Climate.* P. L. Pfeffer (ed.). Pergamon, New York, 1960, p. 71.
42. W. P. Graebel, R. R. Long, and T. Y. Wu, "Report on the

International Conference on Stratified Fluids." *J. Fluid Mech.* **31** (1968), 689.
43. J. Pedlosky, "The Spin Up of a Stratified Fluid." *J. Fluid Mech.* **28** (1967), 463.
44. F. H. Busse and C. R. Carrigan, "Laboratory Simulation of Thermal Convection in Rotating Planets and Stars." *Science* **191** (1976), 81.
45. M. R. Samuels, "Free Convection in a Rotating Medium." *Phys. Fluids* **11** (1968), 1889.
46. G. P. Williams, "Thermal Convection in a Rotating Fluid Annulus: Part 1,2,3." *J. Atm. Sci.* **24** (1967), 144.

Chapter 9

1. A. H. Oort, "The Energy Cycle of the Earth." *Sci. Am.* **223** (Sept. 1970), 54.
2. P. Morel (ed.), *Dynamic Meteorology*. Reidel, Dortrecth, 1973.
3. E. Halley, "An Historical Account of the Trade-Winds and Monsoons Observable in the Seas between and Near the Tropicks with an Attempt to Assign the Physical Cause of Said Winds. *Philos. Trans.* **26** (1686), 153.
4. G. Hadley, "Concerning the Cause of the General Trade-Winds." *Philos. Trans.* **29** (1735), 58.
5. H. W. Dove, *Meteorologische Untersuchungen*. Sandersche Buchhandlung, Berlin, 1837.
6. W. Ferrel, "An Essay on the Winds and the Currents of the Ocean." *Nashville J. Medicine and Surgery* **11** (1856), 287.
7. H. von Helmholtz, "Über atmosphärische Bewegungen." Sitzungsber. Akad. Wiss. Berlin, 1888, 647.
8. V. Bjerknes, "On the Dynamics of the Circular Vortex with Applications to the Atmosphere and Atmospheric Vortex and Wave Motions." *Geofys. Publ.* **2** (1921), No. 4.
9. A. Defant, "Die Zirkulation der Atmosphäre in den gemässigten Breiten der Erde." *Geograf. Ann.* **3** (1921), 209.
10. H. Jeffreys, "On the Dynamics of Geostrophic Winds." *Quart. J. Roy. Soc.* **52** (1926), 85.
11. V. P. Starr, "An Essay on the General Circulation of the Earth's Atmosphere." *J. Met.* **5** (1948), 39.
12. E. N. Lorenz, "The Nature of the Global Circulation of the Atmosphere: A Present View." *Roy. Met. Soc.* (1969), 3.
13. E. Reiter, *Jet Streams*. Doubleday, Garden City, New York, 1967.
14. J. Lighthill and R. P. Pearce (eds.). *Monsoon Dynamics*. Cambridge University Press, 1981.
15. R. E. Newell, "Climate and the Ocean." *Am. Sci.* **67** (1979), 405.
16. P. D. Thompson, *Numerical Weather Analysis and Prediction*. MacMillan, New York, 1961.
17. R. C. Sutcliffe, *Weather and Climate*. Signet Science Library Q3699, 1969.
18. H. Stommel, *The Gulf Stream*. University of California Press, 1960.
19. J. Teal and M. Teal, *The Sargasso Sea*. Little, Brown and Co., Boston, 1975.
20. W. H. Munk, "On the Wind-Driven Ocean Circulation." *J. Met.* **7** (1950), 79.
21. K. Bryan, "A Numerical Investigation of a Nonlinear Model of a Wind-Driven Ocean." *J. Atm. Sci.* **20** (1963), 594.
22. G. Dietrich, K. Kalle, W. Krauss, and G. Siedler, *General Oceanography*. Wiley, New York, 1980.
23. M. J. Lighthill, "Dynamic Response of the Indian Ocean to Onset of the Southwest Monsoon." *Philos. Trans. Roy. Soc. Lond. Ser. A* **265** (1969), 1159, 49.
24. K. O. Emery and G. T. Csanady, "Surface Circulation of Lakes and Nearly Land-Locked Seas." *Proc. Nat. Acad. Sci.* **70** (1973), 93.
25. J. G. Kohl, *Geschichte des Golfstroms und seiner Erforschung*. C. E. Müller, Bremen, 1868.
26. G. A. Maul, "The Annual Cycle of the Gulf Loop Current, Part I: Observations during a One-Year Time Series." *J. Marine Res.* **35** (1977), No.1, 29.
27. G. F. Carrier,"Singular Perturbation Theory and Geophysics." *SIAM Rev.* **12** (1970), 175.
28. B. J. Thompson and G. A. Gotthardt, "Life Cycle of a North Atlantic Eddy." *Abstr. Trans. Am. Geophys. Union* **52** (1971), 241.
29. F. C. Fuglister and L. V. Worthington, "Some Results of a Multiple Ship Survey of the Gulf Stream." *Tellus* **3** (1951), No.1, 14.
30. P. L. Richardson, "Anticyclonic Eddies Generated Near the Corner Rise Seamounts." *J. Marine Res.* **38** (1980), No.4, 673.
31. R. A. Kerr, "Small Eddies Proliferating in the Atlantic." *Science* **213** (1981), 632.
32. "Movement and Acoustic Influenced of Cyclonic Eddies." *Nav. Res. Rev.*, June 1975, 1.
33. H. J. Lugt and P. Uginčius, "Acoustic Rays in an Ocean with Heat Source or Thermal-Mixing Zone." *J. Acoust. Soc. Am.* **36** (1964), 689.
34. B. A. Taft, "Structure of Kuroshio South of Japan." *J. Marine Res.* **36** (1978), No.1, 77.
35. H. Solomon, "Detachment and Recombination of a Current Ring with the Kuroshio." *Nature* **274** (1978), 580.
36. T. Maxworthy and F. K. Browand, "Experiments in Rotating and Stratified Flows: Oceanographic Application." *Ann. Rev. Fluid Mech.* **7** (1975), 273.
37. F. Vettin, "Über den aufsteigenden Luftstrom, die Entstehung des Hagels und der Wirbel-Stürme." *Ann. Phys. und Chem.* (2) Leipzig **102** (1857), 246.
38. F. M. Exner, *Dynamische Meteorologie*. J. Springer, Wien, 1925.
39. R. Pfeffer, G. Buzyna, and W. W. Fowlis, "Synoptic Features and Energetics of Wave-Amplitude Vacillation in a Rotating, Differentially Heated Fluid." *J. Atm. Sci.* **31** (1974), 622.
40. D. J. Baker, "Models of Oceanic Circulation." *Sci. Am.* **222** (1970), 114.
41. J. Charney, R. Fjørtoft, and J. von Neumann, *Tellus* **2** (1950), 237.
42. B. R. Döös, *Numerical Experimentation Related to GARP*. World Meteorological Organization. GARP Publ. Ser. No.6, Sept. 1970.

43. N. A. Phillips, "Models for Weather Prediction." *Ann. Rev. Fluid Mech.* **2** (1970), 251.
44. K. Miyakoda, "Numerical Weather Prediction." *Am. Sci.* **62** (1974), 564.
45. A. Gilchrist, "Numerical Modelling of the Atmosphere." *Rep. Progr. Phys.* **42** (1979), 503.
46. H. H. Hess, "History of Ocean Basins." In *Adventures in Earth History*. P. Cloud (ed.). Freeman, San Francisco, 1970, 277.
47. R. K. Bambach, C. R. Scotese, and A. M. Ziegler, "Before Pangea: The Geographies of the Paleozoic World." *Am. Sci.* **68** (1980), 26.
48. *Continents Adrift*. Readings from Scientific American. Freeman, San Francisco, 1971.
49. S. K. Runcorn, "Towards a Theory of Continental Drift." *Nature* **193** (1962), 311.
50. W. Morgan, *Nature* **230** (1971), 42.
51. J. W. Elder, "Thermal Turbulence and its Role in the Earth's Mantle." In *The Mantles of the Earth and Terrestrial Planets*. S. K. Runcorn, (ed.). Wiley, London, 1967, 525.
52. G. Schubert and D. L. Turcotte, *J. Geophys. Res.* **77** (1972), 945
53. R. B. Gordon, "Diffusion Creep in the Earth's Mantle." *J. Geophys. Res.* **70** (1965), 2413.
54. D. P. McKenzie and F. Richter, "Convection Currents in the Earth's Mantle." *Sci. Am.* **235**, (Nov. 1976).
55. H. G. Wunderlich, *Das neue Bild der Erde—Faszinierende Entdeckungen der modernen Geologie*. Hoffmann und Campe, 1975.
56. B. G. Hunt, "On the Death of the Atmosphere." *J. Geophys. Res.* **81** (1976), 3677.
57. J. Charney, "The Dynamics of Long Waves in a Baroclinic Westerly Current." *J. Meteor.* **4** (1947), 135.
58. E. T. Eady, "The Cause of the General Circulation of the Atmosphere. Centenary Proc. Royal Meteorological Soc. 1950, 156.
59. A. Merz, "Die Deutsche Atlantische Expedition auf dem Vermessungs- und Forschungsschiff "Meteor." *Sitzungsber. Akad. Wiss. Berlin, Math. Phys. Kl.* **31** (1925), 562.
60. R. D. Pingree, P. M. Holligan, and G. T. Mardell, "Phytoplankton Growth and Cyclonic Eddies." *Nature* **278**, (1979), 245.
61. M. Gregg, "The Microstructure of the Ocean." *Sci. Am.* **228**, (1973), 64.

Chapter 10

1. F. Fiedler and H. A. Panofsky, "Atmospheric Scales and Spectral Gaps." *Bull. Am. Met. Soc.* **51** (1970), 1114.
2. B. J. Hoskins and F. P. Bretherton, "Atmospheric Frontogenesis Models: Mathematical Formulation and Solution." *J. Atm. Sci.* **29** (1972), 11.
3. A. Eliassen, *Advances in Earth Science*. P. Hurley (ed.). MIT Press, Cambridge, Mass. 1966.
4. *Bull. Am. Met. Soc.* **46** (1965), No. 8.
5. *Weatherwise* **19** (1966), Feb., No. 1.
6. S. D. Flora, *Tornadoes in the United States*. University of Oklahoma Press, Norman, 1954.
7. A. Wegener, *Wind- und Wasserhosen in Europa*. Vieweg, Braunschweig, 1917.
8. R. K. Smith and L. M. Leslie, "Tornadogenesis." *Q. J. Roy. Met. Soc.* **104** (1978), 189.
9. S. Thorarinsson and B. Vonnegut, "Whirlwinds Produced by the Eruption of Surtsey Volcano." *Bull. Am. Met. Soc.* **45** (1964), 440.
10. B. R. Morton, "Geophysical Vortices." In *Progress in Aeronautical Sciences*. Vol. 7. Pergamon, New York, 1966, p. 145.
11. W. S. Lewellen, "Theoretical Models of the Tornado Vortex." Symposium on Tornadoes, Lubbock, Texas, June 1976. Texas Technical University, 107.
12. L. J. Budney, "Unique Damage Patterns Caused by a Tornado in Dense Woodlands." *Weatherwise* **18** (1965), 74.
13. A. I. Barcilon, "A Theoretical and Experimental Model for a Dust Devil." *J. Atm. Sci.* **24** (1967), 453.
14. T. Maxworthy, "On the Structure of Concentrated, Columnar Vortices." *Astronaut. Acta* **17** (1972), 363.
15. C. M. Reber, "The South Platte Valley Tornadoes of June 7, 1953." *Bull. Am. Met. Soc.* **35** (1954), 191.
16. O. R. Burggraf and M. R. Foster, "Continuation or Breakdown in Tornado-like Vortices." *J. Fluid Mech.* **80** (1977), 685.
17. J. Leighly, "An Early Drawing and Description of a Tornado." *Isis* **65** (1974), 474.
18. R. T. Arnold, H. E. Bass, and L. N. Bolen, "Acoustic Spectral Analysis of Three Tornadoes." *J. Acoust. Soc. Am.* **60** (1976), 584.
19. B. Vonnegut, "Electrical Theory of Tornadoes." *J. Geophys. Res.* **65** (1960), 203.
20. T. T. Fujita, "Proposed Mechanism of Suction Spots Accompanied by Tornadoes." Satellite and Mesometeorology Research Project, Paper 102. Dept. Geophys. Sci., University of Chicago, 1971.
21. R. Rotunno, "A Note on the Stability of a Cylindrical Vortex Sheet." *J. Fluid Mech.* **87** (1978), 761.
22. F. H. Ludlam, "Severe Local Storms: A Review." *Met. Monogr.* **5** (1963), No.27, 1.
23. G. Gillies, G. Withnell, and J. Glass, "Laboratory Production of Tornado-Like Vortices." *J. Atm. Sci.* **31** (1974), 2231.
24. H. Markgraf, "Zur Theorie der Tromben." *Arch. Meteorol. Geophys. Bioklimatol. Ser. A* **12** (1961), 339.
25. J. H. Golden, "The Life Cycle of Florida Keys' Waterspouts." *J. Appl. Met.* **13** (1974), 676.
26. C. A. Wan and C. C. Chang, "Measurement of the Velocity Field in a Simulated Tornado-Like Vortex Using a Three-Dimensional Velocity Probe." *J. Atm. Sci.* **29** (1972), 116.
27. R. P. Davies-Jones, "The Dependence of Core Radius on Swirl Ratio in a Tornado Simulator." *J. Atm. Sci.* **30** (1973), 1427.
28. C. T. Hsu and B. Fattahi, "Mechanism of Tornado Funnel Formation." *Phys. Fluids* **19** (1976), 1853.
29. N. B. Ward, "The Exploration of Certain Features of Tornado Dynamics Using a Laboratory Model." *J. Atm. Sci.* **29** (1972), 1194.

30. C. R. Church, J. T. Snow, and E. M. Agee, "Tornado Vortex Simulation at Purdue University." *Bull. Am. Met. Soc.* **58** (1977), 900.
31. R. L. Ives, "Behavior of Dust Devils." *Bull. Am. Met. Soc.* **28** (1947), 168.
32. A. I. Barcilon and P. G. Drazin, "Dust Devil Formation." *Geophys. Fluid Dyn.* **4** (1972), 147.
33. J. B. Mullen and T. Maxworthy, "A Laboratory Model of Dust Devil Vortices." *Dyn. Atm. Oceans* **1** (1977), 181.
34. P. C. Sinclair, "On the Rotation of Dust Devils." *Bull. Am. Met. Soc.* **46** (1965), 388.
35. A. Wegener, "Staubwirbel auf Island." *Met. Z.* **31** (1914), 199.
36. S. B. Idso, "Tornado or Dust Devil: The Enigma of Desert Whirlwinds." *Am. Sci.* **62** (1974), 530.
37. S. E. Logan, "An Approach to the Dust Devil Vortex." *AIAA J.* **9** (1971), 660.
38. I. Langmuir, "Surface Motion of Water Induced by Wind." *Science* **87** (1938), 119.
39. A. J. Faller, "The Angle of Windrows in the Ocean." *Tellus* **16** (1964), 363.
40. S. Leibovich, "On the Evolution of the System of Wind Drift Currents and Langmuir Circulations in the Ocean." *J. Fluid Mech.* **79** (1977), 715.
41. A. D. D. Craik, "The Generation of Langmuir Circulations by an Instability Mechanism." *J. Fluid Mech.* **81** (1977), 209.
42. A. J. Faller, "Oceanic Turbulence and the Langmuir Circulations." *Ann. Rev. Ecol. Syst.* **2** (1971), 201.
43. R. T. Pollard, "Observations and Theories of Langmuir Circulations and Their Role in Near Surface Mixing." In *A Voyage of Discovery; George Deacon 70th Anniversary Volume.* M. Angel (ed.). Pergamon, Oxford, 1977, p. 235.
44. S. Leibovich, "The Form and Dynamics of Langmuir Circulations." *Ann. Rev. Fluid Mech.* **15** (1983).
45. W. Price, "Cruising Japan's Inland Sea." *Nat. Geogr. Mag.* (Nov. 1953), 619.
46. C. Burland, *North American Indian Mythology.* P. Hamlyn, New Ed. 1968, p. 40.
47. H. Mosby, "Oberflächenströmungen in der Meerenge bei Tromsö." *Arch. Met. Geophys. Bioklimatol. Ser. A* **7** (1954), 378.
48. D. M. Farmer and H. E. Huppert, "The Oceanography of Fjords." *Nature* **280** (1979), 273.
49. C. Berlitz, *The Bermuda Triangle.* Avon, New York, 1974.
50. A. Defant, "Scylla und Charybdis und die Gezeitenströmungen in der Strasse von Messina." *Ann. Hydrogr. Mar. Met.* **68** (1940), 145.
51. P. A. Zahl, "Fishing in the Whirlpool of Charybdis." *Nat. Geogr. Mag.* (Nov. 1953), 579.
52. G. Mazzarelli, "I vortici, i tagli e altri fenomeni delle correnti dello stretto di Messina." *Atti R. Accad. Peloritana* **XL** (1938).
53. K. P. Chopra and L. F. Hubert, "Kármán Vortex Streets in Wakes of Islands." *AIAA J.* **3** (1965), 1941.
54. O. M. Griffin, "Vortex Streets and Patterns." *Mech. Eng.* **104** (March, 1982), 56.
55. E. N. Lorenz, "Barotropic Instability of Rossby Wave Motion." *J. Atm. Sci.* **29** (1972), 258.
56. P. H. Stone, "On Non-geostrophic Baroclinic Stability." *J. Atm. Sci.* **23** (1966), 390; **27** (1970), 721; **29** (1972), 419.
57. J. T. Yen, "Tornado-Type Wind Energy System." Proceedings of the 10th Intersociety Energy Conversion Engineering Conference, University of Delaware, 1975, p. 987.
58. A. J. Faller, "Large Eddies in the Atmospheric Boundary Layer and their Possible Role in the Formation of Cloud Rows." *J. Atm. Sci.* **22** (1965), 176.

Chapter 11

1. W. Keller, *The Bible as History: A Confirmation of the Book of Books.* Morrow, New York, 1964.
2. R. Weyl, "Die geologischen Studien Leonardo da Vincis und ihre Stellung in der Geschichte der Geologie." *Philosophia Naturalis* **1** (1950), 243.
3. *Neptun* **3** (1963), Heft 1, 26.
4. D. M. Ludlum, *Early American Hurricanes 1492–1870.* American Meteorological Society, Boston, 1963.
5. C. J. Neumann, "Trends in Forecasting the Tracks of Atlantic Tropical Cyclones." *Bull. Am Met. Soc.* **62** (1981), 1473.
6. *Nature* **240** (1972), 198.
7. G. R. Flierl and A. R. Robinson, *Nature* **239** (1972), 213.
8. *History of United States Naval Operations in World War II, Vol. XIII: The Liberation of the Philippines.* Little, Brown and Co., Boston, 1959.
9. *Monthly Weather Rev.* (Dec. 1957).
10. I. R. Tannehill, *The Hurricane.* U.S. Dept. Commerce, Weather Bureau, Washington, D.C., 1956.
11. H. Riehl, *Tropical Meteorology.* McGraw–Hill, New York, 1954.
12. P. J. Hebert and G. Taylor, "Hurricanes." *Weatherwise* **32** (1979), Part I, No.2, 61; Part II, No.3, 100.
13. C. O. Erickson, "An Incipient Hurricane near the West African Coast." *Mon. Weather Rev.* (Feb. 1963), 61.
14. *Hurricane.* U.S. Dep. Commerce, ESSA. U.S. Govt. Print. Office 0-336-564, 1969.
15. G. F. Carrier, "The Intensification of Hurricanes." *J. Fluid Mech.* **49** (1971), 145.
16. D. J. Shea, "The Structure and Dynamics of the Hurricane's Inner Core Region." *NOAA N22-65-72(G), Atm. Sci. Paper No. 182, Apr. 1972.*
17. B. I. Miller, "Characteristics of Hurricanes." *Science* **157** (1967), 1389.
18. R. A. Anthes, S. L. Rosenthal, and J. W. Trout, "Preliminary Results from an Asymmetric Model of the Tropical Cyclone." *Mon. Weather Rev.* **99** (1971), 744.
19. P. G. Black and R. A. Anthes, "On the Asymmetric Structure of the Tropical Cyclone Outflow Layer." *J. Atm. Sci.* **28** (1971), 1348.
20. J. S. Malkus. In *The Sea,* Vol. 1. M. N. Hill (ed). Interscience, New York, 1962, 231.
21. B. Ackerman, *The Distribution of Liquid Water in Hurricanes.* Nat. Hurricane Res. Project, Rep. No. 62, Apr. 1963.
22. L. F. Hubert and A. Timchalk, "Estimating Hurricane

Wind Speeds from Satellite Pictures." *Mon. Weather Rev.* **97** (1969), 382.
23. R. W. Fett, *Some Characteristics of the Formative Stage of Typhoon Development: A Satellite Study.* National Conference on the Physics and Dynamics of Clouds, Chicago, March 24–26, 1964.
24. A. K. Mukherjee and T. R. Sivaramakrishnan, "Surface Wind and Sea Waves in a Hurricane Field." *Nature* **267** (May 1977), 237.

Chapter 12

1. F. J. Dyson, "Energy in the Universe." *Sci. Am.* **225** (Sept. 1971), 50.
2. C. Sagan, "The Solar System." *Sci. Am.* **233,** (Sept. 1975), 22.
3. H. Reeves, "The Origin of the Solar System." In *The Origin of the Solar System.* S. F. Dermott (ed.). Wiley, New York, 1978.
4. C. F. von Weizsäcker, "Über die Entstehung des Planetensystems." *Z. Astrophys.* **22** (1943), 319.
5. G. W. Wetherill, "The Formation of the Earth from Planetesimals." *Sci. Am.* **244** (June 1981), 162.
6. J. G. Hills, "Dynamic Relaxation of Planetary Systems and Bode's Law." *Nature* **225** (1970), 840.
7. M. W. Ovenden, "Bode's Law and the Missing Planet." *Nature* **239** (1972), 508.
8. M. Lecar, "Bode's Law." *Nature* **242** (1973), 318.
9. N. Wade, "Three Origins of the Moon." *Nature* **223** (1969), 243.
10. V. P. Starr and P. A. Gilman, "The Circulation of the Sun's Atmosphere." *Sci. Am.* **218** (Jan. 1968), 100.
11. A. Nesis, "Convection and the Solar Granulation." In *Recent Developments in Theoretical and Experimental Fluid Mechanics.* Springer-Verlag, New York, 1979, p. 309.
12. D. J. Galloway, M. R. E. Proctor, and N. O. Weiss, "Formation of Intense Magnetic Fields near the Surface of the Sun." *Nature* **266** (1977), 686.
13. *New Scientist* (Jan. 13, 1977), 77.
14. A. P. Ingersoll, "The Meteorology of Jupiter." *Sci. Am.* **234** (March 1976), 46.
15. R. Gore, "What Voyager Saw: Jupiter's Dazzling Realm." *Nat. Geogr. Mag.* **157** (1980), Jan., 2.
16. G. E. Hunt, "The Voyager Encounters with Jupiter." *Phys. Bull.* **31** (1980), 275.
17. T. Maxworthy and L. G. Redekopp, "New Theory of the Great Red Spot from Solitary Waves in the Jovian Atmosphere." *Nature* **260** (1976), 509.
18. A. P. Ingersoll, Jupiter and Saturn." *Sci. Am.* **245,** (Dec. 1981), 66.
19. J. B. Pollack, "The Rings of Saturn." *Am. Sci.* **66** (1978), 30.
20. J. B. Pollack and J. N. Cuzzi, "Rings in the Solar System." *Sci. Am.* **245,** (Nov. 1981), 78.
21. G. E. Hunt and P. B. James, "Martian Extratropical Cyclones." *Nature* **278** (1979), 531.
22. G. Schubert and C. Covey, "The Atmosphere of Venus." *Sci. Am.* **245** (July 1981), 44.
23. J. L. Tassoul, *Theory of Rotating Stars.* Princeton University Press, New Jersey, 1978.
24. S. Huang, "Life Outside the Solar System." *Sci. Am.* (Apr. 1960).
25. B. J. Bok, "The Birth of Stars." *Sci. Am.* **227,** (Aug. 1972), 48.
26. M. Zeilik, "The Birth of Massive Stars." *Sci. Am.* **238** (Apr. 1978), 110.
27. R. L. Dickman, "Bok Globules." *Sci. Am.* **236** (June 1977), 66.
28. J. P. Ostriker, "The Nature of Pulsars." *Sci. Am.* **224** (Jan. 1971), 48.
29. E. Schatzman, "A Theory of the Role of Magnetic Activity During Star Formation." *Ann. d' Astrophys.* **25** (1962), 18.
30. S. Huang, "Rotational Behavior of the Main-Sequence Stars and its Plausible Consequences Concerning Formation of Planetary Systems." *Astrophys. J.* **141** (1965), 985.
31. R. Sexl and H. Sexl, *White Dwarfs—Black Holes.* Academic, New York, 1979.
32. P. C. Joss and S. A. Rappaport, "Observational Constraints on the Masses of Neutron Stars." *Nature* **264** (1976), 219.
33. K. S. Thorne, "The Search for Black Holes." *Sci. Am.* **231** (Dec. 1974), 32.
34. N. D. Birrell and P. C. W. Davies, "On Falling Through a Black Hole into Another Universe." *Nature* **272** (March 1978), 35.
35. S. W. Hawking, "The Quantum Mechanics of Black Holes." *Sci. Am.* **236** (Jan. 1977), 34.
36. R. Berendzen, R. Hart, and D. Seeley, *Man Discovers the Galaxies.* Science History Publ., Neale Watson Academic Publ, New York, 1976.
37. M. S. Roberts, "The Rotation Curves of Galaxies." Int. Astrophys. Union Symp. No.69, Sept. 1974, p. 331.
38. R. H. Sanders and G. T. Wrixon, "The Center of the Galaxy." *Sci. Am.* **230** (April 1974), 66.
39. J. Pringle, "A Black Hole in the Centre of M89? *Nature* **274** (Aug. 1978), 419.
40. C. C. Lin and W. W. Roberts, "Some Fluid-Dynamical Problems in Galaxies." *Ann. Rev. Fluid Mech.,* **13** (1981), 33.
41. R. B. Larson, "Infall of Matter in Galaxies." *Nature* **236** (1972), 21.
42. H. Gerola and P. Seiden, "Stochastic Star Formation and Spiral Structure of Galaxies." *Astrophys. J.* **223** (1978), 129.
43. A. Toomre and J. Toomre, "Violent Tides between Galaxies." *Sci. Am.* **229** (Dec. 1973), 39.
44. E. J. Groth, P. J. E. Peebles, M. Seldner, and R. M. Soneira, "The Clustering of Galaxies." *Sci. Am.* **237** (Nov. 1977), 76.
45. P. Gorenstein and W. Tucker, "Rich Clusters of Galaxies." *Sci. Am.* **239** (Nov. 1978), 110.
46. E. R. Harrison, *Cosmology.* Cambridge University Press, 1981.
47. A. Webster, "The Cosmic Background Radiation." *Sci. Am.* **231** (Aug. 1974), 26.
48. P. J. E. Peebles, *Physical Cosmology.* Princeton University Press, Princeton, N.J. 1971.

49. E. R. Harrison, "On the Origin of Structure in Certain Models of the Universe." *Mem. Soc. Roy. Sci., Liege* **14** (1967), 15.
50. P. J. E. Peebles, "Origin of the Angular Momentum of Galaxies." *Astrophys. J.* **155** (1969), 393.
51. J. D. Barrow and J. Silk, "The Structure of the Early Universe." *Sci. Am.* **242** (April 1980), 118.
52. G. F. R. Ellis and E. R. Harrison, "Cosmological Principles." *Comm. Astrophys. Space Phys. VI, Numbers 1 and 2* (1974), 23 and 29.
53. C. W. Misner, *Astrophys. J.* **151** (1968), 431.
54. J. D. Barrow, "A Chaotic Cosmology." *Nature* **267** (1977), 117.
55. J. D. Barrow, "Quiescent Cosmology. "*Nature* **272** (1978), 211.
56. G. Gale, "The Anthropic Principle." *Sci. Am.* **245** (Dec. 1981), 154.
57. "Planetary Nebulae." *Nature* **215** (1967), 1438.
58. J. S. Miller, "The Structure of Emission Nebulas." *Sci. Am.* **231** (Oct. 1974), 34.
59. J. D. Barrow and F. J. Tipler, "Eternity is Unstable." *Nature* **276** (1978), 453.

MATHEMATICAL SUPPLEMENT

Notation for the Mathematical Supplement

This mathematical supplement is intended to give the theoretically inclined reader some mathematical justification for the material offered in the main text. A selection of solutions of the basic equations of motion is presented to demonstrate the physical properties of vortex flows.

The mathematical supplement requires background knowledge in fluid dynamics and familiarity with vector analysis and differential equations.

COMMONLY USED SYMBOLS

a	vector field
A	area, angular momentum
b	unit vector in the binormal direction
c, c_p	specific heat, specific heat at constant pressure
D	drag, dissipation
Ek	Ekman number
f	Coriolis parameter
F	external force
F	similarity function
g	gravitational constant
G	conservative thermodynamic variable, similarity function
H	Hamiltonian, height, similarity function
J	kinematic momentum transfer
J_n	Bessel function of the first kind and nth order
k	unit vector in the z direction
k	thermal conductivity
K_0	modified Bessel function of the second kind and zero order
L	length, lift, linear operator
m	source strength
$M\rho$	total angular momentum
n	unit vector normal to a surface
N	Brunt–Väisälä frequency
p	pressure
P	similarity function, $\nabla P = (1/\rho)\nabla p$
r	radius vector
r, φ, z	cylindrical polar coordinates
r, θ, λ	spherical polar coordinates
R	radius
Re	Reynolds number
Ro	Rossby number
s	span of a wing, entropy
t	time
T	temperature
U	phase velocity
v	velocity vector
v_x, v_y, v_z	velocity components in x, y, z coordinates
v_r, v_φ, v_z	velocity components in r, φ, z coordinates
v_r, v_θ, v_λ	velocity components in r, θ, λ coordinates
V	volume, free-stream velocity
x, y, z	Cartesian coordinates
$\mathrm{erf}(x)$	$= \dfrac{2}{\sqrt{\pi}} \int_0^x e^{-s^2}\, ds$
α	thermal expansion coefficient
β	$= \partial f/\partial y$
γ	angle in Biot–Savart law, slope angle of a discontinuity surface
Γ	circulation
δ	boundary-layer thickness, delta function, phase angle
ζ	similarity coordinate

θ	coordinate in (r, θ, λ) system	χ	deviation angle
κ	vortex strength $= \Gamma/2\pi$	ψ	stream function in plane flows
λ	wavelength, coordinate in (r, θ, λ) system	$\tilde{\psi}$	stream function in axisymmetric flows
μ	dynamic viscosity	$\boldsymbol{\omega}$	vorticity vector
ν	kinematic viscosity $= \mu/\rho$	ω	vorticity in z direction $= \omega_z$
ξ	$r/\sqrt{\nu t}$	$\boldsymbol{\Omega}$	angular velocity
ρ	density	Ω	angular velocity in plane motion, absolute value of $\boldsymbol{\Omega}$
φ	coordinate in (r, φ, z) system		
ϕ	latitude $= \pi/2 - \theta$	$*$	dimensionless quantity
Φ	velocity potential		

1M. Some Definitions and Kinematical Aspects

1.1M. SOME DEFINITIONS

A fluid particle may rotate in a plane about a center located a distance r away. The azimuthal (or tangential) velocity is

$$v_\varphi = \Omega r, \tag{1.1}$$

where $\Omega(r, t)$ is the angular velocity defined by the time derivative of the angle φ:

$$\Omega = \frac{d\varphi}{dt}. \tag{1.2}$$

For solid-body rotation, Ω is by definition not a function of r.

In three-dimensional space the following general relation holds:

$$\mathbf{v} = \mathbf{\Omega} \times \mathbf{r}, \tag{1.3}$$

with $\mathbf{v}$ being the velocity vector, $\mathbf{v} = d\mathbf{r}/dt$, $\mathbf{\Omega}$ being the vector of the angular velocity, and $\mathbf{r}$ being the radius vector.

The vorticity vector $\boldsymbol{\omega}$ is defined by

$$\boldsymbol{\omega} = \operatorname{curl} \mathbf{v}. \tag{1.4}$$

This relation yields, for the vorticity components in Cartesian coordinates x, y, z, corresponding to velocity components v_x, v_y, v_z,

$$\omega_x = \frac{\partial v_z}{\partial y} - \frac{\partial v_y}{\partial z}, \quad \omega_y = \frac{\partial v_x}{\partial z} - \frac{\partial v_z}{\partial x},$$
$$\omega_z = \frac{\partial v_y}{\partial x} - \frac{\partial v_x}{\partial y}. \tag{1.5}$$

The vorticity components in cylindrical polar coordinates r, φ, z, which are related to the Cartesian coordinates by $x = r\cos\varphi, y = r\sin\varphi, z = z$, are

$$\omega_r = \frac{1}{r}\frac{\partial v_z}{\partial \varphi} - \frac{\partial v_\varphi}{\partial z}, \quad \omega_\varphi = \frac{\partial v_r}{\partial z} - \frac{\partial v_z}{\partial r},$$
$$\omega_z = \frac{1}{r}\frac{\partial}{\partial r}(rv_\varphi) - \frac{1}{r}\frac{\partial v_r}{\partial \varphi}. \tag{1.6}$$

For solid-body rotation the vorticity is

$$\boldsymbol{\omega} = \operatorname{curl}(\mathbf{\Omega} \times \mathbf{r}) = 2\mathbf{\Omega}, \tag{1.7}$$

that is, twice the angular velocity. For an interpretation of the factor 2 see Reference 21, Chapter 3 and also Reference 1M. Since Ω does not depend on r, ω does not either: it is constant in space. Conversely, constant vorticity does not necessarily imply solid-body rotation. In a parallel shear flow (Fig. 2.6) the vorticity is also constant: From the velocity $v_x = cy$, $v_y = 0$ of the shear flow it follows that $\omega_z = -c$.

The circulation Γ around a closed curve is defined by

$$\Gamma = \oint \mathbf{v} \cdot d\mathbf{r} = \int \boldsymbol{\omega} \cdot \mathbf{n} \, dA, \tag{1.8}$$

where $d\mathbf{r}$ is a line element of the curve and A is the area inside the closed curve; $\mathbf{n}$ is the unit vector normal to this area. For a rotating solid body the circulation is

$$\Gamma = 2\pi \Omega r^2. \tag{1.9}$$

Thus, the circulation of a rotating solid body depends on the distance r.

Potential or irrotational flow is defined by $\omega =$ curl $\mathbf{v} \equiv 0$ except at some singular points. curl $\mathbf{v} = 0$ means that a potential function Φ exists such that

$$v = \nabla\Phi. \quad (1.10)$$

In a simply connected region it follows from Eq. (1.8) that $\Gamma = 0$. In this region every closed line contracts to a point excluding singularities.

1.2M. ROTATION AND CONSERVATION LAW OF MATTER

In a fluid flow region in which no fluid source exists, the amount of fluid must be preserved. For an incompressible fluid (the density ρ of the fluid is assumed constant) the conservation law of matter is

$$\int \mathbf{v} \cdot \mathbf{n} \, dA = \int \text{div} \, \mathbf{v} \, dV = 0, \quad (1.11)$$

where A is the area that surrounds the volume V. For a volume element it follows that

$$\text{div} \, \mathbf{v} = 0. \quad (1.12)$$

In a closed region with impermeable boundaries

$$\mathbf{v} \cdot \mathbf{n} = 0 \quad (1.13)$$

everywhere on the boundary. From this condition and from Eq. (1.12) it follows that the flow must be rotary, or, to be exact, it must have closed streamlines (Section 1.3M).

Example

In a square (Fig. 1.1M) the fluid flow can be described by a Fourier series, the nth term of which is

$$\begin{aligned}(v_x)_n &= C_n \sin nx \cdot \cos ny, \\ (v_y)_n &= -C_n \cos nx \cdot \sin ny,\end{aligned} \quad (1.14)$$

where C_n is an arbitrary constant. Equation (1.14) fulfills the conditions (1.12) and (1.13).

In a simply connected region a potential flow cannot have closed streamlines. Hence, the flow in such a region with closed streamlines must have

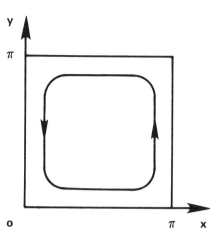

Figure 1.1M. In a simply connected closed region the streamlines must be closed loops. This kinematical statement is here illustrated for a square with the length π and $n = 1$.

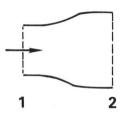

Figure 1.2M. Example of an open region.

vorticity. This can be proven by means of the circulation formula (1.8) with $\Gamma = 0$ for potential flow. Since the velocity vector is tangent to the streamline and does not change sign, the circulation along the closed streamline cannot be zero. For this reason the flows described by Eqs. (1.1) and (1.14) cannot be irrotational.

In an open region conservation of mass requires that the incoming flow must equal the outgoing flow. According to Eq. (1.11) the mass flow in a tube with variable cross section (Fig. 1.2M) is

$$v_1 A_1 = v_2 A_2 = \text{const.} \quad (1.15)$$

Hence, in a widening tube the velocity (averaged over the cross section) decreases.

1.3M. STREAMLINES AND PATHLINES

A streamline is defined by

$$d\mathbf{r} \times \mathbf{v} = 0, \quad (1.16)$$

and the pathline of a particle is computed by

integrating

$$\frac{d\mathbf{r}}{dt} = \mathbf{v}. \qquad (1.17)$$

A streak line is formed by fluid particles that pass a given point in the course of time (Fig. 2.17). Thus, a streak line can be obtained with the aid of pathlines as outlined in Section 2.4.

1.4M. LINES, TUBES, AND FILAMENTS

If **a** is a vector field, then a "line" can be drawn whose tangent at every point is in the direction of the **a** vector through that point. A "tube" is a surface in the fluid formed by all lines through a given closed curve in the fluid (Fig. 4.20). If **v** represents the velocity vector, the lines are called "streamlines" and the tubes "streamtubes." If **a** is the vorticity vector **ω**, "vorticity lines" and "vorticity tubes" can be drawn. If the vorticity field is zero except at a certain vorticity line, this line is called a "vortex line." If a vorticity tube is surrounded by an irrotational flow field, it becomes a "vortex tube." The interior of this tube is a "vortex filament" [2M].

Example 1

The potential vortex $v_\varphi = \kappa/r$ is a straight vortex line with $\omega_z = \infty$ at $r = 0$. In the (r, φ) plane it is a point vortex. The flow region is doubly connected because the center is excluded from the potential flow field.

Example 2

In the Rankine vortex (Fig. 3.10) the core is a vortex tube with $\omega_z = \text{const}$.

2M. Conservation Laws of Viscous Fluids

2.1M. THE NAVIER–STOKES EQUATIONS

Conservation laws of fluid motion exist for mass, momentum, and energy. For their description it is of importance to distinguish between a reference frame moving with the fluid particles (Lagrangian frame) and a reference frame fixed to a prescribed coordinate system (Eulerian frame). The material derivative d/dt is the time change appearing to an observer who moves with the particle; the partial derivative $\partial/\partial t$ is the local time rate of change observed from a fixed point in the Eulerian reference frame. These derivatives are related to each other by

$$\frac{d}{dt} = \frac{\partial}{\partial t} + (\mathbf{v} \cdot \nabla) \qquad (2.1)$$

where $(\mathbf{v} \cdot \nabla)$ represents the convection term. To write the conservation laws as simply as possible for the time being, the fluid is considered as homogeneous and incompressible, that is, the density ρ of the fluid is constant. Then, one obtains in differential form [Reference 39, Chapter 1]

mass: $\qquad \text{div}\,\mathbf{v} = 0, \qquad (2.2)$
(continuity equation)

momentum:

$$\frac{\partial \mathbf{v}}{\partial t} + (\mathbf{v} \cdot \nabla)\mathbf{v} = -\frac{1}{\rho}\nabla p + \nu\nabla^2 \mathbf{v} + \mathbf{F},$$
(Navier–Stokes equation) $\qquad (2.3)$

energy: $\quad \rho c\left[\dfrac{\partial T}{\partial t} + (\mathbf{v} \cdot \nabla)T\right] = k\nabla^2 T + D,$
(heat equation) $\qquad (2.4)$

with the dissipation function according to [1M]

$$D = \mu\left[\omega^2 - 2\,\text{div}(\mathbf{v} \times \boldsymbol{\omega}) + \nabla^2 \mathbf{v}^2\right]. \qquad (2.5)$$

In these equations T, ρ, p, and $\mathbf{F}$ denote the temperature, density, pressure, and external force, respectively. It is also assumed that the fluid is Newtonian, that is, the shear stress is proportional to the rate of deformation, that the heat flux via thermal conduction is proportional to the temperature gradient, and that heat sources or sinks are neglected in Eq. (2.4). It is further assumed that the kinematic viscosity ν, dynamic viscosity μ, thermal conductivity k, and specific heat c are constant. In Eqs. (2.3) and (2.4) ∇^2 is the diffusion term.

A specific problem is defined when appropriate initial and boundary conditions for Eqs. (2.2) through (2.4) are prescribed.

Equations (2.2) and (2.3) do not depend on T (provided that $\mathbf{F}$ does not depend on it). They can, therefore, be solved without information about the temperature field. The heat equation (2.4) can then be solved with the given function $\mathbf{v}$.

The total dissipation D over a volume V with a surface area A is

$$\int D\,dV = \mu\left\{\int \omega^2\,dV + \int\left[\frac{\partial \mathbf{v}^2}{\partial n} - 2(\mathbf{v} \times \boldsymbol{\omega})_n\right]dA\right\}. \qquad (2.6)$$

For the potential flow $\boldsymbol{\omega} = 0$, it is

$$\int D\,dV = \mu\int \frac{\partial \mathbf{v}^2}{\partial n}\,dA, \qquad (2.7)$$

that is, the total dissipation depends only on $\partial \mathbf{v}^2/\partial n$ at the surface.

In cylindrical polar coordinates r, φ, z with the corresponding velocity components v_r, v_φ, v_z, Eqs. (2.2) through (2.5) simplify under the assumption of axisymmetry, $\partial/\partial\varphi \equiv 0$, to

$$\frac{\partial v_r}{\partial t} + v_r \frac{\partial v_r}{\partial r} + v_z \frac{\partial v_r}{\partial z} - \frac{v_\varphi^2}{r}$$
$$= -\frac{1}{\rho}\frac{\partial p}{\partial r} + \nu\left[\frac{\partial^2 v_r}{\partial r^2} + \frac{\partial}{\partial r}\left(\frac{v_r}{r}\right) + \frac{\partial^2 v_r}{\partial z^2}\right] + F_r, \quad (2.8)$$

$$\frac{\partial v_\varphi}{\partial t} + v_r \frac{\partial v_\varphi}{\partial r} + v_z \frac{\partial v_\varphi}{\partial z} + \frac{v_r v_\varphi}{r}$$
$$= \nu\left[\frac{\partial^2 v_\varphi}{\partial r^2} + \frac{\partial}{\partial r}\left(\frac{v_\varphi}{r}\right) + \frac{\partial^2 v_\varphi}{\partial z^2}\right] + F_\varphi, \quad (2.9)$$

$$\frac{\partial v_z}{\partial t} + v_r \frac{\partial v_z}{\partial r} + v_z \frac{\partial v_z}{\partial z} = -\frac{1}{\rho}\frac{\partial p}{\partial z}$$
$$+ \nu\left[\frac{\partial^2 v_z}{\partial r^2} + \frac{1}{r}\frac{\partial v_z}{\partial r} + \frac{\partial^2 v_z}{\partial z^2}\right] + F_z, \quad (2.10)$$

$$\frac{\partial v_r}{\partial r} + \frac{v_r}{r} + \frac{\partial v_z}{\partial z} = 0 \quad (2.11)$$

$$\rho c\left(\frac{\partial T}{\partial t} + v_r \frac{\partial T}{\partial r} + v_z \frac{\partial T}{\partial z}\right)$$
$$= k\left(\frac{\partial^2 T}{\partial r^2} + \frac{1}{r}\frac{\partial T}{\partial r} + \frac{\partial^2 T}{\partial z^2}\right) + \mu\left[\left(\frac{\partial v_\varphi}{\partial r} - \frac{v_\varphi}{r}\right)^2 + \left(\frac{\partial v_r}{\partial r} + \frac{\partial v_r}{\partial z}\right)^2 + \left(\frac{\partial v_\varphi}{\partial z}\right)^2 + 2\left(\frac{\partial v_r}{\partial r}\right)^2 + 2\left(\frac{v_r}{r}\right)^2 + 2\left(\frac{\partial v_z}{\partial z}\right)^2\right]. \quad (2.12)$$

2.2M. THE VORTICITY-TRANSPORT EQUATION

The Navier–Stokes equation (2.3) can be rewritten by applying the curl-operator and using the vorticity vector $\boldsymbol{\omega} = \operatorname{curl} \mathbf{v}$. Then, for an incompressible fluid one obtains the so-called "vorticity-transport equation":

$$\frac{\partial \boldsymbol{\omega}}{\partial t} + (\mathbf{v}\cdot\nabla)\boldsymbol{\omega} - (\boldsymbol{\omega}\cdot\nabla)\mathbf{v} = \nu\nabla^2\boldsymbol{\omega} + \operatorname{curl}\mathbf{F}$$
$$(2.13)$$

with

$$\operatorname{div}\boldsymbol{\omega} = 0. \quad (2.14)$$

The relation (2.14) follows from the definition $\boldsymbol{\omega} = \operatorname{curl}\mathbf{v}$. If $\mathbf{F}$ has a potential, the term $\operatorname{curl}\mathbf{F}$ vanishes. Equations (2.13) and (2.14) simplify for two-dimensional flows in the (x, y) plane to

$$\frac{\partial \omega}{\partial t} + (\mathbf{v}\cdot\nabla)\omega = \nu\nabla^2\omega + (\operatorname{curl}\mathbf{F})_z \quad (2.15)$$

$$\nabla^2\psi = \omega, \quad (2.16)$$

where $\omega_z = \omega$ is the vorticity component normal to the (x, y) plane and ψ is the stream function, defined by

$$v_x = -\frac{\partial \psi}{\partial y}, \qquad v_y = \frac{\partial \psi}{\partial x}, \quad (2.17)$$

or in the (r, φ) plane

$$v_r = -\frac{1}{r}\frac{\partial \psi}{\partial \varphi}, \qquad v_\varphi = \frac{\partial \psi}{\partial r}. \quad (2.18)$$

Note that a single stream function exists for two-dimensional and axisymmetrical flows only.

2.3M. ANALOGY BETWEEN VORTICITY AND HEAT TRANSFER

According to Eq. (2.4) the heat-transfer equation is for vanishing dissipation

$$\rho c\left[\frac{\partial T}{\partial t} + (\mathbf{v}\cdot\nabla)T\right] = k\nabla^2 T. \quad (2.19)$$

For plane flows this equation is analogous to the vorticity-transport equation (2.15), provided the force has a potential. Thus, vorticity in two-dimensional flows is transported in a way similar to that in which heat is transported. The operators $(\mathbf{v}\cdot\nabla)$

and ∇^2 represent convection and diffusion, respectively.

2.4M. DIMENSIONLESS FORM OF THE VORTICITY-TRANSPORT EQUATION

If the dimensionless quantities ω^*, t^*, and $\mathbf{v}^*$ are introduced for ω, t, and $\mathbf{v}$ by relating them to the characteristic length L and the characteristic velocity V through $\omega = \omega^* V/L$, $t = t^* L/V$, and $\mathbf{v} = V\mathbf{v}^*$, Eq. (2.13) yields

$$\frac{\partial \omega^*}{\partial t^*} + (\mathbf{v}^* \cdot \nabla^*)\omega^* = (\omega^* \cdot \nabla^*)\mathbf{v}^* = \frac{1}{\mathrm{Re}} \nabla^{*2}\omega^*, \tag{2.20}$$

where curl $\mathbf{F}$ is omitted. The constant parameter $\mathrm{Re} = LV/\nu$ is called the Reynolds number. It is the ratio of convection to diffusion. Two viscous flows are dynamically similar if their Reynolds numbers are the same.

3M. Vortices in Viscous Fluid Flows

3.1M. SOME CLOSED-FORM SOLUTIONS

In general, solutions of the basic equations of motion (2.2) through (2.4) or (2.13) and (2.14) can be found only by numerical integration with the aid of computers. However, for certain symmetry (or similarity) properties closed-form solutions do exist, although they are quite rare. Some of them, which give information on the characteristics of vortices, will be discussed in the text which follows.

Example 1

Assume that a motion takes place in the (r, φ) plane and is time independent (steady) in the absence of external forces. If $v_r = v_z = 0$, it follows from Eq. (2.9) that

$$v_\varphi = \frac{a}{r} + br \qquad (3.1)$$

with a and b constant and determined by the boundary conditions. For instance, the flow between two concentric cylinders of radii r_1 and r_2, rotating with constant angular velocities Ω_1 and Ω_2, requires

$$a = (\Omega_1 - \Omega_2)\frac{r_1^2 r_2^2}{r_2^2 - r_1^2}, \qquad (3.2)$$

$$b = (\Omega_2 r_2^2 - \Omega_1 r_1^2)\frac{1}{r_2^2 - r_1^2}. \qquad (3.3)$$

For $b = 0$, Eq. (3.1) represents the "potential vortex"

$$v_\varphi = \frac{\kappa}{r} \qquad (3.4)$$

with $a = \kappa$ being the strength of the vortex, which is related to the circulation by $\Gamma = 2\pi\kappa$. The vorticity is zero except at $r = 0$, and thus Eq. (3.4) is a potential-flow solution, hence, the name "potential vortex." The streamlines are circular, a fact which does not contradict the theorem mentioned in Section 1.2M that closed streamlines cannot exist in a potential flow occupying a simply connected region. The field of the potential vortex is doubly connected since the singularity at $r = 0$ must be excluded.

For $a = 0$ in Eq. (3.1) [or if $r_1 = 0$ in Eqs. (3.2) and (3.3)], one obtains the solid-body rotation (1.1) with $b = \Omega$.

From Eq. (2.8) it follows that the pressure difference in a potential vortex between points 1 and 2 is

$$p_1 - p_2 = \rho \frac{\kappa^2}{2}\left(\frac{1}{r_2^2} - \frac{1}{r_1^2}\right), \qquad (3.5)$$

and, for solid-body rotation,

$$p_1 - p_2 = \rho \frac{\Omega^2}{2}(r_1^2 - r_2^2). \qquad (3.6)$$

See Figs. 3.13 and 3.15. The two solutions can be combined to describe the Rankine vortex (Fig. 3.16). According to Eq. (2.12) dissipation for the solid-body rotation is $D = 0$ and for the potential vortex $D = 4\mu\kappa^2/r^4$. The total dissipation of the Rankine vortex with a nonzero core radius r_1 is, according to Eq. (2.7), $4\pi\mu\kappa^2/r_1^2$. The total dissipation of the potential vortex with $r_1 = 0$ is then infinite.

In the rotating-bucket experiment (Fig. 3.17) the water rotates like a solid body. If z is the axis of rotation perpendicular to the earth's surface, the gravitational force acts with $F_z = -g$. Then, Eqs.

(2.8) and (2.10) reduce to

$$\frac{v_\varphi^2}{r} = \frac{1}{\rho}\frac{\partial p}{\partial r}, \quad (3.7)$$

$$-g = \frac{1}{\rho}\frac{\partial p}{\partial z}. \quad (3.8)$$

With $v_\varphi = \Omega r$, Eqs. (3.7) and (3.8) can be written

$$\frac{dp}{\rho} = r\Omega^2\, dr - g\, dz$$

or

$$\frac{p}{\rho} = \tfrac{1}{2}\Omega^2 r^2 - gz + \text{const.} \quad (3.9)$$

At the free surface the pressure p is constant. Hence, the surface is the paraboloid $z = \Omega^2 r^2/2g$ with the origin at the lowest point of the surface.

For a radial flow $v_r \neq 0$ ($v_z = 0$), it follows from Eq. (2.11) that

$$v_r = \frac{m}{r}, \quad m = \text{const}, \quad (3.10)$$

and from Eq. (2.9)

$$v_\varphi = \frac{a}{r} + br^{1+m/\nu}, \quad \frac{m}{\nu} \neq -2,$$

$$v_\varphi = \frac{a}{r} + b\frac{1}{r}\log r, \quad \frac{m}{\nu} = -2. \quad (3.11)$$

This flow can be realized experimentally by two concentric porous cylinders rotating with different angular velocities. Plots of (3.11) are published in Reference 1M.

The assumption $v_z = 0$ can also be relaxed to $v_z = v_z(r)$ (but $v_r = 0$) without affecting Eq. (3.1). Then, Eqs. (2.10) and (2.8) yield

$$v_z = \frac{1}{4\mu}\frac{\partial p}{\partial z}r^2 + A\log r + B \quad (3.12)$$

$$p = \frac{\partial p}{\partial z}z + \rho\int v_\varphi^2\frac{dr}{r} + C, \quad (3.13)$$

where A, B, C, and $\partial p/\partial z$ are arbitrary constants. Equations (3.1), (3.12), and (3.13) describe the helical flow in a circular cylinder. For $v_\varphi = 0$, one obtains with the nonslip condition the well-known Poiseuille solution of a steady laminar flow in a circular pipe (Fig. 6.58).

Example 2

The steady potential vortex is maintained by the work done in the core (Section 3.2). If this source of energy ceases at time $t = 0$, the potential vortex decays. The solution to this problem is given by Lamb [Reference 2M, Chapter 1M]

$$v_\varphi = \frac{\kappa}{r}\left(1 - e^{-r^2/4\nu t}\right), \quad v_r = v_z \equiv 0. \quad (3.14)$$

The corresponding vorticity is

$$\omega_z = \frac{\kappa}{2\nu t}e^{-r^2/4\nu t}, \quad \omega_r = \omega_\varphi \equiv 0. \quad (3.15)$$

A graph of (3.14) is displayed in Fig. 3.12. Near the axis the vortex rotates like a solid body with

$$v_\varphi = \frac{\kappa}{4\nu t}r \quad \text{or} \quad \Omega = \frac{\kappa}{4\nu t} \quad \text{for} \quad \frac{r^2}{4\nu t} \ll 1. \quad (3.16)$$

Thus, the angular velocity Ω decreases with $1/t$. The total vorticity of the vortex is a constant, but the kinetic energy, the angular momentum, and the energy dissipation are infinite in an unbounded flow region.

Example 3

A decaying vortex with different properties can be obtained from Eq. (3.14) by differentiation with respect to t. This solution is due to G. I. Taylor [2M]:

$$v_\varphi = \frac{Mr}{4\pi\nu t^2}e^{-r^2/4\nu t}, \quad v_r = v_z \equiv 0, \quad (3.17)$$

$$\omega_z = \frac{M}{2\pi\nu t^2}\left(1 - \frac{r^2}{4\nu t}\right)e^{-r^2/4\nu t}, \quad \omega_r = \omega_\varphi \equiv 0, \quad (3.18)$$

where M is a constant which, multiplied by ρ, represents the angular momentum:

$$\int_0^\infty 2\pi\rho v_\varphi r\, dr = \rho M. \quad (3.19)$$

The total kinetic energy and the energy dissipation are also finite. Again, near the axis the vortex behaves like a solid body:

$$v_\varphi = \frac{Mr}{4\pi\nu t^2} \quad \text{or} \quad \Omega = \frac{M}{4\pi\nu t^2} \quad \text{for} \quad \frac{r^2}{4\nu t} \ll 1. \tag{3.20}$$

This vortex decays faster than the potential vortex.

Example 4

In order to balance the outward radial diffusion of angular momentum in the decaying potential vortex and to arrive at a steady state, either the core must provide the angular momentum (which results in the steady potential vortex) or angular momentum must be brought in from outside. In Burgers' solution [3M] the inward radial convection of angular momentum just balances the outward diffusion so that the flow is steady:

$$v_\varphi = \frac{\kappa}{r}\left(1 - e^{-ar^2/2\nu}\right), \quad a > 0 \tag{3.21}$$

$$v_r = -ar, \quad v_z = 2az \tag{3.22}$$

$$\omega_z = \frac{a\kappa}{\nu} e^{-ar^2/2\nu}, \quad \omega_r = \omega_\varphi \equiv 0. \tag{3.23}$$

The positive constant a is a measure of the radial influx. The flow field is not only axisymmetric but the half-space $z > 0$ is a mirror of the half-space $z < 0$, with $z = 0$ the plane of symmetry (Fig. 3.1M). If the plane $z = 0$ is considered an impermeable surface (with slip, however), Burgers' solution represents a vortex flow restricted by a plane perpendicular to the vortex axis. v_φ is independent of z, but depends on the influx parameter a. At the center $r = 0$, ω_z has the extremum $a\kappa/\nu$, which is twice the angular velocity of the solid-body rotation of the core:

$$v_\varphi = \frac{a\kappa}{2\nu}r \quad \text{or} \quad \Omega = \frac{a\kappa}{2\nu} \quad \text{for} \quad \frac{ar^2}{2\nu} \ll 1. \tag{3.24}$$

The meridional flow (3.22) is independent of v_φ, that is, the strength κ of the rotation does not affect the influx and outflow. However, the stronger the inflow, the faster is the decay of angular momentum over the distance from the axis. The

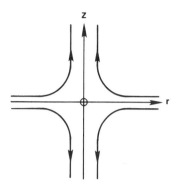

Figure 3.1M. The meridional flow of Burgers' solution.

pressure field is given by

$$p(r,z) = p(0,0) + \rho\frac{a\kappa^2}{4\nu}\int_0^{ar^2/2\nu}\left(\frac{1-e^{-x}}{x}\right)^2 dx$$
$$- \frac{\rho}{2}a^2(r^2 + 4z^2). \tag{3.25}$$

Example 5

The development of the steady-state vortex flow (3.21) and (3.22) was studied by Rott [4M], who found the following remarkable solution:

$$v_\varphi = \frac{\kappa}{r}\left[1 - e^{-\frac{ar^2}{2\nu}(1+be^{-2at})^{-1}}\right], \tag{3.26}$$

$$v_r = -ar, \quad v_z = 2az, \tag{3.27}$$

$$\omega_z = \frac{a\kappa}{\nu}(1 + be^{-2at})^{-1} e^{-\frac{ar^2}{2\nu}(1+be^{-2at})^{-1}}. \tag{3.28}$$

For a meridional influx toward the z axis $a > 0$, it follows that $e^{-2at} \leq 1$ and $1 + b \geq 0$. Burgers' steady-state solution (3.21) is obtained when $t \to \infty$.

If the meridional flow turns away from the z axis, that is, if $a < 0$, then $e^{-2at} \geq 1$ and $1 + b \leq 0$. Diffusion of angular momentum is in the same direction as the convection of angular momentum. The rotation ceases for $t \to \infty$, since $v_\varphi \to 0$. (This is the reason that Burgers' solution is restricted to $a > 0$.)

For a meridional influx $a > 0$, a small initial value of ω_z at $t = 0$ results in the development of a concentrated vortex.

Example 6

For a steady two-cell vortex, characterized by an axial flow reversal near the axis compared to the

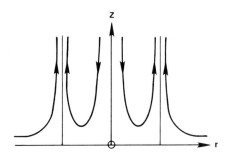

Figure 3.2M. The meridional flow of Sullivan's solution.

outer flow (Fig. 3.2M), Sullivan [5M] found the following solution:

$$v_\varphi = \frac{\kappa}{r} \frac{H(ar^2/2\nu)}{H(\infty)}, \tag{3.29}$$

$$v_r = -ar + \frac{6\nu}{r}\left(1 - e^{-ar^2/2\nu}\right), \tag{3.30}$$

$$v_z = 2az\left(1 - 3e^{-ar^2/2\nu}\right), \tag{3.31}$$

with

$$H\left(\frac{ar^2}{2\nu}\right) = \int_0^{ar^2/2\nu} \exp\left[-x + 3\int_0^x (1 - e^{-s})s^{-1}\,ds\right]dx,$$

$$H(\infty) = 37.905, \tag{3.32}$$

which is tabulated in Reference 6M. The pressure is

$$p(r, z) = p(0, z) \\ - \frac{\rho}{2}\left\{4a^2z^2 + a^2r^2 \\ + \frac{36\nu^2}{r^2}\left[1 - e^{-ar^2/2\nu}\right]^2\right\} \\ + \rho\int_0^r \frac{v_\varphi^2}{r}\,dr. \tag{3.33}$$

Sullivan's solution has features similar to those of Burgers' solution. In fact, for $r \to \infty$ the two solutions approach each other. The most noticeable property of Eqs. (3.29) through (3.31) is the fact that v_φ is decoupled from v_r and v_z, as in the case of Burgers' solution. This means that there is no control of v_r and v_z by v_φ, contrary to known observations in nature.

The unsteady case is discussed by Bellamy-Knights [7M].

Example 7

A closed-form solution of the Navier–Stokes equation also exists for the decay of a vortex grid. If the length of a cell is π in the (x, y) plane (Fig. 1.1M), the solution due to G. I. Taylor [8M] is

$$v_x = +A \cos x \cdot \sin y \cdot e^{-2\nu t}, \tag{3.34}$$

$$v_y = -A \sin x \cdot \cos y \cdot e^{-2\nu t}, \tag{3.35}$$

with the vorticity field

$$\omega_z = -2A \cos x \cdot \cos y \cdot e^{-2\nu t}. \tag{3.36}$$

A is an arbitrary constant.

CONCLUDING REMARKS

Of the seven solutions discussed, four describe a pure diffusion process, since the nonlinear inertial terms are identically zero or are balanced by the pressure gradient (as in the last example). In the steady-state solutions (example 1) energy loss through diffusion is compensated for by energy produced outside and transmitted through the movement of the walls. Unsteady solutions reveal the rate of diffusion which is of the order r_0^2/ν, with r_0 a characteristic radius. Here, Lamb's solution (3.14) is of particular interest since experiments have shown that detached vortices can be described quite well by this solution (e.g., the individual vortices in periodic vortex streets) [Reference 23, Chapter 6].

Solutions with nonvanishing inertial terms demonstrate the interaction between convection and diffusion of the rotating fluid (examples 4, 5, and 6). Their usefulness, however, is limited by the independence of the meridional flow from v_φ and by the free slip on the horizontal plane $z = 0$. Boundary-layer effects, which are caused by the nonslip condition and which are important in rotating flows above a solid surface, are thus not included. The nonslip condition makes the meridional flow strongly dependent on the primary rotating fluid (Section 3.3M).

3.2M. NOTE ON THE OCCURRENCE OF EXTREMAL VALUES INSIDE A VORTICITY FIELD

A vortex is sometimes identified by the occurrence of a local extremum in the vorticity, but this is not necessarily a criterion for the existence of a vortex. In fact, no steady plane motion of an incompressible homogeneous fluid can have an extremum of vorticity inside the fluid unless certain nonconservative forces are acting. This limitation follows from the maximum principle for nonlinear elliptic equations [9M]. The physical explanation for this situation is the fact that vorticity must be fed continuously into the vortex attached to the body to balance dissipation (Fig. 5.18). In unsteady flows, this balance need not exist and the vortex may separate from the body, a process defined by the first occurrence of an extremum.

In a steady three-dimensional flow (without nonconservative forces) the existence of a vorticity extremum requires the convective transport of vorticity into the location of the extremum. This is illustrated for an axisymmetric flow with the assumption that $\omega_r = \omega_\varphi \equiv 0$ and that the meridional flow v_r, v_z is independent of ω_z. Then, the vorticity-transport equation is, according to Eq. (2.13),

$$L(\omega_z) + \frac{\partial v_z}{\partial z}\omega_z = 0, \qquad (3.37)$$

where the linear operator L is

$$L = \nu\left(\frac{\partial^2}{\partial r^2} + \frac{1}{r}\frac{\partial}{\partial r} + \frac{\partial^2}{\partial z^2}\right) - v_r\frac{\partial}{\partial r} - v_z\frac{\partial}{\partial z}. \qquad (3.38)$$

For this linear case a positive maximum or a negative minimum of ω_z is according to Reference 9M not possible if

$$\frac{\partial v_z}{\partial z} \leqslant 0. \qquad (3.39)$$

Conversely, this means that the existence of a vorticity extremum requires the influx $\partial v_z/\partial z > 0$. An example is Burgers' solution (3.21) through (3.23). From (3.22) it follows that $\partial v_z/\partial z = 2a$, which must be positive as a prerequisite for a nonzero solution. A local positive maximum occurs at $r = 0$, as Eq. (3.23) immediately reveals.

For slow motion, for which $\nabla^2\omega_z = 0$, no extremum at all occurs inside the flow region.

3.3M. SIMILARITY SOLUTIONS

The examples of Section 3.1M reveal that closed-form solutions of the Navier–Stokes equation exist only for simple cases. "Simple" means either that the solutions have symmetry properties like $f(z) = f(-z)$ or that they differ at two locations only by a scale factor. Mathematically, such "similarity" properties may be expressed in the form

$$u(x_1, x_2) = f(x_1)g\left[\frac{x_2}{h(x_1)}\right], \qquad (3.40)$$

where the geometric coordinates or the time are designated by x_1 and x_2. The scalar function u stands for the velocity components, and $f, g,$ and h are functions to be determined by the governing equations and boundary conditions. More specifically, f determines the scale of u such that g is an order-one quantity which varies solely with the similarity variable $\zeta = x_2/h(x_1)$. For a given value of ζ, the values of $u(x_1, x_2)/f(x_1)$ are equal at two different locations. All closed-form solutions in Section 3.1M are similarity solutions.

Similarity properties expressed by Eq. (3.40) reduce the original partial differential equations to ordinary differential equations, which are much easier to solve numerically.

Some examples of vortex flows are presented which demonstrate the nature of these similarity solutions, their usefulness, and also their shortcomings.

Example 1

The flow field near a rotating plane (or a rotating disk of infinite radius) was first discussed by von Kármán [Reference 32, Chapter 7] with the following similarity assumptions:

$$v_r = r\Omega F(\zeta),$$

$$v_\varphi = r\Omega G(\zeta),$$

$$v_z = \sqrt{\nu\Omega}\, H(\zeta), \qquad (3.41)$$

$$p = \rho\nu\Omega P(\zeta),$$

where Ω is the angular velocity of the plane and ζ is the dimensionless z component: $\zeta = z\sqrt{\Omega/\nu}$. Then Eqs. (2.8)–(2.11) reduce to the set of ordinary differential equations:

$$2F + H' = 0,$$
$$F^2 + F'H - G^2 - F'' = 0,$$
$$2GF + HG' - G'' = 0, \quad (3.42)$$
$$P' + HH' - H'' = 0,$$

with the following boundary conditions for nonslip at the surface and for the flow at infinity:

$$\zeta = 0: \quad F = 0, G = 1, H = 0, P = 0;$$
$$\zeta = \infty: \quad F = 0, G = 0. \quad (3.43)$$

The boundary-value problem (3.42) and (3.43) has been solved numerically [10M] and is displayed in Fig. 3.3M.

The solution reveals that the fluid layer near the rotating plane is pushed outward through the action of friction and centrifugal force. This fluid is then replaced from an axial flow with velocity v_z toward the rotating plane. This behavior is in agreement with experimental observations.

The similarity properties of the solution are, however, responsible for the following flow characteristics: The thickness of the boundary layer is constant over the plane, that is, it is independent of r. The same is true for v_z and p, as well as for the deviation of the streamlines on the rotating plane from the circular motion, that is,

$$\tan \chi = -\left(\frac{\partial v_r / \partial z}{\partial v_\varphi / \partial z}\right)_{z=0} = 0.838 \quad \text{or} \quad \chi = 39.6°.$$

Moreover, these properties do not depend on a Reynolds number. A closer look at the boundary-value problem reveals that uniqueness of the solution has been enforced by the similarity assumptions, which exclude prescription of boundary conditions far away from the axis of rotation, that is, at $r \to \infty$, $0 < z < \infty$. The fact that these solutions satisfy simultaneously the boundary-layer equations (Section 3.4M) suggests that von Kármán's solution is an asymptotic one for Re $\to \infty$. This suggestion is supported by studies of Schwiderski and Lugt [11M] in which experimental data are also evaluated. More recent investigations have been performed by Dijkstra and Zandbergen [12M].

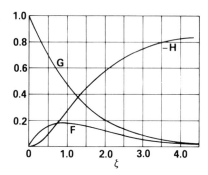

Figure 3.3M. The functions $F(\zeta)$, $G(\zeta)$, and $H(\zeta)$ of von Kármán's problem.

The same similarity assumptions (3.41), except for the pressure, can be applied to the study of solid-body rotation over a fixed flat surface (Bödewadt [Reference 33, Chapter 7]). In this case the direction of the meridional flow is reversed, and the velocity components exhibit an oscillatory behavior in space. Solutions are given in Reference 35, Chapter 1.

Example 2

The study of similarity properties for the potential vortex over a flat surface leads to a solution, contrary to the Kármán–Bödewadt solutions, which is not simultaneously a solution of the boundary-layer equations and which contains a Reynolds number explicitly (Goldshtik [Reference 34, Chapter 7]):

$$v_r = \frac{\kappa}{r} F'(\zeta),$$
$$v_\varphi = \frac{\kappa}{r} G(\zeta),$$
$$v_z = \frac{\kappa}{r}(F - \zeta F'), \quad (3.44)$$
$$p = \frac{\rho}{r^2} \kappa^2 P(\zeta),$$

with $\zeta = \sqrt{\text{Re}}\, z/r$, Re $= \kappa/\nu$. Solutions to this problem have been presented by several investigators with different numerical techniques [References 35 and 36, Chapter 7] (Fig. 3.4M).

The most striking feature of the solution is the existence of an upper bound for the Reynolds number (whose actual value varies with the numerical methods used between 4.75 and 5.5). This means that no similarity solution exists beyond this critical Reynolds number. Another feature not found in the Kármán–Bödewadt flows is the reversal of the

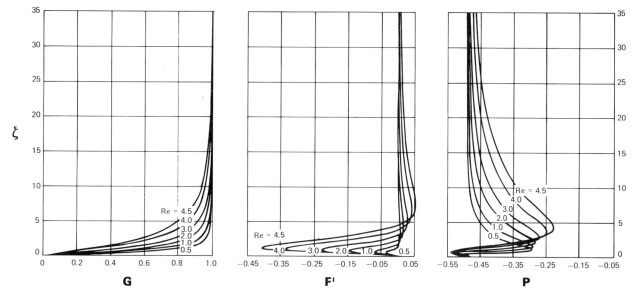

Figure 3.4M. The functions $G(\zeta)$, $F'(\zeta)$, and $P(\zeta)$ of Goldshtik's problem (from G. J. Kidd and G. J. Farris [Reference 35, Chapter 7]).

direction of v_r. The streamlines converge toward the center for small ζ like $z \sim \sqrt{r}$ and diverge for large ζ like $z \sim r^2$. Such behavior is not possible in a boundary-layer solution in which the radial inflow is taken away by a sink at $z = \infty$. The deviation angle χ increases from zero to 90° as Re increases from zero to its critical value.

Schwiderski [Reference 36, Chapter 7] found a solution for the limit Re = 0 that could be constructed only by retaining a quadratic term in the set of ordinary differential equations. This is an interesting exception to the otherwise linear equations for slow motion (Chapter 6M).

3.4M. BOUNDARY-LAYER FLOWS

The Navier–Stokes equations can be simplified considerably by employing Prandtl's boundary-layer idea (Section 4.5). Although this concept has been extremely useful in engineering applications, the mathematical consequences of the truncation are not yet completely clear. Before discussing this further, the boundary-layer equations are presented with the standard order-of-magnitude analysis for two different cases: (1) The boundary layer exists along the solid surface $z = 0$, and (2) the thin core of a vortex forms the boundary layer.

1. With the following estimates of the order of magnitude for each term, that is, with

$$\frac{\partial}{\partial r} \sim O(1), \quad \frac{\partial}{\partial z} \sim O\left(\frac{1}{\delta}\right), \quad \text{Re} \sim O\left(\frac{1}{\delta^2}\right)$$

$$v_r \sim O(1), \quad v_\varphi \sim O(1), \quad v_z \sim O(\delta) \quad (3.45)$$

where $\delta \ll 1$ is the thickness of the boundary layer, the steady-state, dimensionless Navier–Stokes equations (without external forces) (2.8) through (2.11) reduce to (stars are omitted)

$$v_r \frac{\partial v_r}{\partial r} + v_z \frac{\partial v_r}{\partial z} - \frac{v_\varphi^2}{r} = -\frac{\partial p}{\partial r} + \frac{1}{\text{Re}} \frac{\partial^2 v_r}{\partial z^2}, \quad (3.46)$$

$$v_r \frac{\partial v_\varphi}{\partial r} + v_z \frac{\partial v_\varphi}{\partial z} + \frac{v_r v_\varphi}{r} = \frac{1}{\text{Re}} \frac{\partial^2 v_\varphi}{\partial z^2}, \quad (3.47)$$

$$0 \approx \frac{\partial p}{\partial z}, \quad (3.48)$$

$$\frac{\partial v_r}{\partial r} + \frac{v_r}{r} + \frac{\partial v_z}{\partial z} = 0. \quad (3.49)$$

Examples are the Kármán–Bödewadt solutions, which are simultaneously solutions of the full Navier–Stokes equations.

2. If the diameter of the thin core of a vortex is of order δ, then

$$\frac{\partial}{\partial r} \sim O\left(\frac{1}{\delta}\right), \quad \frac{\partial}{\partial z} \sim O(1), \quad \text{Re} \sim O\left(\frac{1}{\delta^2}\right),$$

$$v_r \sim O(\delta), \quad v_\varphi \sim O(1), \quad v_z \sim O(1), \quad (3.50)$$

and

$$\frac{v_\varphi^2}{r} = \frac{\partial p}{\partial r}, \quad (3.51)$$

$$v_r \frac{\partial v_\varphi}{\partial r} + v_z \frac{\partial v_\varphi}{\partial z} + \frac{v_r v_\varphi}{r} = \frac{1}{\text{Re}}\left[\frac{\partial^2 v_\varphi}{\partial r^2} + \frac{\partial}{\partial r}\left(\frac{v_\varphi}{r}\right)\right], \quad (3.52)$$

$$v_r \frac{\partial v_z}{\partial r} + v_z \frac{\partial v_z}{\partial z} = -\frac{\partial p}{\partial z} + \frac{1}{\text{Re}}\left[\frac{\partial^2 v_z}{\partial r^2} + \frac{1}{r}\frac{\partial v_z}{\partial r}\right], \quad (3.53)$$

$$\frac{\partial v_r}{\partial r} + \frac{v_r}{r} + \frac{\partial v_z}{\partial z} = 0. \quad (3.54)$$

Two examples are given below.

Mathematically, the boundary-layer equations are parabolic compared to the original elliptic Navier–Stokes equations. This means that the region of influence is restricted for the boundary-layer equations. For instance, a flat plate positioned parallel to a flow has no upstream influence beyond the plate's edge.

Example 1

The similarity solution by Long [13M] describes a swirling jet emanating from a source at $r = z = 0$ and spreading into an unbounded fluid of constant circulation. There is a strong coupling between swirl and axial flow. Long assumes the following similarity properties:

$$v_r = -\frac{\nu}{r}F(\zeta) + \frac{\nu}{z}F'(\zeta),$$

$$v_\varphi = \frac{\nu}{r}G(\zeta),$$

$$v_z = \frac{\nu}{r}F'(\zeta), \quad (3.55)$$

$$p = -\frac{\nu^2}{z^2}H(\zeta),$$

with $\zeta = r/z$. The boundary-layer equations reduce then to a set of ordinary differential equations that must be solved numerically [13M]. It turns out that a fundamental constant of the flow problem is the kinematic momentum transfer

$$J = \int_0^{2\pi}\int_0^\infty \left(\frac{p}{\rho} + gz + v_z^2\right) r\, dr\, d\varphi \quad (3.56)$$

in addition to the Reynolds number $\text{Re} = \kappa/\nu$. Figure 3.5M shows solutions for three different cases.

Example 2

For the swirling flow with strong axial motion, also originating at $r = z = 0$, Newman [14M] and Batchelor [15M] found a closed-form solution of Eqs. (3.51) through (3.54) under the additional assumption that

$$|v_z - V| \ll V. \quad (3.57)$$

Then, the boundary-layer equations reduce to

$$V\frac{\partial v_\varphi}{\partial z} = \nu\left[\frac{\partial^2 v_\varphi}{\partial r^2} + \frac{\partial}{\partial r}\left(\frac{v_\varphi}{r}\right)\right], \quad (3.58)$$

$$V\frac{\partial v_z}{\partial z} = -\frac{1}{\rho}\frac{\partial p}{\partial z} + \nu\left[\frac{\partial^2 v_z}{\partial r^2} + \frac{1}{r}\frac{\partial v_z}{\partial r}\right], \quad (3.59)$$

whereas Eqs. (3.51) and (3.54) remain unchanged. A similarity solution of Eqs. (3.51), (3.54), (3.58), and (3.59), found by Newman and Batchelor, is

$$v_\varphi = \frac{\kappa}{r}(1 - e^{-\zeta}), \quad (3.60)$$

$$v_z = V - \frac{\kappa^2}{8\nu z}\log\frac{V_z}{\nu}e^{-\zeta}$$

$$+ \frac{\kappa^2}{8\nu z}\{e^{-\zeta}[\log\zeta + \text{ei}(\zeta) - 0.807]$$

$$+ 2\text{ei}(\zeta) - 2\text{ei}(2\zeta)\} - L\frac{V^2}{8\nu z}e^{-\zeta}, \quad (3.61)$$

$$p_0 - p = \frac{\kappa^2 V\rho}{8\nu z}P(\zeta), \quad (3.62)$$

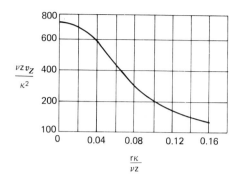

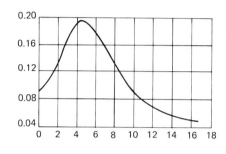

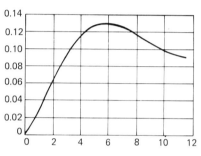

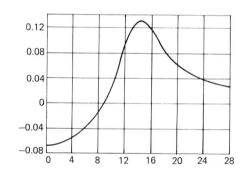

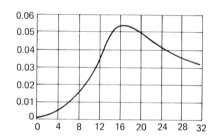

Figure 3.5M. Similarity solutions according to R. R. Long [13M]. (a) The axial velocity v_z is large at the center, expressed by a parameter $B = 10.02$. (b) v_z is small at the center, $B = 1.10$; and (c) v_z is negative at the center, $B = 0.18$.

where $\zeta = Vr^2/4\nu z$,

$$P(\zeta) = \frac{1}{\zeta}(1 - e^{-\zeta})^2 + 2\mathrm{ei}(\zeta) - 2\mathrm{ei}(2\zeta), \tag{3.63}$$

and

$$\mathrm{ei}(\zeta) = \int_\zeta^\infty \frac{e^{-s}}{s}\,ds. \tag{3.64}$$

L is a constant which must be determined by additional conditions at $r = z = 0$. The azimuthal velocity has a structure similar to that of Eq. (3.14) if t is replaced by z/V. An evaluation of v_z and p reveals that during the spreading of the swirl, v_z diminishes while p increases.

CONCLUDING REMARKS

Aside from the obvious shortcomings of the similarity and boundary-layer solutions, whose boundary conditions are unrealistic, in general, they may serve as local approximations to real flow problems. Lewellen [Reference 11, Chapter 10] has divided the vortex flow over a solid surface (as in the case of a tornado) into four regions, as sketched in Fig. 3.6M.

In region I similarity solutions of the type of Burgers, Sullivan, or Long may be considered, of which only Long's solution exhibits a strong interaction between swirl and meridional flow. In region II boundary-layer flows can be assumed of which a whole group has been discussed by Lewellen [Reference 11, Chapter 10]. Region III is the most complex one. Goldshtik's similarity solution is valid there, but only for very small Reynolds numbers. Otherwise, the complete Navier–Stokes equations without similarity assumptions and with possible inclusion of turbulence models must be solved. Region IV depends strongly on the total vortex flow. Ideally, this region may be simply represented by a potential vortex. In reality, as in the case of a tornado, strong interactions with the environmental weather situation occur that to a large extent maintain the vortex (Section 10.3).

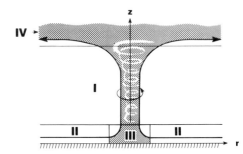

Figure 3.6M. The four regions of vortex flow over a flat surface, exemplified by a tornado, according to W. S. Lewellen [Reference 11, Chapter 10].

All regions interact with each other, and adjustments at their borders must be made. Perturbation techniques, which facilitate the matching, are recorded by Rott and Lewellen [16M], Granger [17M], and Van Dyke [18M].

4M. Vorticity and Circulation Theorems Of Inviscid Fluids

4.1M. HELMHOLTZ'S THEOREMS

If viscous forces are neglected, the Navier–Stokes equation (2.3) reduces to the Euler equation. Moreover, if in this chapter external forces are neglected, the vorticity-transport equations (2.13) and (2.14) simplify to

$$\frac{\partial \omega}{\partial t} + (\mathbf{v} \cdot \nabla)\omega = (\omega \cdot \nabla)\mathbf{v}, \quad (4.1)$$

$$\text{div } \omega = 0. \quad (4.2)$$

These equations are valid for homogeneous fluids with constant density. Relation (4.2), which is unchanged from Eq. (2.14) and is thus generally valid for viscous as well as for inviscid fluids, means that ω is a solenoidal vector field. This may be expressed in a different form by integrating over a certain volume V enclosed by the area A according to Gauss' theorem:

$$\int \text{div } \omega \, dV = \oint \omega \cdot \mathbf{n} \, dA = 0. \quad (4.3)$$

The flow of vorticity out of any closed area at any instant of time is zero. This is Helmholtz's first theorem. Equation (4.3) is applied to a vorticity tube with the cross sections A_1 and A_2. If the ω components normal to the areas A_1 and A_2 are ω_1 and ω_2, and if they are constant or represent average values over the cross sections, then,

$$\omega_1 A_1 = \omega_2 A_2. \quad (4.4)$$

ω behaves exactly analogous to the velocity field $\mathbf{v}$ of an incompressible fluid for which the continuity equation div $\mathbf{v} = 0$ holds (Fig. 1.2M). It may be pointed out that the vorticity lines and the streamlines can spiral through the tube without affecting the validity of Eqs. (4.4) and (1.15) because only the components of ω and $\mathbf{v}$ normal to the surface area appear (Fig. 5.71).

To interpret Eq. (4.1) physically it is of advantage to use the material derivative (2.1). Then, Eq. (4.1) assumes the form:

$$\frac{d\omega}{dt} = (\omega \cdot \nabla)\mathbf{v} \quad (4.5)$$

This is Helmholtz's second theorem. It states that a fluid element without vorticity will remain without it (see, however, discussion in Section 4.6). For two-dimensional flow Eq. (4.5) simplifies to

$$\frac{d\omega}{dt} = 0. \quad (4.6)$$

Thus, vorticity is carried along with a fluid element in a two-dimensional flow. To find a corresponding relation in three-dimensional flow, one may consider a vorticity tube with the infinitesimal cross section dA. Then, according to Reference 1M it is

$$\frac{d}{dt}(\omega \cdot \mathbf{n} \, dA) = \left[\frac{\partial \omega}{\partial t} - \text{curl}(\mathbf{v} \times \omega)\right] \cdot \mathbf{n} \, dA, \quad (4.7)$$

and hence

$$\frac{d}{dt}(\omega \cdot \mathbf{n} \, dA) = 0. \quad (4.8)$$

The quantity $\omega \cdot \mathbf{n} \, dA$ is carried with a fluid particle and is preserved. Vorticity can, therefore, change due to stretching as long as $\omega \cdot \mathbf{n}\, dA$ remains constant.

4.2M. KELVIN'S CIRCULATION THEOREM

The results of the previous section can be expressed in a different form that permits another physical interpretation. For an inviscid fluid, the material acceleration is, according to Eq. (2.3),

$$\frac{d\mathbf{v}}{dt} = -\frac{1}{\rho}\nabla p. \qquad (4.9)$$

If one assumes a barotropic fluid defined by $\rho = f(p)$, $\nabla P = (1/\rho)\nabla p$, then

$$\frac{d\mathbf{v}}{dt} = -\nabla P. \qquad (4.10)$$

The material change of circulation over a closed curve moving with the fluid is

$$\frac{d\Gamma}{dt} = \oint_c \frac{d\mathbf{v}}{dt}\cdot d\mathbf{r} + \oint_c \mathbf{v}\cdot d\mathbf{v}$$

$$= \oint_c \frac{d\mathbf{v}}{dt}\cdot d\mathbf{r} \qquad (4.11)$$

and because of Eq. (4.10)

$$\frac{d\Gamma}{dt} = 0. \qquad (4.12)$$

This is Kelvin's theorem, which states that the circulation along a fluid curve remains constant with time. Equation (4.12) is more general than Eq. (4.8) because it holds for barotropic fluids. For a baroclinic fluid with $\nabla\rho \times \nabla p \neq 0$, however, the material derivative of Γ is, according to Eq. (4.11),

$$\frac{d\Gamma}{dt} = -\oint \frac{\nabla p}{\rho}\cdot d\mathbf{r}, \qquad (4.13)$$

$$= \int \frac{1}{\rho^2}\nabla\rho \times \nabla p \cdot \mathbf{n}\, dA, \qquad (4.14)$$

and the right-hand term of Eq. (4.14) does not vanish. Equation (4.14) is called Bjerknes' theorem.

4.3M. EXTENSIONS

For the vorticity field the following theorems are presented. Extension of the continuity equation (2.2) and Helmholtz's second theorem (4.5) to compressible fluids yield

$$\operatorname{div}\mathbf{v} = -\frac{1}{\rho}\frac{d\rho}{dt}, \qquad (4.15)$$

$$\frac{d\boldsymbol{\omega}}{dt} = (\boldsymbol{\omega}\cdot\nabla)\mathbf{v} - \boldsymbol{\omega}\operatorname{div}\mathbf{v} + \frac{1}{\rho^2}\nabla\rho \times \nabla p, \qquad (4.16)$$

which combined give

$$\frac{d}{dt}\left(\frac{\boldsymbol{\omega}}{\rho}\right) = \left(\frac{\boldsymbol{\omega}}{\rho}\cdot\nabla\right)\mathbf{v} + \nabla\rho \times \frac{\nabla p}{\rho^3}. \qquad (4.17)$$

(Silberstein's theorem, 1896). For a barotropic fluid this equation reduces to

$$\frac{d}{dt}\left(\frac{\boldsymbol{\omega}}{\rho}\right) = \left(\frac{\boldsymbol{\omega}}{\rho}\cdot\nabla\right)\mathbf{v} \qquad (4.18)$$

whose form is analogous to the Helmholtz equation (4.5) when $\boldsymbol{\omega}$ is replaced by $\boldsymbol{\omega}/\rho$.

With the relation

$$\nabla T \times \nabla s = -\nabla\left(\frac{1}{\rho}\right) \times \nabla p, \qquad (4.19)$$

where s is the entropy, Eq. (4.17) can be written

$$\frac{d}{dt}\left(\frac{\boldsymbol{\omega}}{\rho}\right) = \left(\frac{\boldsymbol{\omega}}{\rho}\cdot\nabla\right)\mathbf{v} + \frac{1}{\rho}\nabla T \times \nabla s. \qquad (4.20)$$

(Vazsonyi's theorem, 1945) [Reference 39, Chapter 1].

For a scalar quantity G which is carried along with a fluid particle, that is,

$$\frac{dG}{dt} = 0, \qquad (4.21)$$

Eq. (4.17) can be extended to

$$\frac{d}{dt}\left(\frac{\boldsymbol{\omega}}{\rho}\cdot\nabla G\right) = \frac{1}{\rho}\nabla G \cdot \nabla p \times \nabla\frac{1}{\rho}. \qquad (4.22)$$

This is Ertel's theorem (1942) in which G may be the salinity or entropy of the fluid. If G is a conservative thermodynamic variable, the right-hand side of Eq. (4.22) is zero whether the fluid is barotropic or baroclinic:

$$\frac{d}{dt}\left(\frac{\boldsymbol{\omega}}{\rho}\cdot\nabla G\right) = 0. \qquad (4.23)$$

Entropy, for instance, is such a conservative thermodynamic variable. The quantity in parentheses, which is carried along with the fluid particle, is called "potential vorticity."

For an incompressible fluid layer of variable height H in the z direction of a Cartesian coordinate system, an important conservation law can be derived with the aid of the continuity equation which takes the form [2M]

$$\text{div } \mathbf{v} = \frac{1}{H}\frac{dH}{dt} + \frac{\partial v_x}{\partial x} + \frac{\partial v_y}{\partial y} = 0. \qquad (4.24)$$

With this relation the vorticity-transport equation for the vertical vorticity component $\omega_z = \omega$ can be written as

$$\frac{d}{dt}\left(\frac{\omega}{H}\right) = 0. \qquad (4.25)$$

In a column of fluid moving in a stream the quantity ω/H must be preserved. Stretching causes an increase in ω (Fig. 4.21).

4.4M. SOME SOLUTIONS

Example 1

In three-dimensional flow, $\boldsymbol{\omega}$ is not preserved on a fluid particle but may change its direction and magnitude according to Eqs. (4.5) and (4.7). A simple example [3M] illustrates the change of the initial direction of $\boldsymbol{\omega}$ and, hence, the production of secondary vorticity. Assume a steady swirling motion

$$rv_\varphi = f(z) \qquad (4.26)$$

Then, Eq. (4.5) yields

$$\frac{v_\varphi}{r}\frac{\partial \omega_\varphi}{\partial \varphi} + \frac{\omega_r v_\varphi}{r} = \omega_r \frac{\partial v_\varphi}{\partial r}. \qquad (4.27)$$

From Eqs. (4.26) and (4.27) it follows

$$\frac{\partial \omega_\varphi}{\partial \varphi} = -2\omega_r, \qquad (4.28)$$

so that

$$\Delta\omega_\varphi = -2\omega_r \Delta\varphi, \qquad (4.29)$$

where $\Delta\varphi$ is a small finite change in azimuth. Therefore, secondary vorticity $\Delta\omega_\varphi$ in the original streamwise direction is produced by the tilting of the vorticity lines.

Example 2

On the stagnation-point flow $v_r = -ar$, $v_z = 2az$ is superposed a solid-body rotation, which follows from

$$\frac{\partial v_\varphi}{\partial t} + v_r\frac{\partial v_\varphi}{\partial r} + v_z\frac{\partial v_\varphi}{\partial z} + \frac{v_r v_\varphi}{r} = 0 \qquad (4.30)$$

to be of the form

$$v_\varphi = r\Omega e^{2at} \qquad (4.31)$$

and

$$\omega_z = 2\Omega e^{2at}. \qquad (4.32)$$

In a stagnation-point flow for which $a < 0$ an existing solid-body rotation $\omega_z = 2\Omega$ at $t = 0$ will be suppressed with time. For $a > 0$ the meridional flow away from the surface will spin up exponentially. One may compare this solution with that of Rott, example 5 in Section 3.1M.

Example 3

As a consequence of Eq. (4.6), that is, as vorticity ω is carried along with a fluid particle, vorticity in a steady motion is conserved on a streamline. Thus, Eq. (2.16) can be written

$$\nabla^2 \psi = \omega(\psi), \qquad (4.33)$$

where ω is an arbitrary function of ψ. The simplest case is solid-body rotation with $\omega = \text{const}$. If ω is restricted to a finite circular region, one obtains the

Rankine vortex with potential flow outside the region of vorticity. Another special case is $\omega = -k^2\psi$, where k is a constant. Equation (4.33) yields in polar coordinates

$$\frac{\partial^2\psi}{\partial r^2} + \frac{1}{r}\frac{\partial\psi}{\partial r} + \frac{1}{r^2}\frac{\partial^2\psi}{\partial\varphi^2} + k^2\psi = 0. \quad (4.34)$$

A special solution to this equation is the flow past a vortex pair that is confined in a circular streamline of radius R (Fig. 5.78, when this picture is considered two-dimensional). $\psi = 0$ on the circular boundary:

$$\psi = CJ_1(kr)\sin\varphi, \qquad r \leq R \quad (4.35)$$

$$\psi = V\left(r - \frac{R^2}{r}\right)\sin\varphi, \quad r \geq R \quad (4.36)$$

with

$$J_1(kR) = 0 \quad \text{and} \quad C = \frac{2V}{kJ_0(kR)}, \quad (4.37)$$

where V is the constant velocity of the parallel flow and J_0 and J_1 are Bessel functions of the first kind. The smallest value of k satisfying $J_1(kR) = 0$ is $kR/\pi = 1.2197$. The second condition of Eq. (4.37) follows from the requirement that the tangential velocity at the circle $r = R$ is continuous for both solutions (4.35) and (4.36).

Regions of vorticity in an irrotational environment other than circular have been considered in a number of papers. The oldest goes back to Kirchhoff who studied the behavior of elliptic regions with constant ω. Such a region rotates in a fluid at rest with angular velocity $ab\omega/(a + b)^2$, where a and b are the semiaxes of the ellipse. This angular velocity assumes in the limit $a = b$ the value $\omega/4$, which one obtains from perturbation theory.

More recent studies include those by Saffman and Szeto [4M], Pierrehumbert [5M], Burbea [6M], and Zabusky [7M]. The movement of regions of vorticity can also be studied in Fig. 3.31.

Example 4

For axisymmetric flows the solutions corresponding to Eqs. (4.35) and (4.36) are found in the following way. The analogue to Eq. (4.6) in axisymmetric flow is

$$\frac{d}{dt}\left(\frac{\omega}{r}\right) = 0. \quad (4.38)$$

Then, it is in cylindrical polar coordinates

$$\left(\frac{\partial^2}{\partial r^2} - \frac{1}{r}\frac{\partial}{\partial r} + \frac{\partial^2}{\partial z^2}\right)\psi = -r^2 f(\psi) \quad (4.39)$$

with $f(\psi) = \omega/r$. For $f = A$ a constant, one arrives at Hill's spherical vortex solution:

$$\psi = \frac{A}{10}r^2(R^2 - r^2 - z^2), \quad r^2 + z^2 \leq R^2, \quad (4.40)$$

$$\psi = -\frac{V}{2}r^2\left(1 - \frac{R^3}{(r^2 + z^2)^{3/2}}\right), \quad r^2 + z^2 \geq R^2, \quad (4.41)$$

with

$$A = \frac{15}{2R^2}V. \quad (4.42)$$

The streamlines are displayed in Fig. 5.78.

It is interesting, as was pointed out by O'Brien [8M], that Hill's spherical vortex solution is also a solution of the Navier–Stokes equations. However, in considering this solution for a viscous fluid, one must keep in mind that this solution has a discontinuity at the spherical boundary $r^2 + z^2 = R^2$: The shear stress is not continuous there. Physically, one may imagine that Eqs. (4.40) and (4.41) describe two fluids of slightly different densities that do not mix.

Other steady vortex rings, in particular those of ellipsoidal shape, have been investigated by O'Brien [8M], Norbury [9M], and Durst and Schönung [10M], among others.

5M. Vortices in Potential Flow

This chapter presents solutions of the Laplace equation for potential flows that neglect dissipation, that is, the potential flow is defined by $\omega \equiv 0$ and $\nu = 0$. Vortices are then considered only in the form of singularities like point vortices, discontinuity sheets, and vortex filaments, rather than as extended regions of vorticity as described in the previous chapter. Such solutions give insight into the behavior of vortex systems.

5.1M. POINT VORTICES

A point vortex in the (x, y) plane, located at x_1, y_1, is described by $v_\varphi = \kappa_1/r$ with the stream function $\psi = \kappa_1 \log r$ where $r^2 = (x - x_1)^2 + (y - y_1)^2$. v_φ is a solution of the Laplace equation $\nabla^2 \Phi = 0$ with $v_\varphi = (1/r)\partial \Phi/\partial \varphi$. The corresponding Cartesian velocity components are

$$v_x = -\kappa_1 \frac{y - y_1}{r^2}, \qquad v_y = \kappa_1 \frac{x - x_1}{r^2}. \quad (5.1)$$

In the presence of another vortex with strength κ_2 the first vortex moves with the velocity, which the other vortex has at the location of the first vortex. Vice versa, the second vortex moves under the influence of the first vortex. Thus, the velocity components of the two vortices are:

$$(v_x)_1 = \frac{dx_1}{dt} = -\kappa_2 \frac{y_1 - y_2}{r^2},$$
$$(v_y)_1 = \frac{dy_1}{dt} = \kappa_2 \frac{x_1 - x_2}{r^2}, \quad (5.2)$$

and correspondingly

$$(v_x)_2 = \frac{dx_2}{dt} = \kappa_1 \frac{y_1 - y_2}{r^2},$$
$$(v_y)_2 = \frac{dy_2}{dt} = -\kappa_1 \frac{x_1 - x_2}{r^2}, \quad (5.3)$$

with $r^2 = (x_1 - x_2)^2 + (y_1 - y_2)^2$.

If Eq. (5.2) is multiplied by κ_1, and Eq. (5.3) by κ_2 and the two sets of equations are added, one obtains

$$\kappa_1 \frac{dx_1}{dt} + \kappa_2 \frac{dx_2}{dt} = 0, \qquad \kappa_1 \frac{dy_1}{dt} + \kappa_2 \frac{dy_2}{dt} = 0$$

or

$$\kappa_1 x_1 + \kappa_2 x_2 = \text{const},$$
$$\kappa_1 y_1 + \kappa_2 y_2 = \text{const}. \quad (5.4)$$

Dividing by $\kappa_1 + \kappa_2$ gives

$$x_s = \frac{\kappa_1 x_1 + \kappa_2 x_2}{\kappa_1 + \kappa_2}, \qquad y_s = \frac{\kappa_1 y_1 + \kappa_2 y_2}{\kappa_1 + \kappa_2}. \quad (5.5)$$

The point (x_s, y_s) is always at rest during the motion and is called the "center of the vortex system." It is analogous to the center of gravity when κ is replaced by the mass [1M].

Subtraction of Eq. (5.3) from Eq. (5.2) yields

$$\frac{d}{dt}(x_1 - x_2) = -(\kappa_1 + \kappa_2)\frac{y_1 - y_2}{r^2}, \quad (5.6)$$

$$\frac{d}{dt}(y_1 - y_2) = (\kappa_1 + \kappa_2)\frac{x_1 - x_2}{r^2}. \quad (5.7)$$

268 Vortices in Potential Flow

Multiplication of Eq. (5.6) by $x_1 - x_2$ and of Eq. (5.7) by $y_1 - y_2$ and summation, results in $r = $ const. Hence, the distance between the two vortices remains unaltered. From this fact and from Eq. (5.5) it follows that the two vortices revolve around (x_s, y_s) with constant r and constant angular velocity $(\kappa_1 + \kappa_2)/r^2$, where r is the distance between the two vortices (Figs. 3.25b and 3.27).

In the special case $\kappa_1 = -\kappa_2$ the center of the vortex system is at infinity. If one places the two vortices at a distance $s = x_2 - x_1$ on the axis $y = 0$, then

$$(v_x)_1 = (v_x)_2 = 0, \quad (v_y)_1 = (v_y)_2 = \frac{\kappa_1}{s}. \quad (5.8)$$

This means that the two vortices move parallel to the y axis (Fig. 3.25a).

The discussion is now extended to n point vortices [1M]. From the velocity components of the ith vortex

$$\frac{dx_i}{dt} = -\sum_{k=1}^{n} \kappa_k \frac{y_i - y_k}{r_{ik}^2},$$
$$\frac{dy_i}{dt} = \sum_{k=1}^{n} \kappa_k \frac{x_i - x_k}{r_{ik}^2}, \quad (5.9)$$

with $r_{ik}^2 = (x_i - x_k)^2 + (y_i - y_k)^2$, one derives in a way analogous to that used for the problem of two vortices the following constants of motion, which are independent of time [2M]:

Center of vortex system:

$$x_s = \frac{\sum_{1}^{n} \kappa_i x_i}{\sum_{1}^{n} \kappa_i}, \quad y_s = \frac{\sum_{1}^{n} \kappa_i y_i}{\sum_{1}^{n} \kappa_i}. \quad (5.10)$$

Moment of vortex system:

$$M = 2\pi \sum_{1}^{n} \kappa_i (x_i^2 + y_i^2). \quad (5.11)$$

Angular momentum of vortex system:

$$A = \frac{1}{2} \sum_{i} \sum_{k} \kappa_i \kappa_k r_{ik}^2. \quad (5.12)$$

"Stream function" of vortex system:

$$H = \sum_{i} \sum_{k} \kappa_i \kappa_k \log r_{ik}, \quad i \neq k. \quad (5.13)$$

Except for a few special cases (two examples are given below) the calculation of a system of vortices for $n > 2$ requires the aid of computers. The general cases of $n = 3$ and $n = 4$ are already quite complicated [3M, 4M].

Example 1

For an infinite straight row of equidistant point vortices, each of strength κ, a closed-form solution exists [Reference 2M, Chapter 1M]:

$$v_x = -\frac{\kappa}{2a} \frac{\sinh(2\pi y/a)}{\cosh(2\pi y/a) - \cos(2\pi x/a)},$$
$$v_y = \frac{\kappa}{2a} \frac{\sin(2\pi x/a)}{\cosh(2\pi y/a) - \cos(2\pi x/a)}. \quad (5.14)$$

For $y = \pm\infty$: $v_x = \mp\kappa/2a$, $v_y = 0$. The row of point vortices thus approximates a discontinuity line with the velocity jump κ/a (Fig. 5.1M).

Example 2

The simple circular configuration of point vortices shown in Fig. 5.2M also permits a closed-form solution for the rotational velocity of this system. If the circle contains n vortices, each with the strength $\kappa_i = \kappa/n$, the velocity of the ith vortex at the point $(R, 0)$ is

$$\frac{dx_i}{dt} = 0, \quad \frac{dy_i}{dt} = \frac{\kappa}{2n} \frac{n-1}{R}. \quad (5.15)$$

Hence, the vortices revolve around the center with the velocity

$$v_\varphi = \frac{\kappa}{2R} \frac{n-1}{n}. \quad (5.16)$$

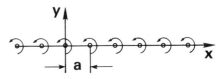

Figure 5.1M. Row of point vortices with distance a between each vortex.

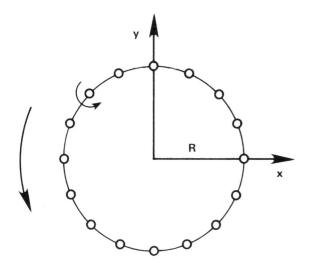

Figure 5.2M. Circular arrangement of n vortices of equal strength and equal direction.

For $n = \infty$ the tangential velocity is $v_\varphi = \kappa/2R$.

Computer-generated solutions of more complex problems are given in Reference 39, Chapter 3; Reference 54, Chapter 5; and 5M.

5.2M. POINT VORTICES AS A HAMILTONIAN SYSTEM

Kirchhoff [6M] found that Eq. (5.9) can be written in the form of Hamilton's canonical equations of mechanics with H as defined in Eq. (5.13) being the Hamiltonian:

$$\kappa_i \frac{dx_i}{dt} = -\frac{\partial H}{\partial y_i}, \qquad \kappa_i \frac{dy_i}{dt} = \frac{\partial H}{\partial x_i}. \qquad (5.17)$$

From the invariance (or symmetry) properties of H the constants of motion (5.10) through (5.13) can be derived. The existence of the Hamiltonian opens the door for using the theory of Hamiltonian mechanics including its statistical treatment [7M]. In particular, the system of vortices behaves according to Liouville's theorem like an incompressible fluid, that is, the size of the area occupied by the vortices does not change with time. Flow regimes with solid boundaries are treated by Lin [8M].

5.3M. SPIRAL DISCONTINUITY LINES

The development of a spiral discontinuity line, resulting from the abrupt start of a thin plate and representing the roll up of a vortex layer at high Reynolds numbers (Fig. 5.34), may be treated as a singular initial and boundary-value problem for the Laplace equation. However, the formulation of this problem at $t = 0$ is not known; even the existence of a solution of such an initial and boundary problem is debated (Section 4.6). To overcome this deficiency, Prandtl hypothesized that near $t = 0$ the discontinuity line must be similar at two instants t_1 and t_2. The resulting "self-similar" shape of the discontinuity line is then valid for $t = \varepsilon \ll 1$, but $\varepsilon \neq 0$.

This similarity property may also be expressed in the way that the discontinuity line is always similar to itself behind a semiinfinite plate. If the plate has a finite width, the discontinuity lines at various times are only similar as long as the influence of the other edge is not felt.

Before a limit is given when this influence is observed, the similarity hypothesis for the roll up of a discontinuity line is as follows:

$$L_1 : L_2 = (t_1 : t_2)^n \qquad (5.18)$$

where L_1 and L_2 are the distances between two points of a discontinuity line that remain in similar positions at the times t_1 and t_2; n is to be determined.

The solution of the potential flow about a semiinfinite plate yields, after some analysis, for the velocity of the two points,

$$V_1 : V_2 = (L_2 : L_1)^{1/2} = (t_1 : t_2)^{-n/2}. \qquad (5.19)$$

The relation $dL/dt = V$ then determines n to be $\frac{2}{3}$. Kaden [9M] found a first approximation for the shape of the spiral close to the center:

$$r = \text{const}\left(\frac{t}{\varphi}\right)^{2/3}. \qquad (5.20)$$

This spiral has infinitely many windings when $r \to 0$. Better approximations must be sought if the spiral is to originate at the edge of the plate. Wedemeyer [Reference 27, Chapter 5] computed such spirals for the semiinfinite plate and for the finite plate. A comparison shows that deviations of the two solutions start at about the dimensionless time $tV/D = 0.25$ with D the width of the finite plate.

More recently, Pullin [10M] made extensive computer studies on the spiral form of discontinuity lines for various wedge angles.

In reality, of course, the core of a vortex rotates like a solid body (Section 3.1M). Models that match the spiral discontinuity line with the viscous core by means of a thickened shear layer between them are given by Moore and Saffman [Reference 29, Chapter 5].

5.4M. VORTEX LINES AND TUBES

The point-vortex model may be imagined in three-dimensional space as the cross section of a cylindrical vortex consisting of an infinitely long and straight vortex line (perpendicular to the plane). To compute the flow characteristics of vortex lines of finite length, straight or curved, an exact analogy in the theory of electromagnetism, called the "Biot–Savart law," can be used. Applied to potential flow it is

$$\mathbf{v} = -\frac{\Gamma}{4\pi} \int \frac{\mathbf{r} \times d\mathbf{s}}{r^3}, \quad (5.21)$$

and for an element ds it follows (Fig. 5.3M) that

$$dv = \frac{\Gamma}{4\pi r^2} \sin\gamma \, ds. \quad (5.22)$$

For a straight vortex line from A to B one obtains (Fig. 5.4M):

$$v = \frac{\Gamma}{4\pi r} \int_{\pi - \gamma_1}^{\gamma_2} \sin\gamma \, d\gamma$$
$$= \frac{\Gamma}{4\pi r}(\cos\gamma_1 + \cos\gamma_2), \quad (5.23)$$

where the velocity v at point C is perpendicular to the plane $A - B - C$. For an infinitely long line $\gamma_1 = \gamma_2 = 0$, one recovers the "point vortex" $v_\varphi = \Gamma/2\pi r$.

Figure 5.3M. Illustration of the Biot–Savart law.

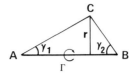

Figure 5.4M. Vortex line A-B.

For a horseshoe vortex of circulation Γ, representing a wing of finite span with the bound vortex line $A - B$ of length s and the two semiinfinite tip vortices emanating from A and B, the velocity field at a point x, y, z (Fig. 5.5M) is with

the bound vortex: $(v_x)_B; (v_y)_B = 0; (v_z)_B;$

and the tip vortices: $(v_x)_T = 0; (v_y)_T; (v_z)_T:$

(5.24)

$$v_x = (v_x)_B$$
$$v_y = (v_y)_T \quad (5.25)$$
$$v_z = (v_z)_B + (v_z)_T$$

where

$$(v_x)_B = \frac{\Gamma}{4\pi} \frac{z}{x^2 + z^2} \left[\frac{\frac{s}{2} - y}{r(x, -y, z)} + \frac{\frac{s}{2} + y}{r(x, y, z)} \right],$$

$$(v_y)_T = \frac{\Gamma}{4\pi} \left\{ \frac{z}{z^2 + \left(\frac{s}{2} + y\right)^2} \left[1 + \frac{x}{r(x, y, z)}\right] \right.$$
$$\left. - \frac{z}{z^2 + \left(\frac{s}{2} - y\right)^2} \left[1 + \frac{x}{r(x, -y, z)}\right] \right\},$$

$$(v_z)_B = \frac{\Gamma}{4\pi} \frac{-x}{x^2 + z^2} \left[\frac{\frac{s}{2} - y}{r(x, -y, z)} + \frac{\frac{s}{2} + y}{r(x, y, z)} \right],$$

$$(v_z)_T = \frac{\Gamma}{4\pi} \left\{ \frac{-\left(\frac{s}{2} + y\right)}{z^2 + \left(\frac{s}{2} + y\right)^2} \left[1 + \frac{x}{r(x, y, z)}\right] \right.$$
$$\left. - \frac{\frac{s}{2} - y}{z^2 + \left(\frac{s}{2} - y\right)^2} \left[1 + \frac{x}{r(x, -y, z)}\right] \right\}$$

(5.26)

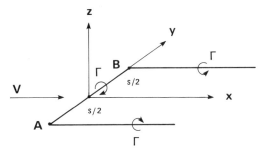

Figure 5.5M. Horseshoe vortex as a wing model with finite span.

with $r(x, y, z) = (x^2 + (s/2 + y)^2 + z^2)^{1/2}$ [11M].

In wing theory the velocity field at the bound-vortex position $x = z = 0$ is of interest. Here, the bound vortex itself does not contribute, and the trailing vortices cause a vertical velocity $(v_z)_T$ downward only, called "downwash":

$$v_z = \frac{-\Gamma}{4\pi}\left(\frac{1}{\frac{s}{2}+y} + \frac{1}{\frac{s}{2}-y}\right). \quad (5.27)$$

This concept will be taken up in Section 5.5M.

The model of the horseshoe vortex assumes rigid vortex lines, which do not change their relative positions. In general, however, straight or curved vortex lines and tubes influence each other by "inducing" a velocity component perpendicular to the plane of curvature (if the curve lies in a plane) or in the binormal direction.

An example of this is the movement of a circular vortex filament, which is a thin vortex ring. The solution to this problem can also be found from the Biot–Savart law [Reference 2M, Chapter 1M]. In addition to the occurrence of an "induced" velocity perpendicular to the plane of the circle, that is, a translational velocity V with which the vortex ring travels, the solution also shows that a curved vortex filament must have a nonzero core radius a:

$$V = \frac{\Gamma}{4\pi R}\left(\log\frac{8R}{a} - c\right), \quad (5.28)$$

where R is the radius of the vortex ring and c is a constant which depends on the vorticity distribution inside the core. For constant vorticity it is $c = \frac{1}{4}$, and for a hollow core $c = \frac{1}{2}$ [Reference 2M, Chapter 1M]. It is readily seen that if $a \to 0$, the velocity $V \to \infty$.

A core of elliptic cross section with constant vorticity will rotate with the period $\pi^2(a+b)^2/\Gamma$.

Moore [12M] has shown that this period is short compared to the time taken for the vortex ring to move a distance of the order of its radius. The average translational velocity $\overline{V}$ is (with a and b being the semiaxes of the ellipse):

$$\overline{V} = \frac{\Gamma}{4\pi R}\left(\log\frac{16R}{a+b} - \frac{1}{4}\right). \quad (5.29)$$

This velocity is a little smaller than that of a vortex ring with a circular core if cores of equal areas are compared. Detailed discussions of vortex rings are given in Reference 8, Chapter 2 and Reference 26, Chapter 3.

For arbitrarily curved vortex tubes numerical methods must be applied [13M]. As a first approximation, derived by Arms from the Biot–Savart law by a Taylor expansion and called the "localized induction equation" [14M], the following formula holds for the induced velocity $\mathbf{V}$:

$$\mathbf{V} = \mathbf{b}\frac{\Gamma}{4\pi R}\left[\log\frac{1}{a} + O(1)\right], \quad (5.30)$$

where R is the local radius of the curve and $\mathbf{b}$ is the unit vector in the binormal direction.

5.5M. VORTEX CONCEPT IN WING THEORY

The horseshoe vortex of the previous section serves as the simplest model for a wing of finite span. The trailing vortices cause a downward deflection of the fluid flow whose vertical velocity component is the downwash (5.27).

In the general case of a wing with circulation distribution $\Gamma(y)$, the downwash at a point y of a span section dy located at η is

$$dv_z = \frac{d\Gamma(\eta)}{4\pi(y-\eta)} \quad (5.31)$$

and the total downwash of the wing with span s is

$$v_z(y) = \frac{1}{4\pi}\int_{-s/2}^{+s/2}\frac{1}{y-\eta}\frac{d\Gamma}{d\eta}\,d\eta. \quad (5.32)$$

For discrete sections $\Delta y = \Delta s$, the system of horseshoe vortices is depicted in Fig. 5.6Mb ("lifting-line" model).

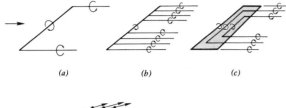
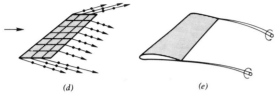

Figure 5.6M. Vortex models of wings with increasing complexity: (a) horseshoe vortex, (b) lifting-line vortex, (c) wing-sheet model, (d) vortex lattice, and (e) vortex tubes for trailing vortices (from Reference 54, Chapter 5).

Lift L and circulation Γ are related to each other by the Kutta–Joukowsky formula for a wing section dy:

$$dL = \rho V \Gamma(y)\, dy, \qquad (5.33)$$

and, since lift and drag are connected by the downwash through

$$\frac{dD}{dL} = \frac{v_z}{V}, \qquad (5.34)$$

the drag of a wing section dy is

$$dD = \frac{1}{V}\frac{dL}{dy} v_z(y)\, dy = \rho \Gamma v_z\, dy. \qquad (5.35)$$

The Γ distribution over the wing depends on the geometry of the wing and the angle of attack.

Refinements of the lifting-line model can be made by considering a nonzero chord (Fig. 5.6Mc) or by using a "vortex-lattice" method in which the wing and the trailing vortices are approximated by cells consisting of vortex lines (Fig. 5.6Md). The trailing vortices themselves can be modeled by vortex tubes so that the interaction of long vortex wakes behind aircraft can be studied (Fig. 5.6Me) [11M].

6M. Vortices in Slow Motion

Solutions of the fluid-flow equations for Re = 0 are of practical interest in the areas of steady and unsteady slow motion (in reality for Re < 1) and in the final stage of decay of flows at initially larger Reynolds numbers.

Neglect of the convection terms renders the Navier–Stokes equations linear. Superposition of solutions is now possible, but the single modes are independent of each other and no coupling through nonlinearity can occur.

For two-dimensional flows in the (x, y) plane the slow-motion equations in the form of the vorticity-transport equation reduce to

$$\frac{\partial \omega}{\partial t} = \nu \nabla^2 \omega, \tag{6.1}$$

or, in form of the stream function,

$$\frac{\partial \nabla^2 \psi}{\partial t} = \nu \nabla^2 \nabla^2 \psi. \tag{6.2}$$

The diffusion equation (6.1) cannot be solved easily because meaningful boundary conditions cannot be prescribed for ω. Thus, it is advantageous to solve Eq. (6.2) for which boundary conditions can be given. Vortices for Re = 0 are called "Stokes vortices."

6.1M. PLANE VORTICES

For the study of vortices solutions of Eq. (6.2) are sought in polar coordinates r, φ. From the complete sets of separable solutions $\psi = h(\varphi)G(r, t)$ found in References 1M and 2M, the following examples are selected.

Example 1

For unsteady fluid motion in a bounded circular domain with nonslip on the circle $r = R$ (Lamb's problem), one obtains the following eigenfunctions:

$$\psi_{mn} = A_{mn}\cos(n\varphi + \delta_{mn})$$
$$\times \left[J_n(\lambda_{mn}r) - J_n(\lambda_{mn}R)\left(\frac{r}{R}\right)^n\right] \tag{6.3}$$
$$\times \exp(-\lambda_{mn}^2 \nu t),$$

with some amplitude A_{mn} and phase δ_{mn}. The J_n are Bessel functions with $n = 0, 1, 2, \cdots$. The eigenvalues λ_{mn} ($m = 1, 2, 3, \cdots$) are determined for given n by the condition

$$J_{n+1}(\lambda_{mn}R) = 0. \tag{6.4}$$

Some data are given for $\lambda_{mn}R$:

	$m = 1$	$m = 2$
$n = 0$:	3.832	7.016
$n = 1$:	5.136	8.417
$n = 2$:	6.380	9.761

It is easily seen that Eq. (6.3) describes the motion of $2mn$ vortices that decay exponentially with advancing time. Their centers lie on the intersections of m coaxial circles and $2n$ radial lines. The damping coefficient $\lambda_{mn}^2 \nu$ increases with m and n from small to large values. Therefore, the vortices of low total orders $(m + n)$ decay very slowly, whereas those of high orders disappear rapidly. The spatial extent of the vortices is relatively small for high $(m + n)$ and large for small $(m + n)$. This analysis explains the mechanism of a decaying circular discontinuity line in its final stage as illustrated in Fig. 6.8 for $m = 1$, and $n = 4$ and $n = 3$.

Example 2

Other classes of solutions of vortical motions are found by differentiating or integrating Eq. (6.3). As an example, solutions for an infinite domain can be found by integrating Eq. (6.3) for arbitrary λ over a continuous spectrum of λ. Closed-form solutions are obtained with the special weighting function λ^{n+1}:

$$G_n = A_n \int_0^\infty \lambda^{n+1} J_n(\lambda r) e^{-\lambda^2 \nu t} d\lambda \quad (6.5)$$

$$= A_n \frac{r^n}{(2\nu t)^{n+1}} e^{-r^2/4\nu t}.$$

These eigenfunctions can be superposed, for instance, on the decaying potential vortex (3.14) to include secondary vortices as in example 1.

Separable solutions of (6.2) for sectorial regions also reveal vortices between the solid walls of a dihedral angle as shown in Fig. 6.1M. These vortices have been studied in References 2M–4M. Note that such corner vortices become visible in numerical calculations or in experiments with better resolution of the flow field. Typical examples are seen in Figs. 5.27 and 7.27.

Vortices can also occur in slow motions around solid bodies (external flows). This is somewhat surprising, since the flow past a sphere or even past an infinitely thin disk normal to a flow does not exhibit flow separation at Re = 0. However, for lens-shaped bodies with concave surfaces, flow separation can occur. An analytical solution was constructed by Dorrepaal [Reference 12, Chapter 5] (Fig. 5.17).

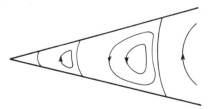

Figure 6.1M. Stokes vortices in a corner. Near the apex, infinitely many vortices occur.

6.2M. AXISYMMETRIC VORTICES

Separable solutions for three-dimensional flows can be constructed in a way similar to those described in the preceding section for plane motions. For instance, eigenfunctions for conical regions are discussed in References 5M and 6M. The structure of vortices is similar in sectorial and conical regions, and the same is true for circular and axisymmetric spherical regions [Reference 2M, Chapter 1M].

The properties of axisymmetric vortices in an infinite region are now briefly discussed. These vortices can occur in the final state of decay of turbulent flows [7M].

For axisymmetric slow motion, expressed in spherical polar coordinates r, θ, λ with $x = r \sin\theta \cos\lambda$, $y = r\sin\theta\sin\lambda$, $z = r\cos\theta$, the basic equation (6.2) must be replaced by

$$D^2 \left(D^2 - \frac{1}{\nu} \frac{\partial}{\partial t} \right) \tilde{\psi} = 0, \quad (6.6)$$

$$\left(D^2 - \frac{1}{\nu} \frac{\partial}{\partial t} \right) (r\sin\theta \, v_\lambda) = 0, \quad (6.7)$$

where

$$D^2 \equiv \frac{\partial^2}{\partial r^2} + \frac{1}{r^2} \frac{\partial^2}{\partial \theta^2} - \frac{\cot\theta}{r^2} \frac{\partial}{\partial \theta}, \quad (6.8)$$

and

$$v_r = -\frac{1}{r^2 \sin\theta} \frac{\partial \tilde{\psi}}{\partial \theta}, \quad v_\theta = \frac{1}{r\sin\theta} \frac{\partial \tilde{\psi}}{\partial r} \quad (6.9)$$

under the axisymmetry condition $\partial/\partial\lambda \equiv 0$.

Phillips [7M] found a solution of Eq. (6.6) which represents a decaying vortex ring with a dipole as initial condition at $t = 0$:

$$\tilde{\psi} = \text{const} \frac{\sin^2\theta}{r} \left[\sqrt{\pi} \, \text{erf}\left(\frac{\xi}{2}\right) - \xi e^{-\xi^2/4} \right], \quad (6.10)$$

with $\xi = r/\sqrt{\nu t}$. The center of the vortex ring is at $\xi = 3.02$, $\theta = \pi/2$.

Differentiation of Eq. (6.10) with respect to time results in another decaying vortex according to Lugt [8M],

$$\tilde{\psi} = \text{const} \frac{r^2}{t^{5/2}} \sin^2\theta \, e^{-\xi^2/4} \qquad (6.11)$$

with the center of the vortex ring at $\xi = 2$, $\theta = \pi/2$. It is interesting that the locations of extremal vorticity for the vortices (6.10) and (6.11) do not coincide with the centers of the vortex ring (centers of nested closed streamlines), contrary to common conception of vortices.

Solutions of Eq. (6.7) represent decaying spherical shells. Of more practical interest is a steady-state solution of Eq. (6.7), which describes a rotating sphere of radius R in a fluid at rest:

$$v_\lambda = \Omega \frac{R^3}{r^2} \sin\theta,$$
$$v_r = v_\theta \equiv 0, \qquad (6.12)$$

where Ω is the angular velocity of the sphere [Reference 2M, Chapter 1M]. The streamlines are concentric circles around the axis of rotation parallel to the equatorial plane. A comparison with experiments shows that the formula for the torque $T = 8\pi\mu R^3 \Omega$ computed from Eq. (6.12) is valid for Re < 10 [Reference 79, Chapter 6]. The important feature of Eq. (6.12) is the absence of any secondary flow v_r, v_θ. However, a perturbation analysis for larger Re will yield a secondary flow as depicted in Fig. 6.35.

7M. Rotating Fluid Systems

7.1M. CENTRIFUGAL AND CORIOLIS FORCES

The equations of motion presented in Chapter 2M are based on Newton's second law and are valid only in an inertial system. In a rotating system, such as one involving the movements of bodies on the surface of the earth, it is often more advantageous to have the reference frame fixed to that rotating system. Then, the acceleration term in Newton's second law must be augmented:

$$\left(\frac{d\mathbf{v}}{dt}\right)_{\text{inert}} = \left(\frac{d\mathbf{v}}{dt}\right)_{\text{rot}} + \mathbf{\Omega} \times (\mathbf{\Omega} \times \mathbf{r}) + 2\mathbf{\Omega} \times \mathbf{v} \tag{7.1}$$

(with $\mathbf{\Omega}$ = const). The two terms on the right are the centripetal and the Coriolis accelerations, respectively. With a change in sign, they may be considered as fictitious forces, that is, centrifugal and Coriolis forces in a rotating system. The Navier–Stokes equations are then

$$\frac{\partial \mathbf{v}}{\partial t} + (\mathbf{v} \cdot \nabla)\mathbf{v} + 2\mathbf{\Omega} \times \mathbf{v} + \mathbf{\Omega} \times (\mathbf{\Omega} \times \mathbf{r})$$
$$= -\frac{1}{\rho}\nabla p + \nu \nabla^2 \mathbf{v}, \tag{7.2}$$

$$\text{div } \mathbf{v} = 0. \tag{2.2}$$

Since the centrifugal force has a potential, the vorticity-transport equation in a rotating frame does not contain the contribution from this force:

$$\frac{\partial \boldsymbol{\omega}}{\partial t} + (\mathbf{v} \cdot \nabla)\boldsymbol{\omega} - [(\boldsymbol{\omega} + 2\mathbf{\Omega}) \cdot \nabla]\mathbf{v} = \nu \nabla^2 \boldsymbol{\omega}, \tag{7.3}$$

or

$$\frac{\partial}{\partial t}(\boldsymbol{\omega} + 2\mathbf{\Omega}) + (\mathbf{v} \cdot \nabla)(\boldsymbol{\omega} + 2\mathbf{\Omega})$$
$$- [(\boldsymbol{\omega} + 2\mathbf{\Omega}) \cdot \nabla]\mathbf{v} \tag{7.4}$$
$$= \nu \nabla^2 (\boldsymbol{\omega} + 2\mathbf{\Omega}),$$

which has the same form as Eq. (2.13) in an inertial frame.

Introduction of dimensionless quantities into Eq. (7.3), as has been done in Section 2.4M, results with $t = t^*/\Omega$ and $\mathbf{\Omega} = \Omega \mathbf{k}$ in:

$$\frac{\partial \boldsymbol{\omega}^*}{\partial t^*} + \text{Ro}[(\mathbf{v}^* \cdot \nabla^*)\boldsymbol{\omega}^* - (\boldsymbol{\omega}^* \cdot \nabla^*)\mathbf{v}^*]$$
$$- 2\mathbf{k} \cdot \nabla^* \mathbf{v}^* \tag{7.5}$$
$$= \text{Ek } \nabla^{*2}\boldsymbol{\omega}^*,$$

where the Rossby number Ro = $V/L\Omega$ and the Ekman number Ek = $\nu/L^2\Omega$ appear [Reference 40, Chapter 1].

7.2M. HYPERBOLICITY AND TAYLOR–PROUDMAN THEOREM

Neglect of the nonlinear terms in Eq. (7.3), or, equivalently, assumption of small Rossby number in Eq. (7.5), leads to

$$\left(\frac{\partial}{\partial t} - \nu \nabla^2\right)\boldsymbol{\omega} = (2\mathbf{\Omega} \cdot \nabla)\mathbf{v}. \tag{7.6}$$

Application of the operator $(\partial/\partial t - \nu \nabla^2)\nabla^2$

eliminates **v** [1M]:

$$\left(\frac{\partial}{\partial t} - \nu\nabla^2\right)^2 \nabla^2\omega + (2\mathbf{\Omega}\cdot\nabla)^2\omega = 0. \quad (7.7)$$

The first term contains the operators $\partial/\partial t$ and $\partial^2/\partial t^2$, indicating diffusion and the possible existence of waves, respectively; the latter indicates hyperbolic-type behavior despite the fact that the diffusion operator is of highest order in the derivatives of the space coordinates. If the viscous term becomes very small or even vanishes, then Eq. (7.7) reduces to

$$\frac{\partial^2}{\partial t^2}\nabla^2\omega + (2\mathbf{\Omega}\cdot\nabla)^2\omega = 0. \quad (7.8)$$

Although the hyperbolicity of this equation can be analyzed as "inertia waves" (see, e.g., Reference 40, Chapter 1), a simpler example leads directly to the following important geophysical concept of "Rossby wave."

For an inviscid fluid on the earth's surface, where $\omega_z = \omega$ is the vorticity component perpendicular to the surface and $f = 2\Omega\sin\phi$ is the Coriolis parameter, Eq. (7.4) yields

$$\frac{d}{dt}(\omega + f) = 0. \quad (7.9)$$

Since the latitude ϕ may be expressed as $(\phi_0 + y/R)$, where ϕ_0 is some reference latitude, R is the radius of the earth, y is the horizontal coordinate directed northward measured from ϕ_0, the following approximation holds:

$$f = f_0 + \beta y, \quad (7.10)$$

where $f_0 = 2\Omega\sin\phi_0$ and $\beta = 2\Omega\cos\phi_0/R$ are constants. Then, Eq. (7.9) simplifies to

$$\frac{d\omega}{dt} + \beta v_y = 0. \quad (7.11)$$

Use of perturbation method with $v_x = V + v_x'$, $v_y = v_y'$, where V is the constant mean west–east flow velocity, results in a wave motion

$$v_y' = A\sin\frac{2\pi}{\lambda}(x - Ut), \quad (7.12)$$

$$U = V - \frac{\beta\lambda^2}{4\pi^2}, \quad (7.13)$$

where λ is the wavelength, U is the phase velocity, and A is some amplitude. These waves, called "Rossby waves" [2M], are sketched in Fig. 7.9.

An extension of Eqs. (4.25) and (7.9) results in the conservation of the potential vorticity $(\omega + f)/H$:

$$\frac{d}{dt}\left(\frac{\omega + f}{H}\right) = 0. \quad (7.14)$$

A steady, inviscid fluid flow is described according to Eq. (7.6) by

$$(\mathbf{\Omega}\cdot\nabla)\mathbf{v} = 0, \quad (7.15)$$

which is known as the Taylor–Proudman theorem [Reference 40, Chapter 1]. The velocity field is independent of the coordinate parallel to the rotation axis. For instance, if Ω is the angular velocity about the z axis, Eq. (7.15) yields

$$\Omega\frac{\partial}{\partial z}\mathbf{v} = 0, \quad (7.16)$$

where v is a function only of x and y. A physical example is the Taylor–Proudman column in Fig. 7.13.

7.3M. GEOSTROPHIC VORTICES

A fluid motion in which the Coriolis force and the pressure gradient balance each other is called "geostrophic":

$$2\mathbf{\Omega}\times\mathbf{v} = -\frac{1}{\rho}\nabla p. \quad (7.17)$$

When on the earth's surface the vertical component of this equation is approximated by the "hydrostatic equilibrium"

$$\frac{\partial p}{\partial z} = -\rho g, \quad (7.18)$$

the remaining horizontal components are

$$\rho f v_y = \frac{\partial p}{\partial x},$$
$$\rho f v_x = -\frac{\partial p}{\partial y}, \quad (7.19)$$

and the velocity field is called the "geostrophic wind." f is assumed to be constant.

Since according to Eq. (2.17) $v_x = -\partial\psi/\partial y$ and $v_y = \partial\psi/\partial x$, it follows that

$$\nabla\psi = \frac{1}{\rho f}\nabla_H p, \qquad (7.20)$$

where ∇_H refers to the horizontal coordinates only. Streamlines of a geostrophic wind are thus isobars, a result that is of great importance for weather maps.

Stewart [3M] and Morikawa [4M] developed a theory of point vortices based on the geostrophic assumption. With this assumption a perturbation analysis of the conservation law of potential vorticity (7.14) yields

$$\frac{d}{dt}\left(\omega - \frac{fH}{H_0}\right) = 0, \qquad (7.21)$$

where H_0 is a reference height. If the geostrophic stream function is introduced with

$$\psi = \frac{g}{f}H, \qquad (7.22)$$

one arrives at the "geostrophic conservation law"

$$\frac{d}{dt}(\nabla^2 - k^2)\psi = 0, \qquad (7.23)$$

where $k^2 = f^2/gH_0$.

A solution of Eq. (7.23), analogous to the solution of a point vortex in an inertial frame (5.1), exists with many properties of a point vortex. The Helmholtz equation (7.23) in polar coordinates is

$$\left(\frac{\partial^2}{\partial r^2} + \frac{1}{r}\frac{\partial}{\partial r} + \frac{1}{r^2}\frac{\partial^2}{\partial\varphi^2} - k^2\right)\psi = \Gamma\delta(r), \qquad (7.24)$$

where $\delta(r)$ is the delta function, and the desired solution is

$$\psi = \frac{\Gamma}{2\pi}K_0(kr). \qquad (7.25)$$

$K_0(kr)$ is the modified Bessel function of the second kind and zeroth order. Equation (7.25) is the geostrophic vortex that corresponds to $\psi = (\Gamma/2\pi)\log r$ in an inertial frame. In fact, Eq. (7.25) has the following properties:

for $r \to 0$:

$$v_\varphi \sim \frac{1}{r} \qquad (7.26)$$

for $r \to \infty$:

$$v_\varphi \sim \frac{1}{\sqrt{kr}}e^{-kr} \qquad (7.27)$$

$$v_\varphi = \frac{1}{\rho f}\frac{\partial p}{\partial r}. \qquad (7.28)$$

Near $r = 0$ the geostrophic vortex behaves like a point vortex in an inertial frame. The direction of the rotating fluid depends on the sign of the pressure gradient, that is, a high-pressure system ($\partial p/\partial r < 0$) causes anticyclonic rotation ($v_\varphi < 0$), a low-pressure system ($\partial p/\partial r > 0$) causes cyclonic rotation ($v_\varphi > 0$). The vorticity, however, is not restricted to a point but is $\omega = k^2\psi$.

Geostrophic vortices have properties analogous to those of point vortices, as described in Sections 5.1M and 5.2M. In particular, a system of vortices can be analyzed in the same way, and a Hamiltonian function also exists [5M].

7.4M. GRADIENT, INERTIA, AND CYCLOSTROPHIC MOTIONS

For fluid motions with curved streamlines, as for vortices, the "gradient motion" is sometimes a better approximation than the geostrophic flow [6M]. In circular gradient motion, which is an exact solution to the Euler equation, Coriolis, centrifugal, and pressure-gradient forces are in equilibrium:

$$fv_\varphi + \frac{v_\varphi^2}{r} = \frac{1}{\rho}\frac{\partial p}{\partial r}, \qquad (7.29)$$
$$v_r = 0.$$

The solution is [6M]

$$v_\varphi = -\frac{fr}{2} \pm \left(\frac{f^2r^2}{4} + \frac{r}{\rho}\frac{\partial p}{\partial r}\right)^{1/2}. \qquad (7.30)$$

First, the positive root of Eq. (7.30) is considered. As in the case of geostrophic vortices, a high-pressure system causes an anticyclonic motion, a low-pressure system a cyclonic one. For vanishing pressure gradient, the velocity $v_\varphi \to 0$. Imaginary roots, which are possible for high-pressure systems, are excluded. Solutions of Eq. (7.30) for positive and negative $\partial p/\partial r$ with negative roots, although mathematically possible, have either not been observed or are considered anomalous [6M].

The special case of vanishing pressure gradient permits, in addition to the solution $v_\varphi = 0$, another solution with

$$v_\varphi = -fr. \qquad (7.31)$$

This flow is called "inertia motion" and has been observed in the ocean (Fig. 7.7). The frequency is $f/2\pi$, which is twice the frequency of the Foucault pendulum [Reference 10, Chapter 7].

Near the equator, where f is small, Eq. (7.29) reduces to the "cyclostrophic" motion

$$\frac{v_\varphi^2}{r} = \frac{1}{\rho}\frac{\partial p}{\partial r}, \qquad (7.32)$$

which is the same as the equation for a circular motion in an inertial frame. $\partial p/\partial r$ is always positive, no matter which sign v_φ has.

7.5M. THE EKMAN LAYER

The effect of rotation on boundary layers is briefly discussed by means of a closed-form solution of the Navier–Stokes equations under simplifying assumptions.

Consider a steady Ekman layer in which the velocity components v_x, v_y, v_z are functions of the vertical component z only. The continuity equation (2.2) together with the nonslip condition immediately gives $v_z = 0$. Since the horizontal pressure gradient is independent of z and is determined by the geostrophic flow outside the boundary layer, directed, say, in the x coordinate, the horizontal velocity components must satisfy

$$-fv_y = \nu\frac{\partial^2 v_x}{\partial z^2},$$

$$fv_x = fV + \nu\frac{\partial^2 v_y}{\partial z^2}, \qquad (7.33)$$

with the boundary conditions

$$z = 0: \qquad v_x = v_y = 0,$$
$$z = \infty: \qquad v_x = V, \quad v_y = 0. \qquad (7.34)$$

The solution of this boundary-value problem is

$$v_x = V(1 - e^{-az}\cos az),$$
$$v_y = Ve^{-az}\sin az, \qquad (7.35)$$

with $a = (f/2\nu)^{1/2}$. The twisting of the boundary layer due to rotation f is pictured in Fig. 7.8. A time-dependent solution of the Ekman layer is given in Reference 40, Chapter 1.

If the geostrophic velocity V is not a constant but changes in the y direction, that is, $V = V(y)$, then v_z is not zero (although small compared to v_x and v_y) and is

$$v_z = -\frac{1}{2a}\frac{\partial V}{\partial y}\left[1 - e^{-az}(\cos az + \sin az)\right]. \qquad (7.36)$$

The geostrophic shear flow $\partial V/\partial y$ (which is also proportional to the vorticity of the geostrophic flow) creates a vertical flow inside the Ekman layer and hence a mechanism to transport fluid from the Ekman layer to the outer geostrophic flow. The mass flux associated with v_z is, therefore, called "Ekman pumping" [Reference 2M, Chapter 4M].

8M. Vortices in Stratified Fluids

8.1M. THE BOUSSINESQ APPROXIMATION

Fluid motion with variable density not only eliminates the decoupling of the momentum equation from the energy equation of incompressible homogeneous fluids (Section 2.1M), it also requires consideration of the equation of state. A useful perturbation approximation, named for Boussinesq, was given by Spiegel and Veronis [1M] and is based on the equation of state

$$\rho = \rho_0(1 - \alpha T'), \tag{8.1}$$

where $T' = T - T_0$; α is the thermal expansion coefficient and is considered as a small constant; the terms involving α are considered higher order except in the buoyancy term. The subscript 0 denotes a reference state.

The Boussinesq approximation for a steady axisymmetric flow in a compressible atmosphere is then

$$v_r \frac{\partial v_r}{\partial r} + v_z \frac{\partial v_r}{\partial z} - \frac{v_\varphi^2}{r}$$
$$= -\frac{1}{\rho_0} \frac{\partial p'}{\partial r} + \nu \left[\frac{\partial^2 v_r}{\partial r^2} + \frac{\partial}{\partial r}\left(\frac{v_r}{r}\right) + \frac{\partial^2 v_r}{\partial z^2} \right], \tag{8.2}$$

$$v_r \frac{\partial v_\varphi}{\partial r} + v_z \frac{\partial v_\varphi}{\partial z} + \frac{v_r v_\varphi}{r}$$
$$= \nu \left[\frac{\partial^2 v_\varphi}{\partial r^2} + \frac{\partial}{\partial r}\left(\frac{v_\varphi}{r}\right) + \frac{\partial^2 v_\varphi}{\partial z^2} \right], \tag{8.3}$$

$$v_r \frac{\partial v_z}{\partial r} + v_z \frac{\partial v_z}{\partial z} = -\frac{1}{\rho_0} \frac{\partial p'}{\partial z} + \alpha g T'$$
$$+ \nu \left[\frac{\partial^2 v_z}{\partial r^2} + \frac{1}{r} \frac{\partial v_z}{\partial r} + \frac{\partial^2 v_z}{\partial z^2} \right], \tag{8.4}$$

$$\frac{\partial v_r}{\partial r} + \frac{v_r}{r} + \frac{\partial v_z}{\partial z} = 0, \tag{8.5}$$

$$v_r \frac{\partial T'}{\partial r} + v_z \left(\frac{\partial T'}{\partial z} + \gamma \right)$$
$$= \frac{k}{c_p \rho_0} \left[\frac{\partial^2 T'}{\partial r^2} + \frac{1}{r} \frac{\partial T'}{\partial r} + \frac{\partial^2 T'}{\partial z^2} \right] \tag{8.6}$$

with

$$p' = p + \rho_0 g z \quad \text{and} \quad \gamma = \frac{g}{c_p} \tag{8.7}$$

the adiabatic lapse rate. A closed-form solution of this system of equations will be presented in the next section.

8.2M. VISCOUS-FLOW SOLUTIONS

As for homogeneous fluid motions, general solutions of the Navier–Stokes equations using the Boussinesq approximation must be obtained numerically with computers. However, a closed-form similarity solution exists that is similar to Sullivan's two-cell solution, Eqs. (3.29) through (3.33). According to Kuo [2M], with the assumption

$k/c_p = \nu$, this solution is

$$v_\varphi = \frac{\kappa}{r} \frac{1}{\sqrt{2}} \left[\tilde{H}\left(\frac{Nr^2}{4\nu}\right) \Big/ \tilde{H}(\infty) \right], \quad (8.8)$$

$$v_r = -\frac{N}{2}r + \frac{4\nu}{r}\left(1 - e^{-Nr^2/4\nu}\right), \quad (8.9)$$

$$v_z = Nz\left(1 - 2e^{-Nr^2/4\nu}\right), \quad (8.10)$$

$$\frac{T'}{T_0} = \frac{N^2}{g}z\left(1 - 2e^{-Nr^2/4\nu}\right), \quad (8.11)$$

with

$$\tilde{H}\left(\frac{Nr^2}{4\nu}\right) = \int_0^{Nr^2/4\nu} \exp\left[-x + 2\int_0^x (1 - e^{-s})s^{-1} ds\right] dx. \quad (8.12)$$

N represents the "Brunt–Väisälä frequency" based on the temperature lapse rate

$$\sqrt{\frac{g}{T_0}\left(\frac{dT_0}{dz} + \frac{g}{c_p}\right)}.$$

The calculation of the pressure is quite cumbersome, and it may suffice here to give the pressure difference between two points 1 and 2 in the outer part of the flow field:

$$p_1 - p_2 = \rho_0 \frac{\kappa^2}{2}\left(\frac{1}{r_2^2} - \frac{1}{r_1^2}\right) \quad \text{for } \frac{Nr^2}{4\nu} > 11. \quad (8.13)$$

This is the pressure field of a potential vortex.

A comparison with Sullivan's solution shows that N plays the role of a, that is, buoyancy has an effect on the flow field similar to that of the meridional inflow. Moreover, as in Sullivan's case, the azimuthal velocity v_φ is decoupled from v_r and v_z, and also from T'/T_0. This means that the downdraft at the center is not enforced by the vortex but occurs simply because the temperature of the core is colder than that of the environment.

Gutman [3M], Kuo [2M], and Franz [4M] found, numerically, solutions of the one-cell flow that resemble Burgers' solution. In these papers, the relevance of the solutions to tornadoes and dust devils is discussed.

8.3M. INVISCID-FLOW SOLUTIONS

Properties of vortices in a stratified fluid are studied with the assumption that the fluid is inviscid and incompressible. For two-dimensional flows Eq. (4.33) can be extended to include changes in the density. With the transformation

$$d\psi' = \sqrt{\frac{\rho}{\rho_0}}\, d\psi \quad (8.14)$$

Yih [5M] arrived at

$$\left(\frac{\partial^2}{\partial r^2} + \frac{1}{r}\frac{\partial}{\partial r} + \frac{1}{r^2}\frac{\partial^2}{\partial \varphi^2}\right)\psi' + \frac{g}{\rho_0}\frac{d\rho}{d\psi'}z = h(\psi'), \quad (8.15)$$

where ρ_0 is a reference density and $z = r\sin\varphi$ is the coordinate in the vertical direction.

Closed-form solutions are found for the special case

$$\frac{g}{\rho_0}\frac{d\rho}{d\psi'} = A \quad \text{and} \quad h(\psi') = -k^2\psi', \quad (8.16)$$

where A and k are constants. Then, Eq. (8.15) yields

$$\left(\frac{\partial^2}{\partial r^2} + \frac{1}{r}\frac{\partial}{\partial r} + \frac{1}{r^2}\frac{\partial^2}{\partial \varphi^2} + k^2\right)\psi' + Az = 0. \quad (8.17)$$

In a way analogous to Eqs. (4.35) and (4.36) for homogeneous fluids, a solution of Eq. (8.17) is sought for a flow with a circular streamline $\psi' = 0$ of radius R. A solution is

$$\psi' = -\frac{A}{k^2}z + CJ_1(kr)\sin\varphi, \quad r \leq R \quad (8.18)$$

with Eq. (4.36) unchanged for $\rho = \rho_0$. The conditions corresponding to Eq. (4.37) are

$$-\frac{AR}{k^2} + CJ_1(kR) = 0,$$

$$2V = \frac{A}{k^2}\left[-1 + \frac{kRJ_1'(kR)}{J_1(kR)}\right]. \quad (8.19)$$

Equations (8.21) and (4.36), with the conditions (8.19), describe a parallel flow with constant V and ρ_0 about a pair of stratified vortices with $\rho = \rho_0$ on $r = R$. The circular bubble is neutrally buoyant.

Other two-dimensional solutions of Eq. (8.17) and those for axisymmetric vortices are presented in Reference 5M.

Like flows in rotating systems, where the constraint of rotation makes it possible for the fluid to produce waves, stratified fluid motion is also capable of wave making.

An example of an exact mathematical analogy between the two systems is the linearized (perturbed) inertial flow

$$\frac{\partial v_x}{\partial t} - fv_y = 0,$$

$$\frac{\partial v_y}{\partial t} + fv_x = 0, \quad (8.20)$$

with the frequency $\pm f/2\pi$, and the perturbation of the hydrostatic equilibrium

$$\frac{\partial v_z}{\partial t} + \frac{g}{\rho_0}\rho' = 0,$$

$$\frac{\partial \rho'}{\partial t} + \frac{d\rho_0}{dz}v_z = 0 \quad (8.21)$$

with the frequency

$$N = \pm\sqrt{\frac{-g}{\rho_0}\frac{d\rho_0}{dz}}, \quad (8.22)$$

which is the Brunt–Väisälä frequency [Reference 40, Chapter 8].

Finally, an estimate is given for inviscid-fluid vortices in stratified, rotating fluid systems.

Consider the geostrophic motion (7.18) and (7.19) with the flow in the y direction:

$$\rho fv_y = \frac{\partial p}{\partial x},$$

$$\rho g = \frac{\partial p}{\partial z}, \quad (8.23)$$

and two fluids of densities ρ_1 and ρ_2, separated by a discontinuity surface with the slope angle γ:

$$\frac{dz}{dx} = \tan\gamma. \quad (8.24)$$

Continuity of pressure across the discontinuity gives, with Eq. (8.23),

$$\tan\gamma = -\frac{f}{g}\left[\frac{\rho_1(v_y)_1 - \rho_2(v_y)_2}{\rho_1 - \rho_2}\right]. \quad (8.25)$$

This relation is called the "Witte–Margules equation" [Reference 10, Chapter 7]. It explains the situation presented in Fig. 8.32 far away from the center. Helland-Hansen and Sandström [Reference 22, Chapter 9] considered a continuously stratified medium.

References for the Mathematical Supplement

Chapter 1M

1M. E. R. Lindgren, "Vorticity and Rotation." *Am. J. Phys.* **48** (1980), 465.

2M. H. Lamb, *Hydrodynamics*, 6th ed. Dover, New York, 1945.

Chapter 2M

1M. K. Oswatitsch, Physikalische Grundlagen der Strömungslehre. In *Handbuch der Physik*, Bd. VIII/1. S. Flügge (ed.), Springer, 1959, p. 1.

Chapter 3M

1M. C. DuP. Donaldson and R. D. Sullivan, *Examination of the Solutions of the Navier-Stokes Equations for a Class of Three-Dimensional Vortices*. AFOSR TN 60-1227, Oct. 1960.

2M. G. I. Taylor, *Aero. Res. Comm. R & M* No. 598 (1918).

3M. J. M. Burgers, "A Mathematical Model Illustrating the Theory of Turbulence." *Adv. Appl. Mech.* **1** (1948), 197.

4M. N. Rott, "On the Viscous Core of a Line Vortex." *ZAMP* **9** (1958), 543; **10** (1959), 73.

5M. R. D. Sullivan, "A Two-Cell Vortex Solution of the Navier-Stokes Equations." *J. Aer. Sci.* **26** (1959), 767.

6M. F. W. Leslie and J. T. Snow, "Sullivan's Two-Celled Vortex." *AIAA J.* **18** (1980), 1272.

7M. P. G. Bellamy-Knights, "An Unsteady Two-Cell Vortex Solution of the Navier-Stokes Equations." *J. Fluid Mech.* **41** (1970), 673.

8M. G. I. Taylor, *Philos. Mag.* **46** (1923), 671.

9M. M. H. Protter and H. F. Weinberger, *Maximum Principles in Differential Equations*. Prentice-Hall, Englewood Cliffs, N.J., 1967.

10M. W. G. Cochran, "The Flow Due to a Rotating Disk." *Proc. Cambridge Philos. Soc.* **30** (1934), 365.

11M. E. W. Schwiderski and H. J. Lugt, "Rotating Flows of von Kármán and Bödewadt." *Phys. Fluids* **7** (1964), 867.

12M. D. Dijkstra and P. J. Zandbergen, "Some Further Investigations on Non-Unique Solutions of the Navier-Stokes Equations for the Kármán Swirling Flow." *Arch. Mech.* **30** (1978), 411.

13M. R. R. Long, "A Vortex in an Infinite Viscous Fluid." *J. Fluid Mech.* **11** (1961), 611.

14M. B. G. Newman, "Flow in a Viscous Trailing Vortex." *Aeronaut. Q.* **10** (1959), 149.

15M. G. K. Batchelor, "Axial Flow in Trailing Line Vortices." *J. Fluid Mech.* **20** (1964), 645.

16M. N. Rott and W. S. Lewellen, "Boundary Layers and their Interactions in Rotating Flows." *Prog. Aeronaut. Sci.* **7** (1966), 111.

17M. R. Granger, "Steady Three-Dimensional Vortex Flow." *J. Fluid Mech.* **25** (1966), 557.

18M. M. Van Dyke, *Perturbation Methods in Fluid Mechanics*. Academic, New York, 1964.

Chapter 4M

1M. A. Sommerfeld, *Mechanik der deformierbaren Medien*. Akademische Verlagsges., Leipzig, 1949.

2M. J. Pedlosky, *Geophysical Fluid Dynamics*. Springer, New York, 1979.

3M. J. H. Horlock, "Vorticity Transfer and Production in Steady Inviscid Flow." *J. Basic Eng.* (March 1968), 65.

4M. P. G. Saffman and R. Szeto, "Equilibrium Shapes of a Pair of Equal Uniform Vortices." *Phys. Fluids* **23** (1980), 2339.

5M. R. T. Pierrehumbert, "A Family of Steady, Translating Vortex Pairs with Distributed Vorticity. *J. Fluid Mech.* **99** (1980), 129.

6M. J. Burbea, "On Patches of Uniform Vorticity in a Plane of Irrotational Flow." *Arch. Rat. Mech. Anal.* **77** (1981), 349.

7M. N. J. Zabusky, "Recent Developments in Contour Dynamics for the Euler Equations." *Ann. N.Y. Acad. Sci.* **373** (Oct. 1981), 160.

8M. V. O'Brien, "Steady Spheroidal Vortices—More Exact

Solutions To the Navier-Stokes Equation." *Q. Appl. Math.* **19** (1961), 163.

9M. J. Norbury, "A Family of Steady Vortex Rings." *J. Fluid Mech.* **57** (1973), 417.

10M. F. Durst and B. Schönung, "Computations of Steady, Ellipsoidal Vortex Rings With Finite Cores." *Comput. Fluids* **10** (1982), 87.

Chapter 5M

1M. N. E. Kochin, I. A. Kibel, and N. V. Roze, *Theoretical Hydromechanics.* Wiley, New York, 1964.

2M. D. M. F. Chapman, "Ideal Vortex Motion in Two Dimensions: Symmetries and Conservation Laws." *J. Math. Phys.* **19** (1978), 1988.

3M. H. Aref and N. Pomphrey, "Integrable and Chaotic Motions of Four Vortices." *Proc. Roy. Soc. Lond. Ser. A* **380** (1982), 359.

4M. H. Aref, "Integrable, Chaotic and Turbulent Vortex Motion in Two-Dimensional Flows." *Ann. Rev. Fluid Mech.* **15** (1983), 345.

5M. R. R. Clements and D. J. Maull, "The Representation of Sheets of Vorticity by Discrete Vortices." *Prog. Aerospace Sci.* **16** (1975), 129.

6M. G. Kirchhoff, "Vorlesungen über mathematische Physik." *Mechanik*, p. 255.

7M. L. Onsager, "Statistical Hydrodynamics." *Nuovo Cimento, Suppl.* **6** (1949), 279.

8M. C. C. Lin, "On the Motion of Vortices in Two Dimensions." *Appl. Math. Series No. 5.* University of Toronto Press, 1943.

9M. H. Kaden, "Aufwicklung einer unstabilen Unstetigkeitsfläche." *Ing.-Arch.* **2** (1931), 140.

10M. D. I. Pullin, The Large-Scale Structure of Unsteady Self-Similar Rolled-Up Vortex Sheets." *J. Fluid Mech.* **88** (1978), 401.

11M. T. von Kármán and J. M. Burgers, "General Aerodynamic Theory—Perfect Fluids." In *Aerodynamic Theory*, Vol. II. W. F. Durand (ed.). Dover, New York, 1963.

12M. D. W. Moore, "The Velocity of a Vortex Ring with a Thin Core of Elliptical Cross Section." *Proc. Roy. Soc. Lond. Ser. A* **370** (1980), 407.

13M. A. Leonard, "Numerical Simulation of Interacting, Three-Dimensional Vortex Filaments." *Lecture Notes in Physics*, Vol. 35. Springer, New York, 1975, p. 245.

14M. F. R. Hama, "Progressive Deformation of a Curved Vortex Filament by its Own Induction." *Phys. Fluids* **5** (1962), 1156.

Chapter 6M

1M. H. J. Lugt and E. W. Schwiderski, "Birth and Decay of Vortices." *Phys. Fluids* **9** (1966), 851.

2M. H. J. Lugt and E. W. Schwiderski, "Flows Around Dihedral Angles." *Proc. Roy. Soc. Lond. Ser. A* **285** (1965), 387.

3M. W. R. Dean and P. E. Montagnon, *Proc. Cambridge Philos. Soc.* **45** (1949), 389.

4M. H. K. Moffat, "Viscous and Resistive Eddies Near a Sharp Corner." *J. Fluid Mech.* **18** (1964), 1.

5M. E. W. Schwiderski, H. J. Lugt, and P. Uginčius, "Axisymmetric Viscous Fluid Motions Around Conical Surfaces." *J. SIAM* **14** (1966), 191.

6M. K. N. Ghia and A. G. Mikhail, "Axisymmetric Stokes Flow Past Cones Including the Case of the Needle." *J. Appl. Mech.* **42** (1975), 569.

7M. O. M. Phillips, "The Final Period of Decay of Non-Homogeneous Turbulence." *Proc. Cambridge Philos. Soc.* **52** (1956), 135.

8M. H. J. Lugt, "Multipole Decomposition of Solutions of the Vector Diffusion Equation." *SIAM J. Appl. Math.* **39** (1980), 264.

Chapter 7M

1M. J. A. Johnson, "The Diffusion of a Viscous Vortex Ring in a Rotating Fluid." *J. Fluid Mech.* **24** (1966), 753.

2M. C. G. Rossby *et al.*, "Relation between Variations in the Intensity of the Zonal Circulation of the Atmosphere and the Displacements of the Semi-Permanent Centers of Action." *J. Marine Res.* **2** (1939), 212.

3M. H. J. Stewart, "Periodic Properties of the Semi-Permanent Atmospheric Pressure Systems." *Q. Appl. Math.* **1** (1943), 262.

4M. G. K. Morikawa, "Geostrophic Vortex Motion. *J. Met.* **17** (1960), 148.

5M. G. K. Morikawa and E. V. Swenson, *Interacting Motion of Rectilinear Geostrophic Vortices.* New York University, Courant Inst. Math. Sci., NR 062-160, IMM 364, Jan. 1968.

6M. S. L. Hess, *Introduction to Theoretical Meteorology.* Holt, Rinehart and Winston, New York, 1959.

Chapter 8M

1M. E. A. Spiegel and G. Veronis, "On the Boussinesq Approximation for a Compressible Fluid." *Astrophys. J.* **131** (1960), 442.

2M. H. L. Kuo, "On the Dynamics of Convective Atmospheric Vortices." *J. Atm. Sci.* **23** (1966), 25.

3M. L. N. Gutman, "Theoretical Model of a Waterspout." *Izv. Akad. Nauk, USSR Bull. (Geophys. Series).* [Pergamon Press trans.] **1** (1957), 87.

4M. H. W. Franz, "On the Cellular Structure of Steady Convective Vortices." *Beitr. Phys. Atmosphäre* **42** (1969), 36.

5M. C. S. Yih, "Vortices and Vortex Rings of Stratified Fluids," *SIAM J. Appl. Math.* **28** (1975), 899.

Name Index

Abell, C. J., 231
Achenbach, E., 105, 233
Ackeret, J., 55, 230
Ackerman, B., 210, 240
Acrivos, A., 234
Adamson, T. C., 230
Agee, E. M., 195, 240
Aggarwal, M., 232
Aiton, E. J., 227
Amphlett, M. B., 229
Anaxagoras, 6, 7
Anaximander, 6, 7
Anaximenes, 6
Anderson, S. B., 230
Andrade, E. N. da C., 35, 229
Angel, M., 240
Anthes, R. A., 240
Anwar, H. O., 228, 229
Archimedes, 60
Archytas, 106
Aref, H., 284
Aristophanes, 8
Aristotle, 8–10, 12, 129, 169, 183, 212, 227
Arius Didymus, 24
Arms, J., 271
Arnold, R. T., 239
Asanuma, T., 233
Aschoff, J., 149
Ashton, R. A., 229
Atkinson, S. A., 88, 89, 232
Auchterlonie, L. J., 234
Avery, D. G., 236
Avicenna, 9

Baade, W., 221
Bach, J. S., 17
Bacon, F., 184
Baker, C. J., 231
Baker, D. J., 183, 238
Baker, G. R., 27, 228
Bambach, R. K., 239
Barcilon, A. I., 191, 239, 240
Barkla, H. M., 234
Barrow, J. D., 226, 242

Bass, H. E., 239
Batchelor, G. K., 88, 121, 229, 260, 283
Beard, K. V., 230
Beér, J. M., 236
Belcher, R., 236
Bellamy-Knights, P. G., 256, 283
Bellhouse, B. J., 11, 227
Bénard, H., 15, 105, 159, 233, 237
Benjamin, T. B., 88
Benton, E. R., 235
Berendzen, R., 241
Berg, H. C., 229
Berger, S. A., 88, 232
Bergeron, T., 189
Berkeley, G., 13, 34, 130, 131
Berlitz, C., 240
Bernoulli, J., and D., 13, 14, 96
Bertelsen, A., 233
Betchov, R., 232
Bethke, R. J., 228
Betz, A., 55, 230
Bilanin, A. J., 231
Binnie, A. M., 35, 229
Birkhoff, G., 231
Birrell, N. D., 241
Bishop, R. L., 232
Bjerknes, J. and V., 15, 171, 189, 238, 264
Black, P. G., 240
Blackwelder, R. F., 233, 235
Blanchette, 36
Blevins, R. D., 234
Bode, 214
Bödewadt. U. T., 142, 236, 258
Bok, B. J., 241
Boldes, U., 232
Bolen, L. N., 239
Boltzmann, L., 14
Bornstein, J., 89
Boscovich, R. G., 13
Bossel, H. H., 89, 232
Bouard, R., 231
Bourot, J. M., 50, 229
Boussinesq, J., 15, 120, 234, 280
Bradley, W. H., 237
Bramwell, A. R. S., 61, 230

Brans, C., 131, 235
Brebbia, C. A., 229
Brennen, C., 229
Brennenstuhl, U., 231
Brenner, H., 229
Bretherton, F. P., 189, 239
Briley, W. R., 236
Brodkey, R. S., 232
Brokaw, C. J., 229
Browand, F. K., 238
Brown, F. N. M., 99, 115, 118, 234
Brunt, D., 237
Bruzek, J., 192
Bryan, K., 177, 183, 238
Buch, H., 173
Budney, L. J., 193, 239
Buffon, G. L. L., 213
Burbea, J., 266, 283
Burgers, J. M., 121, 230, 255–257, 262, 281, 283, 284
Burggraf, O. R., 74, 142, 231, 236, 239
Buridan, 8, 9
Burland, C., 240
Burt, W. V., 237
Busse, F. H., 167, 217, 237, 238
Buzyna, G., 238
Byrne, H. M., 180

Campbell, J., 228
Cantwell, B. J., 235
Carlowitz, O., 236
Carmody, J. M., 233
Carpenter, M. S., 119
Carrier, G. F., 179, 238, 240
Carrigan, C. R., 167, 238
Carusi, E., 10, 227
Cauchy, A. L., 14, 18, 19
Cayley, G., 56
Chamberlin, 213
Chandrasekhar, S., 166, 219, 237
Chang, C. C., 195, 239
Chang, G., 70, 230
Chang, L. M., 157, 158, 237
Chanute, O., 56
Chapman, D. M. F., 284

285

Chapman, D. R., 235
Charney, J., 183, 185, 238, 239
Chen, C. F., 233
Chen, C. J., 157, 158, 237
Chernin, 225
Chevray, R., 227
Childe, V. G., 228
Chong, M. S., 233
Chopra, K. P., 201, 240
Chow, C., 232
Christiansen, J. P., 39, 40, 98, 229
Chun, C. H., 160, 162, 167, 237
Chung, T. J., 229
Church, C. R., 195, 240
Cirlot, J. E., 228
Clark, A., 235
Clark, K., 227
Clarke (Newton's friend), 130
Clements, R. R., 284
Cloud, P., 239
Cochran, W. G., 283
Cohen, K. P., 228
Colón, 206
Columbus, C., 178, 203
Cone, C. D., 237
Connor, J. J., 229
Conrad, J., 211
Cook (Captain), 177
Cook, T. A., 228
Copernicus, N., 11, 129
Corby, 237
Coriolis, G., 13, 131
Cornford, N., 235
Cornhill, J. F., 232
Coutanceau, M., 231
Covey, C., 241
Cox, 213
Cox, R. G., 234
Craik, A. D. D., 197, 240
Crew, H., 237
Criminale, W. O., 232
Crocco, L., 15
Cross, E. J., 231
Csanady, G. T., 178, 238
Curtis, 222
Cuzzi, J. N., 241

d'Alembert, J. L., 14
Dali, S., 17
Daly, B. J., 157, 237
Danckwerts, P. V., 228
Dante, 9, 10, 26, 190, 201
David, King, 106
Davidson, J. F., 35, 229
Davies, E., 236
Davies, J. M., 234
Davies, P. C. W., 241
Davies-Jones, R. P., 195, 239
Dean, W. R., 231, 284
Dechend, H. v., 5, 227
Defant, A., 171, 201, 202, 238, 240
Democritus, 6–8
Dennis, S. C. R., 68, 70, 230

Descartes, R., 12, 13, 20, 22, 129, 130, 149, 213
Dicke, R. H., 131, 214, 235
Dickman, R. L., 241
Didden, N., 122, 232
Dietrich, G., 238
Dijkstra, D., 258, 283
Diogenes Laeritus, 6
Dolzhanskii, F. V., 148, 237
Donaldson, C. DuP., 228, 231, 283
Donnelly, R. J., 229
Döös. B. R., 238
Dorrepaal, J. M., 71, 230, 274
Dove, H. W., 171, 238
Drazin, P. G., 240
DuBuat, 13
Dugas, R., 227
Durand, W. F., 228, 230, 284
Dürer, A., 11
Durst, F., 266, 284
Dussourd, J. L., 236
Dyson, F. J., 241

Eady, E. T., 185, 239
Eaton, K. J., 234
Ebert, C. H. V., 237
Ehrhardt, G., 104, 233
Eichelbrenner, E. A., 74, 84, 231
Eiffel, G., 124
Einstein, A., 13, 131, 235
Ekman, V. W., 133, 235
Elder, J. W., 157, 184, 237, 239
Eliassen, A., 239
Ellis, G. F. R., 226, 242
Emery, K. O., 178, 238
Emmons, H. W., 101, 233
Empedocles, 6–8
Epicurus, 8
Erickson, C. O., 240
Erickson, G. E., 85
Ericsson, L. E., 233
Ertel, H., 15, 264
Escher, M., 17, 66
Escudier, M. P., 89, 232, 236
Eskinazi, S., 233
Euler, L., 14, 24, 54, 130, 250, 263, 27
Exner, F. M., 182, 238

Faller, A. J., 145, 197, 202, 240
Farmer, D. M., 240
Farris, G. J., 142, 236, 259
Fattahi, B., 195, 239
Favaro, A., 10, 227
Favier, D., 233
Fearn, R., 229
Feldman, F., 230
Ferrel, W., 13, 171, 238
Ferreri, J. C., 232
Ferziger, J. H., 235
Fett, R. W., 211, 241
Feynman, R. P., 16, 228
Ffowcs Williams, J. E., 234
Fiedler, F., 239

Fierz, M., 227
Fink, P. T., 44, 229
FitzGerald, G. F., 14
Fjørtoft, R., 183, 238
Flachsbart, O., 227
Flierl, G. R., 204, 240
Flora, S. D., 239
Fohl, T., 232
Foster, M. R., 239
Fowlis, W. W., 238
Franklin, B., 178
Franz, H. W., 281, 284
Freymuth, P., 98, 232
Fromm, J. E., 103, 229, 233
Fuglister, F. C., 179, 180, 238
Fujita, T. T., 239
Fujiwara, 38
Fultz, D., 182, 227, 237
Fung, Y., 71, 230

Gale, G., 242
Galilei Galileo, 11, 12, 129, 218
Galloway, D. J., 241
Gamov, G., 224, 225
Garde, R. J., 228
Gardner, M., 229
Gartshore, I. S., 88
Gaster, M., 104, 233
Genghis Khan, 203
Georgii, W., 163, 237
Gerard of Brussels, 9
Gerbier, 237
Gerola, H., 241
Gerster, G., 161
Ghia, K. N., 231, 284
Ghia, U., 231
Giacomelli, R., 228
Gibbs-Smith, C. H., 230
Gilchrist, A., 239
Gillies, G., 195, 239
Gilman, P. A., 214, 241
Glaberson, W. I., 229
Glass, J., 195, 239
Glatz, H., 146
Glauert, H., 62, 230
Glenn, J. H., 119
Glenny, D. E., 229
Goethe, J. W. v., 16, 17, 226
Gold, M., 229
Goldburg, A., 229
Golden, J. H., 193, 239
Goldshtik, M. A., 142, 236, 258, 262
Goldstein, S., 227, 228
Gordon, M. J. V., 229
Gordon, R. B., 184, 239
Gore, R., 241
Gorenstein, P., 241
Görtler, H., 15, 136, 145, 235, 236
Gotthardt, G. A., 238
Govindaraju, S. P., 228
Grabowski, W. J., 88, 232
Graebel, W. P., 142, 236, 237
Granger, R. A., 229, 262, 283

Grant, E., 227
Greated, C. A., 232
Greenspan, H. P., 148, 228
Gregg, M., 239
Gregory, N., 234
Griffin, O. M., 102, 107, 233, 234, 240
Groth, E. J., 241
Gühler, M., 92, 93
Gustafson, 133
Gutman, L. N., 281, 284

Hadamard, J., 55, 230
Hadley, G., 13, 170, 171, 238
Haeckel, E., 51
Hagen, G., 123, 125
Hall, M. G., 88, 232
Hall, W. D., 70, 230
Halley, E., 170, 238
Halsey (Admiral), 205
Hama, F. R., 233, 284
Hamielec, A. E., 69, 70, 72, 230
Han, T., 231
Happel, J., 229
Happelius, 178
Harlow, F. H., 229, 236
Harrison, E. R., 226, 241, 242
Hart, R., 241
Harvey, J. K., 88, 89, 94, 232
Harvey, R. L., 234
Hashimoto, K., 236
Hasimoto, H., 229
Hassenpflug, H. U., 78, 231
Haussling, H. J., 72, 75, 109, 140, 141, 164, 231, 233, 236, 237
Hawk, N. E., 234
Hawking, S. W., 222, 241
Herbert, P. J., 240
Hegel, G. W. F., 131
Heisenberg, W., 15, 121
Helland-Hansen, 282
Helmholtz, H. V., 14, 15, 54–58, 71, 97, 171, 230, 238, 263
Henry the Seafarer, 178
Heraclitus, 24
Hertel, H., 58, 60, 229
Hesiod, 4, 6
Hess, F., 63, 64, 230
Hess, H. H., 239
Hess, S. L., 284
Hewish, A., 221
Hicks, 14
Hide, R., 137, 182, 235
Hill, M. J. M., 91, 266
Hill, M. N., 240
Hills, J. G., 214, 241
Hinze, J. O, 120, 228
Hippocrates, 8
Hiroshige, 17
Hirschel, E. H., 231
Hocking, L. M., 235
Hoerner, S., 230
Hoffmann, E. R., 228
Hoglund, R. L., 236

Holligan, P. M., 239
Homer, 3, 4, 9, 10, 198, 199
Honji, H., 233
Horlock, J. H., 283
Hoskins, B. J., 189, 239
Hoskins, J., 230
Hoult, D. P., 231
Howarth, L., 121
Hoyle, F., 131, 225, 235
Hoyt, J. W., 235
Hsu, C. T., 195, 239
Huang, M., 232
Huang, S., 220, 241
Hubble, E. P., 16, 222, 223
Hubert, L. F., 201, 240
Humboldt, A. v., 184
Hummel, D., 89, 231
Hunt, B., 229
Hunt, B. G., 185, 239
Hunt, G., E., 241
Hunt, J. C. R., 231, 234
Huppert, H. E., 240
Hurley, P., 239
Hussain, A. K. M. F., 232
Huygens, C., 13

Ibbetson, A., 137, 235, 237
Idso, S. B., 240
Ince, S., 235
Ingersoll, A. P., 217, 241
Ives, R. L., 240
Izumi, K., 232

Jackson, R. G., 236
Jacobs, J., 236
Jacobs, W., 51, 229
Jahn, T. L., 229
Jain, A. K., 228
James, P. B., 241
Jammer, M., 235
Jansen, 215
Jeffrey, D. J., 230
Jeffreys, H., 15, 169, 171, 238
Jenson, V. G., 49, 229
Jeschar, R., 236
John, Saint, 224, 226
Johnson, J. A., 284
Jones, B. M., 230
Jordan, S. K., 103, 233
Joss, P. C., 241
Joubert, P. N., 228
Joukowsky, N., 15, 56, 272

Kaden, H., 269, 284
Kalle, K., 238
Kambe, T., 232
Kant, I., 13, 131, 213, 222
Kao, T. W., 105, 233
Kármán, T. v., 15, 101, 105, 106, 108, 120, 121, 142, 228, 230, 236, 257, 258, 284
Kaylor, R. E., 145
Keck, J. C., 231

Keller, H. B., 70, 230
Keller, W., 240
Kelly, D. L., 237
Kelvin, Lord (W. Thomson), 14–16, 41, 97, 100, 136, 227, 229, 232, 235, 264
Kepler, J., 9, 11–13, 29, 129, 130, 178, 213
Kerr, R. A., 238
Kibel, I. A., 284
Kidd, G. J., 142, 236, 259
Kimura, R., 237
King, R., 234
Kircher, A., 198
Kirchhoff, G., 266, 269, 284
Kirk, G. S., 227
Kirkpatrick, D. L. I., 89
Kiya, M., 233
Klein, F., 55, 230
Klein, H., 229
Klöckner, 163
Knisely, C., 111, 234
Kochin, N. E., 284
Kohl, J. G., 238
Kolmogorov, A. N., 15, 120, 234
Koopmann, G. H., 233
Köppen, W., 181
Koptsik, V. A., 228
Kordulla, W., 231
Korpi, 17
Koschmieder, H., 163, 237
Kovasznay, L. S. G., 121
Koyré, A., 227
Krafft, C. F., 17, 228
Kraichnan, R. H., 121
Krauss, W., 238
Kreith, F., 235
Kreplin, H. P., 231
Krishnamurti, R., 160, 237
Krutzsch, C., 94, 232
Kry, P. R., 235
Kublai Khan, 203
Küchemann, D., 231
Kühn, H., 227
Küttner, J., 163
Kullenberg, 133
Kundt, A., 106, 233
Kuo, H. L., 237, 280, 281, 284
Kutta, W., 15, 56, 74, 75, 96, 272

Lagerstrom, P. A., 27, 228, 232
Lagrange, J. L., 14, 24, 250
Lamar, J. E., 231
Lamb, H., 254, 256, 273, 283
Lanchester, F., 15, 56, 57, 61, 62, 96, 108
Landau, L. D., 16, 42, 119, 228, 234
Landweber, L., 233
Langbein, W. B., 232
Langley, S., 56
Langmuir, I., 197, 240
Laplace, P. S. de, 13, 14, 213
Larson, R. B., 223, 241
Laufer, J., 235

288 Name Index

Lavan, Z., 73, 75, 228, 231, 232
Le Bris, J.-M., 56
Lecar, M., 214, 241
Le Clair, B. P., 69, 70, 230
Lee, D. A., 233
Lee, J., 71, 230
Leibniz, G. W., 13, 14, 129–131
Leibovich, S., 89, 197, 240
Leighly, J., 239
Lemaître, G., 224
Leonard, A., 101, 233, 284
Leopold, L. B., 232
Leslie, F. W., 283
Leslie, L. M., 239
Leucippus, 7, 8
Levi, E., 37, 229
Lewellen, W. S., 236, 239, 262, 283
Liepmann, H. W., 235
Lifshitz, E. M., 228
Lighthill, M. J., 15, 58, 59, 84, 177, 228, 229, 238
Lilienthal (Brothers), 56, 57
Lilley, D. G., 228
Lilly, D. K., 237
Lim, T. T., 233
Lin, C. C., 121, 223, 241, 269, 284
Lin, J. Y., 232
Lindblad, 223
Lindgren, E. R., 283
Linsley, R. M., 229
List, R., 235
Lo, R. K. C., 232
Locke, J., 130
Logan, S. E., 240
Long, R. R., 142, 235–237, 260–262, 283
Lorenz, E. N., 13, 171, 227, 238, 240
Love, A. E. H., 229
Lucan, 9, 196
Lucretius, 8, 9, 10
Ludlam, F. H., 239
Ludlum, D. M., 240
Ludweig, H., 88
Lugt, H. J., 24, 72, 75, 85, 87–90, 100, 112, 140, 141, 227, 228, 230–234, 236–238, 258, 275, 283, 284
Lumley, J. L., 234, 235
Lund, F., 228
Lyne, W. H., 232

Macagno, E. and M., 235
Mach, E., 13, 131, 134, 235
McCutchen, C. W., 228
McEwan, A. D., 237
Mackenzie, D. A., 3, 227
McKenzie, D. P., 184, 239
MacLatchy, C. S., 122, 235
Macleod, G. R., 229
Magarvey, R. H., 122, 230, 232, 235
Magnus, G., 114, 234
Magnus, O., 198, 199
Maier, A., 227
Malkus, J. S., 240

Mandelstam, S., 227
Mangione, B. J., 233
Mannal, C., 236
Mardell, G. T., 239
Maresca, C., 107, 233
Markgraf, H., 239
Martin, B. W., 237
Maskell, E. C., 86, 231
Mason, W. T., 231
Mather, N. W., 236
Matsuda, T., 236
Matta, 17
Mattig, W., 215
Maul, G. A., 238
Maull, D. J., 284
Maupertuis, P. L. M., de, 13, 130
Maxwell, J. C., 14, 118, 125
Maxworthy, T., 92, 94, 122, 138, 191, 194, 217, 232, 234, 235, 238–241
Mazzarelli, G., 240
Meerwein, C. F., 56, 230
Mehta, U. B., 73, 75, 108, 228, 231, 233
Meier, H. U., 231
Meier-Windhorst, A., 234
Meinesz, V., 184
Mercator, 198
Merkli, P., 236
Mermoz, J., 191
Merz, A., 186, 239
Merzkirch, W., 228
Messiter, A. F., 230
Meyer, E., 15
Michalke, A., 99, 232
Mikhail, A. G., 284
Miles, J. W., 237
Miller, B. I., 240
Miller, J. S., 242
Misner, C. W., 228, 242
Miyakoda, K., 239
Moffat, H. K., 284
Mohamed Gad-El-Hak, 233
Mollö-Christensen, E., 145, 233
Montagnon, P. E., 284
Moore, D. W., 75, 231, 235, 270, 271, 284
Morel, P., 238
Morel, T., 231
Morgan, W., 183, 184, 239
Morikawa, G. K., 278, 284
Morkovin, M. V., 233
Morton, B. R., 192, 239
Morton, W. A., 230
Mosby, H., 201, 240
Moulton, 213
Müller, E. A., 234
Mukherjee, A. K., 241
Mullen, J. B., 240
Mullin, T., 232
Munk, W. H., 177, 183, 238

Narain, J. P., 232
Narlikar, J. V., 131, 235
Naudascher, E., 234

Navier, M., 14
Nerem, R. M., 232
Nesis, A., 241
Neumann, C. J., 240
Neumann, E., 227
Neumann, G., 235
Neumann, J. v., 183, 238
Newell, R. E., 174, 238
Newman, B. G., 260, 283
Newton, I., 13, 30, 34, 44, 56, 114, 129–131
Niccols, W. O., 233
Nikuradse, J., 123, 235
Norbury, J., 266, 284

Obasi, G. O. P., 173
O'Brien, V., 237, 266, 283
Obukov, A. M., 121
O'Connor, N. F., 229
Oertel, H., jun., 159
Ohring, S., 112, 234
Okajima, A., 107, 233
Olander, D. R., 236
Olsen, J. H., 229
Onsager, L., 284
Oort, A. H., 170, 238
Oort, J. H., 225, 226
Oppenheimer, R., 221
Oresme, N., 9, 10
Orszag, S. A., 235
Oser, H. J., 136, 235
Oshima, Y., 232
Ostriker, J. P., 241
Oswatitsch, K., 15, 228, 231, 283
Oudart, A., 84, 231
Ovenden, M. W., 214, 241
Ozernoi, 225

Packard, R. E., 229
Palm, E., 158, 237
Palmén, 206
Panofsky, H. A., 164, 237, 239
Pao, H. P., 105, 142, 143, 233, 236
Parkinson, G. V., 230, 234
Parmentier, E. M., 230
Pasmore, 17
Patel, V. C., 231
Pauly, P. J., 227
Payne, R. B., 48, 229
Peake, D. J., 82, 231
Pearce, R. P., 238
Pearson, 14
Pearson, C. E., 148, 236
Peckham, D. H., 88, 89, 232
Pedlosky, J., 238, 283
Peebles, P. J. E., 225, 241, 242
Peixoto, J. P., 173
Penaud, 56
Peng, L., 173
Perry, A. E., 77, 231, 233
Perry, F. J., 94, 232
Peterka, J. A., 231
Petschek, H. E., 230

Pfeffer, R., 238
Phillips, N. A., 239
Phillips, O. M., 237, 274, 284
Philoponus, 9
Pierce, D., 101, 233
Pierrehumbert, R. T., 266, 283
Pingree, R. D., 239
Pistolesi, E., 228
Pitter, R. L., 72, 230
Plaskowski, Z., 230
Plato, 5, 8, 10, 17, 19, 24
Pliny, 9, 10
Poe, E. A., 198
Poe, G. G., 234
Poiseuille, J. L. M., 123, 254
Poisson, S. D., 14
Polhamus, E. C., 82, 231
Pollack, J. B., 241
Pollard, R. T., 197, 240
Pomphrey, N., 284
Ponce de Leon, 178
Poole, S. L., 237
Powell, A., 15, 228, 234
Prandtl, L., 15, 27, 52, 53, 55, 57, 76, 96, 120, 124, 228–231, 236, 259, 269
Price, W., 240
Priestley, C. H. B., 173
Prigogine, I., 15, 125, 228
Pringle, J., 241
Proctor, M. R. E., 241
Protagoras, 8
Protter, M. H., 283
Proudman, J., 136, 235
Pruppacher, H. R., 69, 70, 72, 230
Ptolemy, 178
Pullin, D. I., 77, 231, 270, 284
Pyestock, N., 229
Pythagoras, 8, 10, 17

Queney, P., 163, 237
Quick, M. C., 229

Ragsdale, R. G., 236
Rajagopal, H. Y., 229
Ramberg, S. E., 233
Randall, J. D., 89
Ranga Raju, K. G., 228
Rankin, T. M., 233
Rankine, W. J. M., 31
Rappaport, S. A., 241
Raven, J. E., 227
Rayleigh, Lord, 15, 97, 105, 145, 159, 165, 232, 233, 236, 237
Rayner, J. M. V., 230, 232
Reber, C. M., 192, 194, 239
Rebont, J., 233
Redekopp, L. G., 217, 241
Redfield, 204
Reeves, H., 213, 241
Regge, T., 228
Reiter, E., 238
Reynolds, O., 15, 46, 119, 234
Riabouchinsky, D. P., 15, 228

Richards, C. G., 142, 236
Richardson, H. H., 236
Richardson, L. F., 119, 120, 183
Richardson, P. L., 238
Richter, F., 184, 239
Riehl, H., 174, 206, 235, 240
Rieu, E. V., 4
Riely, J. J., 233
Rimon, Y., 72, 230
Ringleb, F. O., 73, 143, 231
Rittenhouse, L. E., 236
Roache, P., 228
Roberts, K. V., 39, 40, 98, 229
Roberts, M. S., 241
Roberts, P. H., 229
Roberts, W. W., 241
Robins, B., 114, 234
Robinson, A. R., 204, 240
Robinson, J. M., 227
Rochino, A. P., 232
Rockett, J. A., 153, 237
Rockwell, D. O., 111, 233, 234
Roesner, K. G., 148, 236, 237
Rogers, M., 229
Rosen, M., 235
Rosenblat, S., 234
Rosenhead, L., 39, 44, 228, 229
Rosenthal, S. L., 240
Roshko, A., 103, 121, 228, 233
Ross, D., 234
Rossby, C. G., 135, 183, 235, 284
Rott, N., 230–232, 255, 262, 265, 283
Rotunno, R., 194, 239
Rouse, H., 235
Roze, N. V., 284
Rumford, Count, 152, 237
Runcorn, S. K., 184, 239
Rydberg, 5

Saffman, P. G., 27, 75, 228, 231, 266, 270, 283
Sagan, C., 241
St. Venant, B. de, 14
Sallet, D., 92, 93
Samuels, M. R., 167, 238
Sanders, R. H., 241
Sandström, 282
Sarpkaya, T., 89, 228, 232, 233
Sato, H., 98, 232
Saulmon, 13
Savkar, S. D., 234
Sawatzki, O., 113, 114, 146, 148, 234, 236
Schade, H., 99, 232
Schaefer, J. W., 233
Schatzman, E., 220, 241
Schiller, F., 198
Schivley, G. P., 236
Schlichting, H., 227
Schneider, P. E. M., 43, 95, 112, 123, 234, 235
Schönung, B., 266, 284

Schot, J., 227
Schubert, G., 184, 237, 239, 241
Schuchhardt, C., 4, 227
Schwarz, K. W., 229
Schwenk, T., 17, 42, 228
Schwiderski, E. W., 142, 236, 237, 258, 259, 283, 284
Sciama, D. W., 225, 235
Scibor-Rylski, A. J., 231
Scotese, C. R., 239
Sedney, R., 96, 232
Seeley, D., 241
Seiden, P., 241
Seldner, M., 241
Sellers, A. J., 236
Sellers, W. D., 170
Seneca, 9, 10
Serrin, J., 228
Sexl, H. and R., 241
Sforza, P. M., 232
Shacter, J., 236
Shakespeare, W., 204
Shapiro, A. H., 38, 229
Shapley, H., 222
Shea, D. J., 208, 209, 240
Shelton, J., 230
Sherwood, J. D., 230
Shih, H., 142, 143, 236
Shin, C. T., 231
Shlien, D. J., 155, 157, 237
Shoaff, R. L., 233
Shubnikov, A. V., 228
Sibulkin, M., 236
Siedler, G., 238
Siekmann, J., 231
Silberstein, 264
Silk, G., 200
Silk, J., 242
Silliman, R. H., 227
Simon, L. E., 232
Sinclair, P. C., 240
Singh, M. P., 232
Singh, S. N., 138, 235
Sinha, P. C., 232
Sivaramakrishnan, T. R., 241
Skow, A. M., 91, 232
Slettebak, A., 228
Smagorinsky, J., 183
Smith, A. M. O., 234
Smith, E. H., 234
Smith, J. H. B., 229
Smith, J. L., 236
Smith, R. K., 239
Snider, A., 184
Snow, J. T., 195, 240, 283
So, R. M. C., 234
Socrates, 8
Soh, W. K., 44, 229
Solomon, H., 238
Sommerfeld, A., 283
Soneira, R. M., 241
Sonju, O. K., 235
Soong, T. C., 230

Sovran, G., 231
Speiser, A., 227
Spiegel, E. A., 280, 284
Spilhaus, A., 202
Squire, H. B., 88
Starr, V. P., 214, 234, 238, 241
Stein, L. R., 236
Stewart, H. J., 278, 284
Stewartson, K., 134, 235, 236
Stokes, G. G., 14, 18, 19, 65, 71, 273, 274
Stommel, H., 177, 183, 238
Stone, P. H., 240
Stong, C. L., 229
Stork, G. F., 154
Strahler, A. N., 235
Stringfellow, 56
Strouhal, V., 103, 105, 233
Stuart, 5
Stuart, J. T., 234
Sullivan, R. D., 228, 256, 262, 280, 281, 283
Sutcliffe, R. C., 238
Svardal, A., 233
Sverdrup, H. U., 183
Swanson, W. M., 234
Swenson, E. V., 284
Syred, N., 236
Szeto, R., 266, 283
Szewczyk, A. A., 101, 115, 233
Szodruch, J., 231
Szu, H., 41

Ta Phuoc Loc, 231
Tabaczynski, R. J., 78, 231
Taft, B. A., 238
Tagori, T., 231
Tait, 14
Takami, H., 70, 230
Takamoto, M., 232
Takata, H., 233
Takusagawa, Z., 230
Talbot, L., 11, 227
Taneda, S., 74, 101, 104, 107, 231, 233
Tannehill, I. R., 211, 240
Tarbell, J. M., 232
Tassoul, J. L., 241
Tatro, 145
Taylor, E. S., 237
Taylor, G., 240
Taylor, G. I., 15, 32, 120, 136–138, 145–147, 228, 234–236, 254, 256, 276, 277, 283
Teal, J. and M., 238
Telionis, D. P., 95, 232
Teller, E., 225
Tempelmeyer, K. E., 236
Tennekes, H., 234
Ter Linden, A. J., 236
Thales, 6
Thames, F. C., 105
Themon, 9
Theodorsen, T., 101, 233, 234

Theophrastus, 8
Thoman, D. C., 101, 115, 233
Thompson, B. J., 238
Thompson, D. W., 237
Thompson, P. D., 175, 238
Thomson, J. J., 14
Thorarinsson, S., 239
Thorne, K. S., 228, 241
Tietjens, O. G., 228
Tigner, S., 227
Timchalk, A., 240
Timme, A., 102, 233
Ting, L., 232
Tipler, F. J., 226, 242
Titiriga, A., 91, 232
Titius, 213
Tjøtta, S., 233
Tobak, M., 82, 231
Toms, B. A., 126, 235
Toomre, A. and J., 241
Torrance, K. E., 153, 237
Torricelli, 150
Townsend, A. A., 235
Tritton, D. J., 237
Trout, J. W., 240
Truesdell, C., 45, 227, 229
Tsahalis, D. T., 95, 232
Tsai, 89
Tsui, Y. T., 234
Tucker, W., 241
Turcotte, D. L., 184, 239
Turmlitz, O., 37, 229
Turner, J. S., 232, 237

Uginčius, P., 238, 284
Ussher (Bishop), 226

Van Dyke, M., 262, 283
Van Flandern, T. C., 235
Van Gogh, V., 17
Vasseur, P., 234
Vazsonyi, A., 15, 264
Venturoli, 14
Veronis, G., 237, 280, 284
Vettin, F., 182, 238
Viets, H., 228, 232, 233
Vinci, Leonardo da, 10–12, 30, 56, 60, 86, 125, 156, 203, 227
Virgil, 9
Vogel, H. U., 141, 236
Volkov, 221
Von Halle, E., 236
Vonnegut, B., 194, 236, 239
Votaw, C. W., 233
Votta, J. J., 229

Wade, N., 241
Walker, J. D. A., 68, 230
Walker, W. S., 234
Walls, H. A., 236
Walton, I. C., 235
Wan, C. A., 195, 239
Wang, K. C., 85, 231

Ward, N. B., 195, 239
Warren, A., 237
Weber, E. H., 159, 237
Weber, J., 231
Webster, A., 241
Wedemeyer, E., 231, 269
Wegener, A., 184–186, 191, 227, 239, 240
Weinberger, H. F., 283
Weis-Fogh, T., 59, 234
Weiss, N. O., 241
Weizsäcker, C. F. V., 16, 121, 213, 214, 225, 241
Weller, J. A., 229
Werlé, H., 82, 84, 231
Werle, M. J., 95, 232
Weske, J. R., 101, 195, 196, 233
West, M. L., 228
Westley, R., 236
Weston, R. P., 229
Wetherill, G. W., 241
Weyl, H., 229
Weyl, R., 240
Wheeler, J. A., 228
Whitehead, J. A., 158, 237
Whittaker, E., 227
Widnall, S. E., 89, 94, 229
Wieghardt, K., 231
Wilcke, J. C., 13, 14, 195
Williams, G. P., 167, 238
Williams, J. C., 231, 232
Willmarth, W. W., 110, 234
Wilson, R. A. M., 228
Wimmer, M., 146, 236
Windelband, W., 227
Withnell, G., 195, 239
Wolley, 203
Woo, H., 231
Wormley, D. N., 236
Worthington, L. V., 179, 180, 238
Wrixon, G. T., 241
Wu, T. Y., 229, 237
Wunderlich, H. G., 185, 239

Yarmchuk, E. J., 229
Yen, J. T., 202, 240
Yen, Y., 162, 237
Yih, C. S., 237, 281, 284
Yoshioka, I., 230
Yourgrau, W., 227
Yudin, E. Y., 15

Zabusky, N. J., 266, 283
Zahl, P. A., 240
Zaman, K. B. M. Q., 232
Zandbergen, P. J., 258, 283
Zarantonello, E. H., 231
Zehnder, N., 89, 232
Zeilik, M., 241
Ziegler, A. M., 239
Zierep, J., 146, 148, 160, 236, 237
Zipser, E. J., 237
Zwicky, F., 221

Subject Index

Acoustics, 15, 35, 105, 106, 111, 112, 143, 194
Adherence, 7, 14, 45, 64, 87, 96, 123, 167, 256
Aeolian harp, 106
Aircraft, 15, 37, 38, 53, 56, 62, 71, 81, 83, 272. *See also* Airfoil theory
Aircraft carrier, 38, 39
Air-entraining vortex, 34, 35
Airfoil:
   cylindrical, 64, 65
   oscillating, 58, 59, 76, 77, 79, 107–109
   theory, 56–64, 77, 114, 270–272
Air pollution, 70, 152, 164
Amplification, 110, 111, 143
Analogy:
   anthropomorphic, 6
   rotation-stratification, 164, 165, 282
   turbulence, 120
   vorticity-heat transfer, 45, 251
Andromeda nebula, 222, 224
Aneurism, 71
Anisotropic, 148, 225
Anomaly of water, 21, 151
Anticyclonic, 15, 132, 153, 179, 181, 208, 278
Antihail cannon, 92
Aortic valve, 10, 11, 71
Apeiron, 6
Aphrodite, 5
Arteriosclerosis, 70
Asgard's way, 202
Atherosclerosis, 88
Atmosphere:
   planetary, 16, 27, 215–218
   standard, 151
Atom:
   Democritus, 7, 8
   vortex theory (Lord Kelvin), 14–16, 41
Attachment point, concept, 67, 114
Automobile, 83, 96
Autorotation, 61–63, 110, 116–119, 125
Avalanche, 73

Baguio, 203
Baroclinic, 151
Baroclinic instability, 185, 188, 201
Baroclinic wave, 201, 218
Barotropic, 151, 264
Barotropic instability, 188, 201
Barotropic wave, 201
Bathtub vortex, 25, 32, 37, 129, 130, 134, 135, 155, 221. *See also* Discharge vortex
Bénard cell, 21, 91, 97, 125, 158–161, 165, 166, 184, 215
Bergen School, 171, 189
Bernoulli's law, 96
Bible, 4, 203, 226
Bicycle, 60
Big bang, 222, 224–226
Bilge vortex, 83, 85
Bioconvection, 160
Biot-Savart law, 270, 271
Bird, flight, 57–59, 77, 92, 108, 111, 154–156, 163
Bjerknes' theorem, 15, 264
Black hole, 16, 221, 222, 226
Blocking effect, 163, 165
Bödewadt solution, 142, 258, 259
Boiling, 145, 198
Bok globule, 219
Boomerang, 59, 63, 64, 66
Boundary layer:
   atmospheric, 139
   concept, 52, 53, 55
   Ekman, *see* Ekman layer
   equation, 259, 260
   eruption, 191, 194
   instability, 98, 100, 101
   shock, 71
   suction, 126
   turbulent, 121
Bound vortex, 15, 57, 74, 270, 271
Boussinesq approximation, 280
Brunt-Väisälä frequency, 162, 281, 282
Bucket, rotating, 34, 130, 253
Buffeting, 108

Buoyancy, 20, 80, 97, 152, 155, 165, 209
Burgers' solution, 255–257, 281

Cardonazo, 203
Cartesian vortex, 11–14, 130
Cascading vortices, 158, 160
Cat eyes, 100
Cavitation, 33, 143. *See also* Hollow core
Cavity, 3, 4, 72, 73, 111, 160
Cell:
   Bénard, 21, 91, 97, 125, 158–161, 165, 166, 184, 215
   Ferrel, 172
   Hadley, 172, 174, 207, 218
   polar, 172
Center:
   vortex, 23, 31, 35, 41, 42, 75, 91, 96, 122, 125, 254, 259, 270, 271, 281
   vortex system, 38, 267, 268
Centrifugal force, concept, 33, 34, 131, 276
Centrifuge, 15, 26, 33, 143
Chandrasekhar limit, 219, 220
Characteristics, 136, 149. *See also* Hyperbolicity
Charybdis, 3, 4, 9, 10, 198, 199, 201, 202
Chimney, 154, 164
Cilia, 51, 52, 79
Circulation:
   definition, 19, 27, 43, 56, 247
   general, 13, 15, 19, 26, 27, 37, 168–175, 177, 183, 187
   Hadley, 171, 172, 218
   meridional, concept, 138–142
   ocean, 26, 135, 175–178, 183, 188
   potential vortex, 43, 253
   quantized vortex, 42
   secondary, concept, 88, 138
   theorems, 264
   vessel, 138–142, 153, 167. *See also* Dishpan experiment
   Walker, 174

wing, 56–58, 271. *See also* Airfoil theory
Cluster:
  galaxies, 224, 225
  stars, 212, 222
  turbulence, 120
  vortices, 39, 40, 125
Coalescence, 94, 98, 111
Coherent structure, 121. *See also* Turbulence
Combustion chamber, 15, 143, 144
Computer, general, 15, 47, 121, 167, 168, 181, 183, 204, 269, 280
Computer-generated pictures, 23, 24, 39–41, 48–50, 68, 70–75, 78, 83, 98, 100, 101, 104, 105, 108, 109, 112, 113, 116, 138, 140, 141, 153, 157, 164
Conservation laws, 19, 20, 22, 28, 47, 67, 248, 250, 251
Constitutive equation, 28, 126
Container, 138–142, 153, 167
Continental drift, 160, 184–186
Continuity equation, 250, 264
Continuum, 7, 18–20, 27, 28, 42, 47, 48, 64, 212, 226
Control:
  boundary layer, 126
  drag, 123, 124
  flood, 34
  flux, 15, 143
Convection:
  cellular, 158–162. *See also* Bénard cell
  definition, 45, 152, 250, 252
  earth's mantle, 26, 152, 160, 185
  forced, 154
  natural or free, 154
Core of vortex, 23, 31, 35, 41, 42, 75, 91, 96, 122, 125, 254, 259, 270, 271, 281
Coriolis force, concept, 131, 171, 276
Coriolis parameter, definition, 132, 277
Corner vortex, 6, 21, 86, 144, 145
Cosmic background radiation, 224, 225
Cosmic whirl, 5–7, 10, 13, 16, 225
Cosmological principle, 225
Cosmology, 225
Couette flow, 159
Creeping flow, 46, 48–50, 64, 65, 71, 80, 257, 273–275
Crystal growth, 162, 166
Cybernetics, 15, 143
Cyclone:
  concept, 19
  dust, 15, 33, 143, 144. *See also* Centrifuge
  low-pressure system, 26, 132, 135, 169, 171, 174, 185, 187–190, 211, 278
  tornado, 194
  tropical, 203. *See also* Hurricane

Cyclonic, 15, 132, 153, 179, 181, 208, 278
Cyclostrophic motion, 279
Cylinder:
  drag, 124
  flow field, 48, 70, 101, 102
  lift, 115
  oscilliating, 106, 107
  rotating, 96, 112–115, 145–147

d'Alembert's paradox, 56, 66
Decay, *see* Discontinuity line, decay; Potential vortex, decay; Vortex decay
Deformation, 18–20, 27, 28
Deluge, 11, 203, 204
Delta wing, 81, 91
Density effect, 25, 27, 36, 93, 96, 129, 150, 155, 158, 165, 166, 179, 198, 225, 281. *See also* Stratification
Depression, tropical, 207
Descartes' vortex theory, 11–14, 130
Diffusion:
  definition, 45, 152, 252
  equation, 273
  flame, 112
Dine, 6, 8
Dipole, 46, 48, 64, 66, 274
Discharge (drainage) vortex, 34–37, 142, 143
Discontinuity line:
  decay, 53, 54, 100, 101, 273
  description, 39, 53, 98, 122, 189, 268
  invariance, 75
  spiral, 75, 269. *See also* Roll-up
Discus, 64, 65
Dishpan experiment, 167, 168, 182
Disk, 72, 74, 91, 109, 110, 112, 114, 257
Dissipation:
  concept, 32
  function, 250
  total, 250
Double helix:
  DNA, 16
  vortex, 105
Downwash, 271
Drag:
  coefficient, 124
  concept, 29, 123
  control, 123, 124
  induced, 57, 272
  reduction, 126
Dragon, 5
Drift current, 175, 176, 178
Dust cyclone, 15, 33, 143, 144
Dust devil, 9, 26, 135, 188, 195–197, 281

Edda, 5
Eddy:
  general, 19
  large scale, 15, 121
  spectrum, 15
  turbulent, 26, 119, 120, 150, 171, 188
Edge vortex, 81, 86, 96
Eel, 176
Eiffel-Prandtl paradox, 124
Ekman layer:
  concept, 87, 133, 279
  instability, 145, 148, 197
  pumping, 279
Ekman number, definition, 134, 276
Electromagnetism, 14, 27, 97, 270
Elliptic vortex, 41, 266, 271
El Niño, 174
Energy spectrum, 120
Entrainment, 94, 122
Entropy, 124, 212, 213, 226, 264
Epicycle, 8, 129, 130
Eridanus, 4, 5, 17
Ertel's theorem, 15, 264
Euler equation, 263, 278
Eulerian frame, 24, 250
Event-horizon, 221
Evolution, 3, 6, 16, 25, 43, 52, 77, 79, 131, 212–214, 223
Exchange coefficient, 120
Expansion of universe, 131, 224–226
Explosion, 94
Eye of hurricane, 207, 208, 210, 211

Feedback mechanism, 110, 111
Ferrel cell, 172
Filament, 31, 34, 37, 42, 60, 94, 101, 144, 249, 271
Fire ball, 224, 225. *See also* Big bang
Fish, 58, 59, 77, 108
Flagella, 51, 52, 59, 79
Flattening:
  earth, 13, 130
  stars, 218
Flettner rotor, 115
Flight, *see* Airfoil theory: Bird, flight; Insect, flight
Flow separation:
  concept, 14, 20, 23, 67–74
  shock-induced, 71, 82
  three-dimensional, 81–87
Flutter, 108
Foucault's pendulum, 129, 130, 133, 135, 148, 279
Frequency of vortex shedding, 99, 103, 107
Friction factor of pipes, 123–125
Frisbee, 64–66
Front, 53, 58, 172, 174, 185, 188, 189
Froude number, definition, 162, 163
Fujiwara effect, 38

Galaxy, 3, 16, 25–27, 131, 212, 219, 222–226
Galloping, 108

General circulation, 13, 15, 19, 26, 27, 37, 168–175, 177, 183, 187
Geostrophic contour, 137
Geostrophic flow, 132, 133, 137, 172, 277–279, 282
Geostrophic stream function, 278
Geostrophic vortex, 277, 278
Geostrophic wind, 133, 277
Ghost wake, 33
Glauert's criterion, 62
Glide:
    angle, 58
    number, 58
Gliding, 56, 58, 163
Görtler vortex, 15, 145, 148
Goldshtik's solution, 142, 258, 262
Gradient flow, 133, 278
Granulation, sun, 160, 215
Grashof number, definition, 153
Gravitational collapse, 212
Great Red Spot of Jupiter, 27, 137, 216–218
Gulf Stream, 26, 175–181, 185, 187
Gyro, 63, 64, 148

Hadley:
    cell, 172, 174, 207, 218
    circulation, 171, 172, 218
Hailstone, 125
Hamel-Oseen solution, *see* Lamb's solution
Hamiltonian, 269, 278
Heat valve, *see* Aortic valve
Heat:
    conduction, 45
    death, 212, 226
    equation, 250
    exchanger, 144
    specific, 250
Heat-up, 165
Helicopter, 60, 61
Helium, liquid, 16, 26, 42
Helmholtz:
    equation, 278
    instability, 98, 164, 188, 197
    theorems, *see* Vorticity theorems
Hertzsprung-Russel diagram, 218, 219
Hill's spherical vortex, 91, 266
Hilsch tube, 144
Hollow core, 33, 42, 61, 89, 90. *See also* Cavitation
Horseshoe vortex, 57, 81, 85, 110, 271, 272
Hovering, 60, 92
Hub vortex, 60, 86
Hubble's classification, 223
Humming bird, 60, 61
Hurricane, 3, 16, 18, 26, 135, 153, 187, 188, 203–211, 217
Hydrostatic equilibrium, 172, 277
Hydrostatic layer, 165

Hyperbolicity, 135, 164, 276, 277
Hysteresis, 117

Ice, 154, 160, 162, 204
Ideal flow, 14, 54
Impetus theory, 8, 9
Inertia:
    concept, 12, 129–131
    motion, 279
    subrange, 120
    system, 22–24, 55, 149, 276, 278
    vortex, 133, 135
    wave, 140, 141, 148, 277
Insect, flight, 26, 57–59, 77, 92, 108, 109
Instability:
    baroclinic, 185, 188, 201
    barotropic, 188, 201
    boundary layer, 98, 100, 101
    concept, 20, 97
    cosmological, 226
    discontinuity line, 39, 41, 54, 55
    Ekman, 197
    Helmholtz, 98, 164, 188, 197
    rotating fluid, 89, 145
    shear flow, 98
Intake:
    pipe, 34–36
    vortex, 29, 36, 35, 44
Invariance:
    discontinuity line, 75
    pathline and streamline, 22, 23
    vorticity, 53, 55, 95, 96
Inversion, 151, 152, 164
Inviscid flow, 14, 54
Ionic enlightenment, 8
Irrotational flow, *see* Potential flow
Isotropic, 148, 225

Jet:
    flow, 36, 91, 99, 104, 110, 155, 158, 164, 178, 184
    impinging, 110, 111
    propulsion, 80, 81
    rotating, 35, 260
    stream, 167, 172, 173, 181, 185, 188, 189, 191, 206, 218
Jupiter, 16, 27, 47, 214–218
    Great Red Spot of, 27, 137, 216–218

Kaegyihl Depguesk, 198
Kant-Laplace theory, 13, 213
Kamikaze, 203, 211
Kappa, 198
Kármán solution, 142, 257–259
Kármán vortex street, 23, 97, 98, 101–107, 119, 124, 178, 187, 188, 201, 256
Kasper wing, 96
Kelvin's atomic theory, 14–16, 41
Kelvin's theorem, 15, 264
Kepler's laws, 29, 30, 130, 213

Kerr-Newman structure, 221
Kolk, 144, 146, 198
Kolmogorov's hypotheses, 120
Kundt's dust figures, 106
Kuroshio, 177, 181, 186, 187
Kutta condition, 56, 74, 75, 96
Kutta-Joukowsky formula, 15, 56, 272

Lagrangian frame, 24, 250
Lamb's problem, 273
Lamb's solution, 254
Laminar flow, concept, 34, 119, 120
Lanchester-Prandtl hypothesis, 56, 57, 96. *See also* Airfoil theory
Lanchester propeller, 61, 62, 108
Langmuir vortex, 188, 197
Lapland expedition, 13, 130
Large-scale structure, 15, 121. *See also* Turbulence
Latent heat, 194, 195, 207, 209, 210
Lee wave, 58, 163, 164, 188
Lepensky Vir, 144
Lift:
    airfoil, 56–58, 62, 64
    coefficient, 57, 58, 66
    concept, 29
    oscillating body, 58, 59, 108
    rotating body, 114, 115, 118
Lifting-line model, 271, 272
Liouville's theorem, 269
Localized induction equation, 271
Lock-in, 99, 107, 109, 116
Logos, 224, 226
Long's solution, 260–262
Low-pressure system, 26, 132, 135, 171, 174, 188, 189, 190, 211, 278

Mach's principle, 129, 131
Mackerel cloud, 21, 160
Maelstrom, 3, 5, 17, 198, 199
Magnetic field, 12, 168, 214, 220
Magnetohydrodynamics, 144
Magnus:
    effect, 114, 115
    force, 118
Main sequence, 219, 220
Maple seed, 61
Marangoni effect, 159, 160, 162, 166, 167
Materialism, 8
Merging, 40, 94, 98. *See also* Coalescence
Mesorange, 187, 188, 191
Microrange, 187, 188
Milky Way, 222. *See also* Galaxy
Mimicry, 92
Mixing length, 120
Moment of vortex system, 268
Monsoon, 152, 169, 174, 177
Moving-flame experiment, 155
Musical instrument, 106, 111

Naruto straits, 5
Navier-Stokes equations:
    closed-form solutions, 66, 253–257, 266, 279
    description, 64, 250
    different coordinate systems, 251
    history, 14, 15
    numerical solutions, 23, 24, 39–41, 48–50, 68, 70–75, 78, 83, 98, 100, 101, 104, 105, 108, 109, 112, 113, 116, 138, 140, 141, 153, 157, 164
    rotating system, 27, 276
Necklace vortex, 85–87
Neutron star, 16, 42, 221
Newgrange, 16
Newtonian fluid, 44, 64, 126
Newton's laws, 129, 130, 276
Niagara Falls, 144
Noise, 15, 35, 105, 106, 111, 112, 143, 194
Non-Newtonian fluid, 44, 123, 126
Nonslip condition, 7, 14, 45, 64, 87, 96, 123, 167, 256
Nose vortex, 84–86
Nuclear reactor, 35
Numerical experiment, 23, 24, 39–41, 48–50, 68, 70–75, 78, 83, 98, 100, 101, 104, 105, 108, 109, 112, 113, 116, 138, 140, 141, 153, 157, 164

Odyssey, 3, 4, 9, 10
Organization time, 37
Orifice, 73, 89, 90, 157
Orphics, 17
Oscillation:
    body, 35, 52, 58, 59, 76, 80, 92, 96, 98, 106–108, 119, 136, 165
    building, 109
    flow-induced, 15, 77, 79, 105–112
    forced, 76, 79, 107
    in-line, 107
    jet, 110–112
    neutral, 77
    rotary, 79
    translational, 79
    turbulence, 108
    vortex, 40, 41, 43, 143
    vortex-induced, 15, 77, 79, 105–112
Overshoot, 89, 122

Pairing, see Coalescence; Merging
Pamir, 205
Pangaea, 184, 185
Panthalassa, 184
Paradox:
    d'Alembert, 56, 66
    Eiffel—Prandtl, 124
    penny, 43
    Stokes, 65
Parent cloud (tornado), 194, 195
Pathline:
    computation, 248
    definition, 21, 22, 25, 26, 248
    invariance, 22, 23
Pattern recognition, 121
Pendulum, 129, 130, 133, 135, 148, 279
Penny paradox, 43
Periodic system, 226
Perturbation technique, 262, 266, 275, 278, 280, 282
Phaidon, 5, 19
Pipe:
    flow:
        oscillating, 88, 106
        steady, 119, 122–126, 254
    intake, 34–36
    pressure loss, 47, 124, 125
    separation, 70, 71
Planetary nebula, 220, 226
Planetesimals, 213
Plasma, 27
Plate:
    flow field (plate fixed), 23, 46, 50, 72, 78, 81, 99, 104
    lift, 56–59
    oscillating, 58, 59, 79, 80, 109, 110
    rotating, 24, 116–118
Point vortex, see Potential vortex
Poiseuille solution, 254
Polar cell, 172
Polar front, 172
Polar vortex (circumpolar), 174
Polyox, 126
Potential flow:
    definition, 18, 54, 66, 248
    superfluid, 16, 26, 42
Potential vortex:
    circulation, 43, 253
    decay, 32, 88, 102, 254
    definition, 31, 249, 253
    description, 31–33, 42, 122, 267
    dissipation, 32, 253
    generation, 31
    Hamiltonian, 269
    near wall, 142, 258, 259, 262
    pressure, 33, 253
    system, 38–40, 78, 98, 125, 267–269
Potential vorticity, 265, 277, 278
Pothole, 144, 146
Pottery wheel, 16
Prandtl airfoil theory, 56, 57, 96
Prandtl boundary-layer theory, 52, 53, 55
Prandtl Eiffel paradox, 124
Prandtl mixing length, 120
Prandtl number, definition, 153
Precession of equinoxes, 5, 17
Pressure loss in pipe, 47, 124, 125
Primeval vortex, 4, 7, 225
Principia (Newton), 13, 130
Principle:
    anthropic, 226
    cosmological, 225
    Mach's, 129, 131
    maximum, 257
    minimum, 8, 226
    uniformity, 226
Propeller:
    general, 15, 59, 60
    Lanchester, 61, 62, 108
    ship, 61, 83
    Voith-Schneider, 63
    vortex, 60
Pulsar, 221
Pump:
    general, 34
    vortex, 144
Pyramid text, 4

Quantum:
    mechanics, 16, 19, 27, 42, 222
    vortex, 16, 26, 42
Quasar, 223

Raindrop, 69, 70, 93, 96
Rankine vortex, 31–33, 66, 122, 194, 197, 249, 253, 266
Ranque-Hilsch tube, 144
Rayleigh number, definition, 153
Rayleigh-Taylor instability, 155, 188, 197
Real flow, concept, 54
Red giant, 219
Red shift, 224
Reference frame, concept, 129–121. See also Eulerian and Lagrangian frames; Inertia, system; Rotating system
Relativity, theory, 16, 27, 131, 221
Resistance, see Drag
Resonance, 110, 111
Reynolds number, definition, 46, 252
Rigveda, 4
Ro (oar), 66
Roll, 158–160, 165–167. See also Convection
Roll parameter, 114, 115, 125
Roll-up, 39, 41, 55, 75, 76, 78, 81, 86, 98, 99, 101, 110, 121, 125, 143, 269
Rossby number:
    data, 135
    definition, 114, 125, 134, 276
Rossby wave, 135, 137, 165, 174, 177, 185, 188, 277
Rotating system, 129, 149, 276
Rotation:
    absolute, 34, 129–131
    concept, 18, 19
    differential, 214
    translational, 59
Roton, 16, 42. See also Quantum vortex
Rotor:
    airplane, 115
    Flettner, 115

lee wave, 163
ship, 115

Salinity, 152, 175, 179
Salt lake, 161
Sand dunes, 149
Sargasso Sea, 176, 177, 185
Saturn, 27, 47, 214, 217, 218
Schmidt number, 186
Scholasticism, 8, 10, 11, 129
Schooling of fish, 81, 111
Schwarzschild structure, 221
Sculling, 66
Scylla, 4, 199, 202
Sea breeze, 152, 160, 188
Sedimentation, 7, 34, 70
Separation, *see* Flow separation; Vortex separation
Separation point, concept, 67, 84, 114
Shear flow or layer:
   definition, 53
   instability, 98–100. *See also* Instability, discontinuity line
Shells, 3, 43
Ship wave, 33, 83, 86, 135, 136
Shock-induced separation, 71, 81, 82
Shock wave, 71, 76, 78, 81, 82, 135, 136, 219
Silberstein's theorem, 264
Similarity:
   concept, 257, 269
   conical, 81
   dynamic, 46
   geometric, 46
   solutions, 257–262
Simply connected region, 248
Sink flow, 36, 78, 142
Singing of wires, 105
Slip, 14
Slow motion, 46, 48–50, 64, 65, 71, 80, 257, 273–275
Smog, 152
Snow cornice, 73
Soaring, 58, 154, 155, 163
Solar system, 212–214
Solid-body rotation:
   center of vortex, 75, 96, 122, 125, 254, 270
   definition, 30, 247, 253
   description, 30–34, 253
   free surface, 34, 254
   history, 10
   near wall, 142, 258, 259, 265
   occurrence, 36, 52, 87, 131, 138
   pressure, 33, 253
   superfluid, 42
   swirl, 30, 87
Soliton, 137, 217
Sophists, 8
Sound, 15, 35, 105, 106, 111, 112, 143, 194
Source flow, 142

Space:
   absolute, 13, 34, 130, 149
   curvature, 221
   ship, 32, 119
Spectrum:
   eddy, 15
   energy, 120
   vortex, 26, 187
Sphere:
   drag, 65, 124
   flow around, 21, 22, 49, 50, 68, 69, 72, 74, 91, 105
   liquid, 69, 96
   rotating, 113, 114, 145, 275
   in rotating fluid, 137, 138, 142
Spillway, 34
Spin:
   atomic, 27
   airplane, 62, 63
Spin-down, 138, 139, 148
Spin-up, 138, 139, 148, 165
Spiral motif, 3–5, 10, 16, 17, 25, 43
Spraying, 35, 143
Squall line, 188, 190
Squid, 26
Stagnation point:
   concept, 67
   flow, 265
Stall, 62
Standard atmosphere, 151
Star:
   binary, 220, 221
   cluster, 212, 222
   generation, 219
   luminosity, 219
   neutron, 16, 42, 221
   rotating, 33, 218–220
Starting vortex, 57, 74–76
State variable, 18, 27, 28, 43, 172
Stern vortex, 85
Stewartson layer, concept, 134
Stokes law, 65
Stokes paradox, 65
Stokes vortex, 71, 273, 274
Strake vortex, 85
Stratification, 23, 27, 64, 129, 136, 150, 162, 178, 201, 215, 217, 280. *See also* Density effect
Streak line, definition, 25, 26, 248
Stream function:
   definition, 251
   geostrophic, 278
   vortex system, 268
Streamline:
   closed, 22–24, 27, 96, 248, 253, 275
   definition, 21, 25, 248
   invariance, 22, 23
   wavy, 23, 162
Stretching of vortex, 25, 36, 55, 264, 265
Strouhal number, definition, 103, 125
Submarine, 86, 87, 181

Submergence, 34, 35
Suction:
   boundary layer, 126
   mark (tornado), 194
Sullivan's solution, 256, 280, 281
Sun:
   granulation, 160, 215
   rotation, 214
   spots, 27
Superfluid, 16, 26, 42
Supernova, 219, 223
Surface tension, 69, 97, 159, 160. *See also* Marangoni effect
Surge, 42
Suspension, 115. *See also* Sedimentation
Svelgr, 5
Swilkie (svelgr), 198
Swimming, 58, 77, 108
Swirling flow, 19, 35, 87–91, 122, 141, 195, 265
Synchronization, 99, 107, 109, 116
Synoptic range, 187, 188, 191
System:
   closed, 28, 29, 43
   inertial, 22–24, 55, 149, 276, 278
   periodic, 226
   reference, *see* Reference frame
   rotating, 129, 149, 276
   vortex, 38–40, 78, 98, 125, 267–269

Tacoma bridge, 106, 108
Tail fin, 58, 80
Taylor-Proudman column, 136, 137, 163, 277
Taylor-Proudman theorem, 15, 136, 148, 165, 276, 277
Taylor vortex, 119, 125, 145–147, 165
Teacup phenomenon, 7, 8, 13, 141
Temperature lapse rate, 281
Theogony (Hesiod), 4, 6
Thermal conductivity, 250
Thermal diffusivity, 153
Thermocline, 152, 197
Thrust, 59–63, 80
Thunderstorm, 58, 153, 174, 183, 188, 190, 195
Tidal vortex, 4, 5, 9, 145, 188, 198–202. *See also* Maelstrom
Timaeus (Plato), 8, 10
Tip vortex, 38, 57, 61, 81–83, 85, 86, 88, 122, 270, 271
Titius-Bode rule, 213, 214
Tollmien-Schlichting wave, 100
Toms phenomenon, 126
Tornado:
   artifical, 202
   cyclone, 194
   data, 135, 149, 188, 192, 194
   description, 158, 190–197, 262, 281
   history, 13

Subject Index

parent cloud, 194
vortex breakdown, 191, 194
Torpedo, 126
Torque, concept, 29
Trailing vortex, 38, 57, 61, 81–83, 85, 86, 88, 122, 270, 271
Translation, 18, 19
Transmigration, 4, 17
Trombe, 191. *See also* Tornado
Tropical storm, 3, 16, 18, 26, 135, 153, 187, 188, 203–211, 217
Turbine, 15, 35, 53, 143
Turbulence:
  concept, 34, 97, 119
  definition, 120
  eddy, 26, 119, 120, 150, 171, 188
  fine structure, 121
  history, 10, 15, 119, 120
  inertial subrange, 120
  large-scale structure, 15, 121
  rotating fluid, 148
  spot, 101
  theories, 120, 121
  vortex, 34, 121, 122
Two-dimensional constraint, 15, 136, 148, 165, 276, 277
Typhoon, 3, 135, 203–205, 210, 211. *See also* Hurricane

Universal range (turbulence), 120
Universe, expanding, 131, 224–226
Uranium, 143

Väisälä frequency, 162, 281, 282
Vazsonyi's theorem, 15, 264
Velocity, settling or terminal, 50, 124, 158
Vena contracta, 73
Venus, 214, 218
Vessel, 138–142, 153, 167
Vibration, *see* Oscillation
Viscosity, 32, 39, 42, 43, 45, 250
Viscous flow, concept, 54
Voith-Schneider propeller, 63
Volcano, 26, 158, 174, 191
Vortex:
  air-entraining, 34, 35
  alpine, 185
  amplifier, 143
  atom, 14–16, 41
  band, 39, 98, 99
  bathtub, 25, 32, 37, 129, 130, 134, 135, 155, 221
  bilge, 83, 85
  bound, 15, 57, 74, 270, 271
  bow, 86
  breakdown (burst), 81, 88–91, 140, 141, 143, 191, 194, 195
  Cartesian, 11–14, 130
  cascading, 158, 160
  center of vortex system, 38, 267, 268
  chamber, 144
  chain (row), 39, 41, 99, 125, 268
  circular, concept, 19, 23, 20
  circumpolar, 174
  clipped, 111
  cluster, 39, 40, 125
  coalescence, 94, 98, 111
  columnar, 18, 19, 143, 155, 195, 196, 202
  concentrated, 36, 125. *See also* Vorticity extremum
  core (center), 23, 31, 35, 41, 42, 75, 91, 96, 122, 125, 254, 259, 270, 271, 281
  corner, 6, 21, 86, 144, 145
  cosmic, 5–7, 10, 13, 16, 225
  cylindrical, 18, 19, 270
  decay, 32, 34, 43, 55, 88, 91, 92, 100, 102, 108, 111, 122, 155, 157, 158, 164, 254, 255
  definition, 18, 22, 23, 27, 53, 96
  diameter, definition, 31
  discharge or drainage, 34–37, 142, 143
  disklike, 18, 19, 150
  edge, 81, 86, 96
  elliptic, 41, 266, 271
  extraterrestrial, 27, 212–226
  filament, 31, 34, 37, 42, 60, 94, 101, 144, 249, 271
  galactic, 222
  generation (formation), 14, 20, 25, 27, 36, 37, 42, 48, 67, 91, 93, 96, 116. *See also* Roll-up
  geostrophic, 277, 278
  Görtler, 15, 145, 148
  grid, 256
  Gulf Stream, 26, 178–181, 188
  helical, 18, 19, 32, 34, 36, 41, 104, 144
  Hill's, 91, 266
  horseshoe, 57, 81, 85, 110, 271, 272
  hub, 60, 86
  inertial, 133, 135
  intake, 29, 34, 36, 44
  interaction, *see* Coalescence; Merging
  Kármán, 23, 97, 98, 101–107, 119, 124, 178, 187, 188, 201, 256
  Langmuir, 188, 197
  lattice, 272
  lift, 81, 82
  line, 31, 40, 122, 249
  Marangoni, 159, 160, 162, 166, 167
  necklace, 85–87
  nose, 84–86
  oscillation, *see* Oscillation
  periodic, 72, 81, 99, 101–112, 115. *See also* Vortex street
  plane, 18, 30, 87, 273
  potential (point), *see* Potential vortex
  primeval, 4, 7, 225
  propeller, 60
  protoplasmic, 17
  pump, 144
  quantum, 16, 26, 42
  Rankine, 31–33, 66, 122, 194, 197, 249, 253, 266
  reactor, 144
  ring, 14, 15, 20, 26, 41, 42, 68–70, 91–95, 122, 154–158, 266, 271, 274
  roll, 158–160, 165–167
  row, 39, 41, 99, 125, 268
  secondary, 74, 76, 82, 93, 100, 110, 123, 194–196
  separation (shedding), 74–80, 96, 101–111, 116–118
  solid-body, *see* Solid-body rotation
  sound, 15, 112. *See also* Sound
  Spectrum, 26, 187
  Spiral, 18, 19, 32, 34, 36, 41, 105, 144
  starting, 57, 74–76
  stern, 85
  Stokes, 71, 273, 274
  strake, 85
  street, 23, 97, 98, 101–107, 119, 124, 178, 187, 188, 201, 256. *See also* Vortex, periodic
  stretching, 25, 36, 55, 264, 265
  system, 38–40, 78, 98, 125, 267–269
  Taylor, 119, 125, 145–147, 165
  theory, *see* Cartesian vortex; Vortex atom
  tidal, 4, 5, 9, 145, 188, 198–202. *See also* Maelstrom
  tip, *see* Tip vortex
  trailing, 38, 57, 61, 81–83, 85, 86, 88, 122, 270, 271
  tube, 31, 41, 91, 104, 144, 249, 271, 272
  turbulent, 34, 121, 122
  two-cell, 255, 280, 281
  valve, 143, 144
  vibration, 40, 41, 43, 143
  viscous, 253
  wake, 70, 86, 163, 272
  warm air, 154, 155, 157
  wave, 42
  whistle, 144
  wind-generated, 175
Vorticella, 52
Vorticity:
  absolute, 135
  baroclinic, 164, 168
  definition, 18, 19, 27, 29, 247
  dimensionless, 252
  dynamics, 47, 48
  extremum, 27, 69, 75, 76, 78, 96, 164, 257, 275
  generation, 45, 54, 64, 67, 71, 164
  history, 14
  invariance, 53, 55, 95, 96

Subject Index

line, 31, 249
potential, 265, 277, 278
relative, 135
secondary, 265
theorems:
  Bjerknes, 15, 264
  Crocco, 15
  Ertel, 15, 264
  Helmholtz, 14, 15, 54–58, 66, 87, 104, 120, 135, 263, 264
  Oswatitsch, 15
  Silberstein, 264
  Vazsonyi, 15, 264
transport equation, 15, 183, 185, 251, 252, 263, 265, 273, 276
tube, 31, 54–56, 249
Voyagers 1 and 2, 16, 47, 215, 218

Wake, *see* Tip vortex; Vortex, street; Vortex, wake
Walker circulation, 174
Water:
  anomaly, 21, 151
  screw, 60
Waterspout, 26, 37, 188, 190–193, 195

Wave:
  baroclinic, 201, 218
  barotropic, 201
  data, 188
  density, 223
  description, 162, 182
  frozen or trapped, 140, 141
  inertial, 140, 141, 148, 277
  Rossby, 135, 137, 165, 174, 177, 185, 188, 277
  rotating fluid, 89, 135, 164, 277
  ship, 33, 83, 86, 135, 136
  shock, 71, 76, 78, 81, 82, 135, 136, 219
  solitary, 41. *See also* Soliton
  stratified fluid, 162, 164, 282
  streamline, 23, 162
  Tollmien-Schlichting, 100
  trapped, 140, 141
  vortex filament, 41, 42
  vortex ring, 122, 157, 158
Weather prediction, 15, 47, 175, 183
Weir, 86
Weis-Fogh mechanism, 59, 108
Weizsäcker's hypothesis, 214

Westward intensification, 177
Wheel, 4, 7, 16, 59
Whirl, concept, 19
Whirlpool, *see* Bathtub vortex; Intake vortex; Maelstrom; Tidal vortex
Whirlwind, *see* Dust devil; Hurricane; Tornado; Waterspout
White dwarf, 219–221
Willy-willy, 203
Wind mill, 60
Windrow, 197
Wing-sheet model, 272
Wing theory, 56–64, 77, 114, 270–272
Witte-Margules equation, 282
Wobbling, 110
Wokingham model, 195
World:
  age, 10, 17
  mill, 5
  mother, 4

Yin-Yang, 5

Zuni Indians, 4